IMAGE FORMATION FROM COHERENCE FUNCTIONS IN ASTRONOMY

ASTROPHYSICS AND SPACE SCIENCE LIBRARY

A SERIES OF BOOKS ON THE RECENT DEVELOPMENTS OF SPACE SCIENCE AND OF GENERAL GEOPHYSICS AND ASTROPHYSICS PUBLISHED IN CONNECTION WITH THE JOURNAL SPACE SCIENCE REVIEWS

VOLUME 76

PROCEEDINGS

IMAGE FORMATION FROM COHERENCE FUNCTIONS IN ASTRONOMY

PROCEEDINGS OF IAU COLLOQUIUM NO. 49 ON THE FORMATION OF IMAGES FROM SPATIAL COHERENCE FUNCTIONS IN ASTRONOMY, HELD AT GRONINGEN, THE NETHERLANDS, 10–12 AUGUST 1978

Edited by

CORNELIS VAN SCHOONEVELD
Leiden Observatory, Leiden, The Netherlands

D. REIDEL PUBLISHING COMPANY

DORDRECHT : HOLLAND / BOSTON : U.S.A.
LONDON : ENGLAND

Library of Congress Cataloging in Publication Data

CIP

Image Formation from Coherence Functions in Astronomy.
(Astrophysics and space science library ; v. 76)
Includes indexes.
1. Radio astronomy–Congresses. 2. Telescope–Congresses.
3. Coherence (optics)–Congresses.
I. Schooneveld, Cornelis van, 1933- II. Title: International Astronomical Union Colloquium No. 49. III. Series.
QB475.A1I43 522'.682 79-11787
ISBN 90-277-0987-4

Published by D. Reidel Publishing Company,
P. O. Box 17, Dordrecht, Holland.

Sold and distributed in the U.S.A., Canada and Mexico
by D. Reidel Publishing Company, Inc.
Lincoln Building, 160 Old Derby Street, Hingham,
Mass. 02043, U.S.A.

Printed in The Netherlands

TABLE OF CONTENTS

PART III - PROCESSING TECHNIQUES AND DISPLAY METHODS

PART IV - OPTICAL INTERFEROMETRIC METHODS

PART V - MAXIMUM ENTROPY IMAGE RECONSTRUCTION

PART VI - OTHER IMAGE IMPROVEMENT METHODS

INTRODUCTION

The IAU Colloquium No.49, on the formation of images from spatial coherence functions in astronomy, was held at Groningen, the Netherlands, during the period 10-12 August 1978. The colloquium was attended by 108 participants from 14 countries (U.S.A. 29, the Netherlands 20, U.K. 19, Germany 10, France 7, Australia 5, Canada 5, Japan 4, India 2, New Zealand 2, Sweden 2, Argentina 1, Belgium 1, Israel 1). It was sponsored by the Netherlands Foundation for Radio Astronomy, the International Astronomical Union, the Department of Education and Sciences, the Union Radio-Scientifique Internationale, the Leiden Kerkhoven-Bosscha Foundation and the State University at Groningen.

This volume contains 36 of the 37 papers presented. Nearly all papers are followed by a summary of the discussion that took place after their presentation. A few papers, published in full elsewhere, are given only as abstracts.

The majority of the papers are related to aperture synthesis in radio astronomy; a small number deal with optical astronomy and with applications in acoustics and medicine. The presentations are divided in 7 groups: aperture synthesis and its deficiencies, the problem of limited or missing phase information, techniques for processing and data display, optical interferometric methods, maximum entropy image reconstruction, other image improvement methods, and a survey of image formation from projections. Each group contains one or two invited lectures (see Table of Contents), intended as surveys of particular areas; on the average they occupy twice as many pages as the other papers.

Looking at the contents as a whole, one is struck by the fact that relatively little is said about the classic method of image formation through straightforward fourier inversion of the visibility data. On the other hand, many proposals are made for non-fourier techniques. They are prompted by the wish to cope with the phase problem or by the desire to achieve improved resolution in the image. The key to many of these methods is a simple one: the exploitation of previously neglected a priori knowledge of the brightness distribution and of the measurement errors. In spite of this simplicity we are confronted by a profusion of non-fourier image formation schemes. Some 20 different methods are listed in the Subject Index under the heading "Image reconstruction". A comparison of the relative performance of these methods - and of their combinations - is at present hampered by the fact that practically none of them has a closed form solution. More work is needed in the future to clarify the situation. Progress in this area is not only expected through astronomically oriented investigations but also through contributions from geophysics, acoustics, statistical spectrum analysis and picture processing, where similar problems exist.

The editor gratefully acknowledges the cooperation of all authors in preparing the typescripts and expresses his gratitude to Mrs. M.H. van Haaster for her assistance with editing the book.

C. van SCHOONEVELD

ORGANIZING COMMITTEE

SCIENTIFIC

W.N. Brouw (Netherlands)
E. Blum (France)
M.H. Cohen (U.S.A.)
T.W. Cole (Australia)
R.D. Ekers (Netherlands)
E.B. Fomalont (U.S.A.)
M. Ishiguro (Japan)
G.G. Pooley (U.K.)

LOCAL

E. Raimond (Netherlands)
J.P. Hamaker (Netherlands)
R.H. Harten (Netherlands)
R.S. le Poole (Netherlands)
U.J. Schwarz (Netherlands)

EDITOR

C. van Schooneveld (Netherlands)

PART I

APERTURE SYNTHESIS METHODS

FUNDAMENTALS AND DEFICIENCIES OF APERTURE SYNTHESIS

(Invited paper)

Edward B. Fomalont
National Radio Astronomy Observatory
Green Bank, West Virginia 24944 U.S.A.

1. INTRODUCTION

Aperture synthesis is the method used by astronomers to determine the accurate brightness distribution of the radio sky with a resolution much better than that possible with a single large antenna. The technique, now over a decade old, utilizes a large number of connected radio antennas, some of them physically moveable, to follow a region of sky for many hours or days in order to sample the spatial coherence function of the radiation field over a sufficiently large area and with a reasonable filling factor. Landmark references for aperture synthesis are McCready et al. (1947), Stanier (1950), Christiansen and Warburton (1955), Lequeux et al. (1962), Read (1961) and Ryle and Hewish (1960).

In Section 2 the relationship between the spatial coherence function and the brightness distribution, essentially the van Cittert-Zernike Theorem, together with necessary assumptions about the radiation field, will be discussed. Aperture synthesis is a straightforward application of this theorem. The deficiencies in aperture synthesis are many and these are summarized in Section 3. A major part of this colloquium will deal with methods to correct the adverse effects of these deficiencies. Some are fundamental to the technique--e.g., the spotty sampling of the spatial coherence function. Some are related to practical considerations in the design of the array and the speed of map making at the expense of some accuracy--e.g., the use of the Fast-Fourier transform algorithm. Some are related to non-stationary effects during the course of the observations--e.g., tropospheric phase fluctuations. In Section 4 a brief description of problems in aperture synthesis that strike the author as important or annoying as well as a *potpourri* of other ideas will be mentioned.

C. van Schooneveld (ed.), Image Formation from Coherence Functions in Astronomy, 3–18.

2. FUNDAMENTALS OF APERTURE SYNTHESIS

2.1. Coherence Properties of the Radiation Field

Let $E(\bar{u},t)$ represent the radiation field at any point in space, $\bar{u}$, at any time, t. Although electromagnetic radiation is described by a vector field which satisfies Maxwell's Equations, for most astronomical applications an approximate description of the field by a scalar wave function (for example, the transverse component of the electric field) is adequate. The field fluctuates rapidly and behaves ergodically; that is, the time-average properties are well-defined and any measurement of an average property is typical of all such similar measurements.

While it is now possible at radio frequencies to follow the detailed fluctuations of the field, there is little additional information in them and most characteristics of the radiation are embodied in various average functions of the field. The intensity of the field can be easily measured using a suitable probe which responds with a signal $S(\bar{u},t)$ proportional to $E(\bar{u},t)$. The signal is first "detected" by measuring the power and then averaged over some time interval T which is long compared with the time-scale of the fluctuations. The intensity $I(\bar{u})$ is thus

$$\lim_{T\to\infty} \frac{1}{2T} \int_{-T}^{T} S^*(\bar{u},t)\, S(\bar{u},t)\, dt \qquad (2.1)$$

and is usually written in short-hand notation as

$$I(\bar{u}) = \langle S^*(\bar{u},t)\, S(\bar{u},t)\rangle \ . \qquad (2.2)$$

As is customary when dealing with signals associated with the wave equation, a complex number, called the analytic signal, is used in place of the real signal (e.g., Steel 1967, p.22).

When two or more sources of radiation are superimposed the intensity in the region of superposition can vary. This phenomenon is called interference. In general the radiations from different sources are incoherent; i.e. their fluctuations are completely uncorrelated and, thus, they do not interfere over typical averaging time scales $t > 1/\Delta\nu$ where $\Delta\nu$ is the bandwidth of the radiation. But if the radiation from *one* source is separated into two beams, by sampling two points of the wavefront or by splitting the radiation at some point, the fluctuations in the two beams are generally correlated and they will interfere when combined.

The measure of the correlation in a radiation field between two points $\bar{u}_1$ and $\bar{u}_2$ at a time difference τ is given by

$$\Gamma(\bar{u}_1,\bar{u}_2,\tau) = \langle S^*(\bar{u}_1,t) \quad S(\bar{u}_2, t+\tau)\rangle \tag{2.3}$$

and is denoted as the *mutual coherence function* of the field. For example if we combine the signals sampled at two points $\bar{u}_1$ and $\bar{u}_2$, separated by time τ, the time-average intensity of the signal $S_s = S(\bar{u}_1,t) + S(\bar{u}_2,t)$ would be

$$\langle S_s^* \; S_s\rangle = I(\bar{u}_1) + I(\bar{u}_2) + 2 \text{ Real part } \Gamma(\bar{u}_1,\bar{u}_2,\tau). \tag{2.4}$$

The measured intensity could have any value between 0 and 4I depending on the phase and amplitude of the mutual coherence function. The last term in equation (2.4) can be isolated by multiplying, rather than adding, the two signals; the imaginary part can be obtained by introducing a 90-deg phase shift in one of the signals. Thus the mutual coherence function of a radiation field can be easily measured.

2.2. The Van Cittert-Zernike Theorem

The arrays used in radio astronomy sample many parts of the wavefront of the radiation from a source. Thus, the sum (or product) of signals from any pair of antennas is related to $\Gamma(\bar{u}_1,\bar{u}_2,\tau)$ where $\bar{u}_1$ is the position of one antenna, $\bar{u}_2$ is the position of the other antenna and τ is the time difference between the samples measured with respect to a wavefront from some arbitrary direction. If the two points coincide, $\Gamma(\bar{u},\bar{u},\tau)$ is called the *autocorrelation function* of the field at $\bar{u}$. By the Wiener-Khinchin Theorem the autocorrelation function is the Fourier transform of the power spectrum of the field (e.g., Steel 1967, p. 44). On the other hand for the special case when $\tau=0$, $\Gamma(\bar{u}_1,\bar{u}_2,0)$ is called the *spatial coherence function* of the field.

The relationship between the spatial coherence function and the brightness distribution of an extended source is given by the van Cittert-Zernike Theorem. A derivation is given by Born and Wolf (1964, p. 510) or Steel (1967, p. 47). The theorem applies only if the radiation from the source is incoherent; that is the fluctuations of the radiation from different elements of the source are statistically independent.

For most astronomical applications, a radio source can be considered in the far field and the van Cittert-Zernike Theorem reduces to

$$\Gamma(\bar{u},\tau) = \int_{\nu=0}^{\infty} \left\{ \exp(i2\pi\nu\tau) \int_{\bar{\ell}} I(\bar{\ell},\nu)\exp(i2\pi\frac{\nu}{c}\,\bar{u}\cdot\bar{\ell})d\bar{\ell} \right\} d\nu \tag{2.5}$$

where $\bar{u}$ = the separation of the sampled field points = $\bar{u}_1-\bar{u}_2$, independent of origin
$\bar{\ell}$ = the direction cosine to an element of the source
ν = the frequency of the radiation
c = the speed of light
$I(\bar{\ell},\nu)$ = the brightness distribution of the source
$\Gamma(\bar{u},\tau)$ = the mutual coherence function.

The actual measurement of the spatial coherence function $\Gamma(\bar{u},0)$ in typical radio astronomy applications is complicated by the following: 1) the radiation probes (antennas) sample the field over a finite area with relative sensitivity $a(\bar{u})$, usually called the grading of the antenna; 2) the bandwidth of radiation accepted is finite with a relative sensitivity given by $B(\nu)$; 3) all signals must be combined with delay τ=0 with respect to some arbitrary direction, usually denoted as the phase center or map center. With these practical considerations, the measured response of the correlation of the signals separated by $\bar{u}$ is called the *visibility function* and is given by

$$V(u,v,w) = \int_{\nu=0}^{\infty} B(\nu) \left\{ \int_{\ell,m} \frac{1}{n} A(\ell,m,\nu)\ I(\ell,m,\nu)\ \exp\left\{ i2\pi\frac{\nu}{c}\ [u\ell+vm+w(1-n)]\right\} d\ell dm \right\} d\nu \quad (2.6)$$

We have resolved $\bar{\ell}$, the direction cosine to the source with respect to the phase center and $\bar{u}$, the separation of the two samples of radiation into the usual components

$\bar{\ell}$ = (ℓ,m,n) where the n-axis is parallel to the phase center direction, the ℓ- and m-axes define the plane parallel to n, ℓ directed to the east and m directed to the north. [Because the radio source is confined to the unit sphere, $n^2 = 1-\ell^2-m^2$.]

$\bar{u}$ = (u,v,w) where $u\|\ell$, $v\|m$, and $w\|n$.

The effect of the antenna grading is the multiplicative factor $A(\ell,m,\nu)$, called the antenna field pattern. If conventional radio telescopes are used as the radiation probe, A is near zero beyond a radius ℓ_{max} so that little information about $I(\ell,m,\nu)$ is measured beyond a *field of view* of radius ℓ_{max}.

2.3. Solution for the Brightness Distribution

It is clear from equation (2.6) that the visibility function measures something like a Fourier component of the brightness distribution. Aperture synthesis is, essentially, a methodical technique whereby a large number of samples in (u,v,w) are measured by 1) correlating the signals from all pairs (or at least all non-redundant pairs) of antennas in the multi-element array, 2) physically moving antennas to change the spacings between pairs of antennas, and 3) utilizing the rotation of the Earth to change the source-array aspect. With sufficient sampling, it is then possible to estimate the brightness distribution from the measured set of visibility functions.

However, an explicit expression for the brightness distribution can only be obtained with two further simplifications. In equation (2.6) the frequency ν and the angular coordinate $\bar{\ell}$ are coupled in the exponential term and cannot in general be separated. A quasi-monochromatic approximation applies when the change in phase of the exponential term is negligible over the radiation bandwidth $\Delta\nu$, which gives the condition

$$\frac{\Delta\nu}{c}\, u_{max}\, \ell_{max} << 1 \qquad (2.7)$$

where u_{max} = maximum baseline of the array
ℓ_{max} = maximum angular extent of the emission, usually limited by the field pattern of the antenna.

We may then define an *apparent brightness distribution* $I'(\ell,m)$ given by

$$I'(\ell,m) = \int_{\nu} B(\nu)\, \frac{1}{n}\, A(\ell,m,\nu)\, I(\ell,m,\nu)\, d\nu. \qquad (2.8)$$

The apparent brightness distribution is now a function of the array parameters as well as the radiation from the sky. In order to make any Fourier inversion of equation (2.6) meaningful the apparent brightness distribution must be invariant for all antenna pairs and be unchanged for the duration of the observations. This is accomplished, in part, by building an array with identical antennas and supporting electronics. Stringent mechanical and electrical tolerances are also necessary to achieve the necessary stability of I' to permit accurate aperture synthesis. Particular problems will be discussed in Section 4.

The explicit expression of the apparent brightness distribution in terms of the visibility function can now be written

$$I'(\ell,m)\, \delta(n-n') = \int V(u,v,w) \exp\left\{-i2\pi\frac{\nu_0}{c}\,(u\ell+vm+w[1-n])\right\} du dv dw \qquad (2.9)$$

where $n' = (1-l^2-m^2)^{\frac{1}{2}}$ so that the emission is confined to the unit sphere, δ is the Dirac-delta function and ν_0 is the average frequency of I'.

The assumptions made in obtaining equation (2.9) are:

(1) The radiation is ergodic
(2) The radiation is incoherent across the source
(3) The radio source is in the far field
(4) The signals of each pair are correlated with $\tau=0$
(5) The quasi-monochromatic approximation applies
(6) $I'(\ell,m)$ is well-defined and stationary.

The breakdown of assumptions 1 and 2 would be surprising except for non-natural signals. The limitation imposed by the far field assumption can be lessened if we know the distance to the radio source and can correct for the effects of the spherical wavefront in the measured visibility. Many of the problems of high resolution, high frequency aperture synthesis arise from a breakdown of assumption 4 because of the variable differential delay across an array caused by the propagation of radiation in the troposphere of the Earth. If assumption 5 is not valid, then there is a loss of information about the brightness distribution at large angular separations from the phase center. The degree of validity of assumption 6 is a large measure of the ultimate accuracy in which the brightness distribution can be recovered from the visibility function.

The apparent brightness distribution and the visibility function are three-dimensional Fourier pairs; however, since I' is only defined on the unit sphere, it is often possible to reduce the dimensionality of the inversion formula to two. For many aperture synthesis applications the term $w(1-n)\ \nu_0/c \ll 1$ anywhere in the field of view and can be neglected. The three dimensional inversion can also be avoided if the sampled visibility function is coplanar over the entire set of observations. In this case $w=au+bv$ where a and b are arbitrary constants and a redefinition of ℓ and m gives the relation

$$I'(\ell',m') = \int V(u,v,w)\exp\left\{-i2\pi\frac{\nu_0}{c}\ (u\ell'+vm')\right\} dudv \qquad (2.10)$$

with ℓ' and m' now not quite direction cosines with respect to the phase center. Brouw (1971) was the first to show that only an east-west array (or an array on a plane perpendicular to the rotation axis of the earth) describes a coplanar aperture under earth-rotation synthesis. Other ground-based arrays may, at any instant, sample the visibility function on a plane and may use a two-dimensional transform for data taken over a short duration. However, the combination of maps over many hours of earth-rotation synthesis require a continual redefinition of ℓ and m. For ground-based arrays which are not east-west and larger than several kilometers, the curvature of the earth invalidates the coplanar property of the array.

The measurement of the polarization characteristics of the radiation can be obtained by using probes which are sensitive to the various components of the field. No additional problems are encountered in aperture synthesis as long as the entire set of observations are consistent in their polarization properties.

The mutual coherence function may be described as a second-order correlation function of the field since it involves two signals. A generalization to higher-order correlators is possible (Glauber 1963). An example is the intensity interferometer (Brown and Twiss 1954). The instantaneous signal at two points is *first* detected and *then* multiplied together to give the correlation. The response R is

$$R(\bar{u}_1,\bar{u}_2,\tau) = \langle S^*(\bar{u}_1,t)\ S(\bar{u}_1,t)\ S^*(\bar{u}_2,t+\tau)\ S(\bar{u}_2,t+\tau)\rangle \qquad (2.11)$$

and can be shown to be equal to (Mandel 1963)

$$R(\bar{u}_1,\bar{u}_2,\tau) = I(\bar{u}_1)I(\bar{u}_2) + \left|\Gamma(\bar{u}_1,\bar{u}_2,\tau)\right|^2. \qquad (2.12)$$

Thus, only the amplitude of the mutual coherence function is obtained.

3. DEFICIENCIES OF APERTURE SYNTHESIS

3.1. Undersampling and the Two Basic Reconstruction Methods

Perhaps the most annoying and fundamental deficiency in aperture synthesis is the undersampling of the visibility function in the (u,v) plane. (For simplicity equation (2.10), the two-dimensional inversion formula will be used as the explicit solution for I', realizing that equation (2.9) should be used if necessary.) The determination of I' explicitly demands the value of the visibility function measured continuously within some region. Instead, it is sampled at discrete points within a radius of $\sim u_{max}$, along an elliptical track (one for each pair of antennas) in the (u,v) plane as the source moves in a diurnal path, with noticable omissions in coverage such as missing short spacings, a missing angular wedge, and other coverage anomalies such as perfectly-scaled tracks or regions of particularly dense sampling.

The conventional method of determining I' is the direct Fourier Transform (DFT) in which equation (2.10) is summed only for the P samples of the visibility function.

$$I''(\ell,m) = \sum_p \alpha(u_p,v_p)\ V(u_p,v_p)\exp\left\{-2\pi\frac{\nu_0}{c}\,(u_p\ell+v_pm)\right\} \quad 1<p<P. \qquad (3.1)$$

The *reconstructed image* I" is not in general a particularly good representation of I' but it does have the property that

$$I''(\ell,m) = I'(\ell,m) * B(\ell,m) \tag{3.2}$$

with

$$B(\ell,m) = \sum_p \alpha(u_p,v_p)\exp\left\{-i2\pi\frac{\nu_0}{c}(u_p\ell+v_pm)\right\} \quad 1<p<P \tag{3.3}$$

where $B(\ell,m)$ is the response to a point source (point-spread function or dirty beam) and $*$ means convolution. The apodizing function or weights $\alpha(u_p,v_p)$ can be chosen arbitrarily and are used to change some of the characteristics of I" within certain limits. For arrays composed of a line of antennas equation (3.1) can be conveniently summed in the radial and then azimuthal coordinates in the (u,v) plane (Kenderdine 1974). This is equivalent to obtaining a one-dimensional strip brightness distribution of the radio source and then summing the strip distributions over angle. Algebraic reconstruction techniques for this problem have been studied in electron microscopy (e.g., Gordon 1974).

The restoration of I' from I" is a problem of deconvolution (Högbom 1974). The image or map I" and the beam B contain ripples, called *sidelobes*, produced by spotty sampling of the visibility function. The sidelobes come in all shapes and sizes and the major features can be traced to particularly large gaps or noticable periodicities in the (u,v) coverage.

Non-Fourier techniques utilize equation (2.6), with I' defined by equation (2.8), to determine an I' which reproduces the observed visibility function at the sampled points. For example, model-fitting techniques (Fomalont and Wright 1974) start with the assumption that I' can be approximated by a small number of discrete components, usually Gaussian-shaped. The parameters which define the strength, position and size of the components are then adjusted using a search technique, (the simplex method has been found useful in this regard) to better fit the observed data (e.g., Nelder and Mead 1965). In the maximum entropy method (MEM; Ables 1974) a solution of I', usually represented in a pixel form, is obtained which satisfies equation (2.6) and maximizes the entropy of I'. The use of non-Fourier techniques, especially MEM, is becoming more prevalent as more efficient algorithms are being developed and as the properties of these methods are better understood. In addition, certain classes of errors associated with the measurement of the visibility function or with the addition of *a priori* knowledge of the nature of I' cannot be incorporated well in the standard Fourier methods.

3.2. Wide Bandwidths

The basic equations of aperture synthesis were derived for the quasi-monochromatic case of relatively narrow bandwidths. However, this constraint is not satisfied for many large arrays which use wide bandwidths to increase sensitivity. In general there is no way of recovering I' from V [using measurements of the spatial coherence function] without unrecoverable loss of resolution. The mutual coherence function (visibility function at many delay lags) must be sampled from which $I'(\ell,m,\nu)$ can be determined in narrow frequency bands, each of which satisfy the quasi-monochromatic assumption. The inversion formulae would be similar to equation (2.9) with the addition of another dimension of time t with the (u,v,w) domain and frequency ν with the (ℓ,m,n) domain.

Application of equation (3.1) in the wide bandwidth case leads to a reconstructed image which is the result of summing the apparent brightness distribution at each frequency but with the angular scale multiplied by ν/ν_0. This produces a radial smearing in I" apart from the other beam convolution effects. The extent of the radial smearing is proportional to the distance of the source from the phase center and the bandwidth. However, I" is still a convolution of the point source response

$$I'' = I'(\text{radially smeared}) * B \tag{3.4}$$

and standard deconvolution techniques are still useful. In a sense, no information is lost when wide bandwidths are observed; only the radial resolution degrades at angular separations far from the phase center.

3.3. Practical Considerations in Fourier Inversions

The DFT is extremely time consuming to calculate in a digital computer. Significant computation labor is saved by use of the Fast-Fourier Transform algorithm (FFT) which requires the visibility function to be sampled on an equi-spaced grid in the (u,v) plane (Cochran et al. 1967). The transformation of the originally sampled visibility function (by a convolution in the (u,v) plane and then a further sampling at the appropriate grid points) further distorts the reconstructed map (e.g., Fomalont 1973) as compared with that obtained with the DFT algorithm. Perhaps, the most annoying effect of the FFT is the *aliasing* caused by the periodic sampling. If I' contains significant response outside of the area in the field of view of the FFT (either real emission or sidelobes of I') this response will be reflected back into the FFT field of view with an amplitude related to the convolution function. Also, the errors in the reconstructed map will not be as uniformly distributed as with the DFT.

It is possible to use an optical system to perform the Fourier transform and a possible scheme has been investigated (ERIM report to VLA 1977). The visibility function is encoded on a piece of film, coherent light is then passed through the film and focussed by a lens onto an array of sensors. The method works but the accuracy of inversion is far surpassed by digital techniques.

3.4. Propagation Effects

A major source of error in aperture synthesis (at least ground-based aperture synthesis) is produced by changes in the propagation properties of the troposphere and ionosphere (Basart, Miley and Clark 1970, Hamaker 1978). For baselines longer than $\sim$ 1 km the radiation wavefront is continually distorted over minutes and hours by refraction from the dry-air component and the water-vapor component of the troposphere. At frequencies above 2 GHz the resulting differential delay of the radiation between the antenna pairs causes large phase errors in the measured visibility function. At frequencies above about 10 GHz and for baselines longer than 20 km, the phase of the visibility function may be virtually useless in some weather conditions. Generally, the visibility amplitude is unaffected except at frequencies above 5 GHz where water vapor absorption becomes important. At frequencies below about 0.3 GHz, the ionosphere of the earth produces similar fluctuations in the measured phase.

With increasing emphasis on high frequency, high resolution aperture synthesis reconstruction and restoration techniques which can handle data with large phase errors are necessary. Recent advances and understandings of various techniques will be reported during this colloquium.

3.5. Errors and Noise

All of the observations contain random short term errors caused by receiver noise and radiation from the sky and ground. These errors have the usual statistical behavior and their effects are easily analyzed in terms of the sensitivity limits of the observations.

Other errors produce variations of I' (see assumption 6 in Section 2.3) between the various antenna pairs and temporal changes caused by various instrumental imperfections. Obviously the design considerations for an array attempt to balance a loss of accuracy in determining I with various technical and economical considerations.

As a general rule, errors in the measured visibility function which vary slowly in time cause more grief than somewhat larger errors which have shorter characteristic time scales. For example, a pointing error in an antenna of an amount X of short term duration caused by gusty wind conditions would not have as large an effect on the accuracy of a radio map as a gradually changing pointing error much less than X lasting over the duration of the observations. It is often difficult to estimate the limitation in the brightness distribution accuracy caused by these non-random errors.

4. OTHER TOPICS

The previous sections have dealt with fundamentals and deficiencies of aperture synthesis. The main emphasis of this colloquium, however, is to discuss and determine optimum methods of arriving near the true brightness distribution in the presence of imperfections. Only one imperfection is really fundamental to aperture synthesis--the spotty sampling of the visibility function. Others are caused by imperfections in the array and the effect of the troposphere on the radiation.

The exalted, untouchable position of the Fourier solution in equation (2.10) in aperture synthesis has lessened over the last few years for the following reasons: the increased use of high resolution, high frequency observations has prompted a harder look at the general phase problem and its effect on the derived brightness distributions; the sensitive systems now in use enable maps to be made at sensitivity levels for which the desired brightness distribution is comparable or less than the sidelobes of bright sources in the field of view; and *a priori* knowledge about the brightness distribution (positive definiteness, small angular extent, large regions of no emission) are difficult to incorporate into the Fourier method.

4.1. Deconvolution Techniques

The deconvolution technique called clean (Högbom 1974) has been used successfully (with an element of danger) at most radio observatories. The algorithm is able to decompose a radio map into a sum of point-source responses with moderate efficiency, good convergence and now (Schwarz 1978) some understanding of its properties. At the present time, however, very little work has been done on the use of other deconvolution techniques.

The clean algorithm complements the Fourier-inversion solution since the radio map is a true convolution of the apparent brightness distribution and the point-source response. The FFT method and the use of the two-dimensional transform does perturb the exact convolution property; however, these deficiencies can be lessened if necessary.

Clean also works reasonably well for extended sources as long as the loop gain (the percentage of the maximum peak subtracted at each iteration) is set to about 10 percent. Unfortunately, computer time necessary for such cleaning is excessive. Convergence might be faster with more accurate results, if the clean algorithm could deconvolve with a variety of responses of sources with varying angular sizes. First use the broadest beam to clean out the more extended structure and gradually work down to the real point-source response.

4.2. The Phase Problem

The errors of phase of the visibility function have been the main impetus for investigations of non-Fourier reduction methods. By its nature, the standard Fourier inversion treats the amplitude and phase with the same weight and poor maps are generally obtained when non-random phase errors are larger than 20 deg. Other reconstruction methods offer the possibility of using the phase information with less (or no) weight.

The phase problem has long been encountered in VLB observations and optical interferometry. However, until recently these techniques acquired limited sampling of the visibility function and observed preferentially strong sources with relatively small angular diameters, so that the modest technique of model fitting the visibility amplitude was adequate. With increased VLB capability and correlation interferometers now working at optical wavelengths, there is also more pressure to develop efficient algorithms.

Observational and reduction schemes which help reduce the effects of phase errors are

(1) Frequent calibration on an unresolved point source near the radio source
(2) Monitoring the amount of water emission in the direction of the radio source
(3) Use of phase closure to measure accurately the sum of visibility function phase triplets (Fort and Yee 1976)
(4) Use of redundant spacings to separate the effects of the troposphere from the visibility phase
(5) Fourier maps of the square of the visibility amplitude (Baldwin and Warner 1978)
(6) Analytical analysis of the amplitude-spacing data (Bates 1969)
(7) Model fitting the visibility amplitude data
(8) The maximum entropy method
(9) Use of a point source, if any, in the field of view as a phase calibrator.

The major disadvantage to all of the schemes except (1), (2) and (9) is that there must be sufficient signal to noise (> 3 to 1) for *each* visibility function measurement before the techniques work well. Scheme (1) is limited because observations of the calibrator and source are not simultaneous and the calibrator and source are separated in the sky. Scheme (2) does not appear to work very well for a variety of reasons.

Scheme (9), the self calibration of the phase by a point source in the field of view, is promising and has been used in some special cases. The main trick in this method is to isolate the response of the point source by removing the effects of the other emission in the field of view. This might be done in the following way. First, clean the source in the usual way. Next, subtract all of the cleaned components, except the ones associated with the putative calibrator, from the visibility function. With this modified data remake the map which should now include only the point source with sidelobes characteristic of the (u,v) coverage *and* the amplitude and phase fluctuations during the course of the observations. Finally, clean the original map using this new point-source response and iterate.

It is not clear if this particular method would converge to anything useful. This hybrid scheme, I suspect, will be typical of the kind of schemes that will probably be required to make self calibration work.

Closure phase data has been used recently for VLB mapping using a hybrid model-fitting mapping technique (Readhead and Wilkinson 1978). The closure phase probably cannot be used for poor signal-to-noise data because of the 360 degree lobe ambiguities in the sum of the phase of the three visibilities.

4.3. The Three-Dimension Inversion

There is no fundamental problem in aperture synthesis with a non-east/west baseline. The three-dimensional transform of equation (2.9) should be used, if necessary, for skew-baselines. Inadvertent use of the two-dimension inversion would produce 'U'-shaped sources at large angular distances from the phase center. Unlike the bandwidth distortion, however, the convolution property of the map and beam are destroyed in this case.

The w sampling is sparse compared with that in the u and v directions so that the FFT algorithm in all three dimensions would lead to severe aliasing in the n direction. At the VLA the (u,v) inversion will be done using the FFT, the w inversion will be done using the DFT.

4.4. Nomenclature

The nomenclature and symbols used in contemporary aperture synthesis are not particularly uniform, but neither is there much confusion. In this article the term *brightness distribution* has been used for the radio emission I. This is certainly better than the older form of brightness temperature. Perhaps angular power distribution would be more descriptive. There is some confusion in the use of the appropriate sky coordinates. Here and in the general literature (ℓ,m,n) are used for the direction cosines. It is common in radio astronomy to use (x,y,z) except that z usually represents $(1-n)$. For the two-dimensional approximation of the sky, $(\alpha\cos\delta,\delta)$ are used. The triad (u,v,w) are commonly used as the ground coordinates associated with aperture synthesis. Here the units of absolute length were used in order to keep the explicit dependence of frequency in the equations. The units of wavelengths are more commonly used. The trend to use units of delay or time for (u,v,w) is confusing. The direction conversions of (u,v,w) and (ℓ,m,n) given in Section 2.3 are standard with north up and east to the left.

The representation of radio maps is now uniform. Some confusion was caused in the early days when Cambridge displayed maps in which the beams were circularized but the maps were squashed. One bone of contention is the common use of *Jansky per beam area* as an intensity unit. This is the most natural unit from the inversion process, however, some relationship to more externally natural units would be useful. Unfortunately temperature units or Jansky per steradian are not particularly appealing either.

4.5. Map Weighting Terminology

There is confusion among the observatories concerning the meaning of specific weighting (apodizing) schemes in the map making process. *Natural weighting* usually means that the weight of a visibility function sample is proportional to the duration of the sample. The DFT leads to this kind of weighting which gives optimum signal-to-noise for detection of an unresolved source. *Uniform weighting* should mean that the weight of a sample is proportional to the length of the (u,v) track during the period pertaining to the sample. This weighting gives about the highest resolution for the (u,v) coverage. *Smooth weighting* means that the sum of the weights per unit area is reasonably smooth over the entire aperture. This weighting tends to give lowest sidelobe levels. In addition, Gaussian weighting (taper) can be applied.

REFERENCES

Ables, J.G., 1974, Astron. Astrophys. Suppl., 15, 383.
Baldwin, J.E. and Warner, P.J., 1978, Astron. Astrophys. (in press).
Basart, J.P., Miley, G.K. and Clark, B.G., 1970, IEEE Trans., AP-18, 375.
Bates, R.H.T., 1969, Mon. Not. R. Astron. Soc., 142, 413.
Born, M. and Wolf, E., 1964, Principles of Optics, 2nd ed., Oxford: Pergamon.
Brouw, W.N., 1971, Doctoral dissertation, Leiden University.
Brown, R.Hanbury and Twiss, R.Q., 1954, Phil. Mag. (7), 45, 663.
Christiansen, W.N. and Warburton, J.A., 1955, Aust. J. Phys., 8, 474.
Cochran, W.T. et al., Proc. IEEE, 55, 1967.
ERIM, 1977, Study for an Optical Processor System for VLA Radio Telescope System.
Fomalont, E.B., 1973, Proc. IEEE, 61, 1211.
Fomalont, E.B. and Wright, M.C.H., 1974, Galactic and Extragalactic Radio Astronomy, Chap. 10, New York: Springer.
Fort, D.N. and Yee, H.K.C., 1976, Astron. Astrophys., 50, 19.
Glauber, R.J., 1963, Phys. Rev., 130, 2529.
Gordon, R., 1974, IEEE Trans. Nucl. Science, NS-21, 78.

Hamaker, J.P., 1978, Radio Science (in press).
Högbom, J.A., 1974, Astron. Astrophys. Suppl., 15, 417.
Kenderdine, S., 1974, Astron. Astrophys. Suppl., 15, 413.
Lequeux, J., 1962, Ann. Astrophys., 25, 221.
Mandel, L., 1963, Progress in Optics, 2, 181, Amsterdam: North-Holland.
McCready, L.L., Pawsey, J.L. and Payne-Scott, R., 1947, Proc. Roy. Soc., A-190, 357.
Nelder, J.A. and Mead, R., 1965, Computer Journal, 7, 308.
Read, R.B., 1961, IRE Trans. Ant. Prop., AP-9, 31.
Readhead, A.C.S. and Wilkinson, P.N., 1978, Astrophys. J. (in press).
Ryle, M. and Hewish, A., 1960, Mon. Not. R. Astron. Soc., 120, 220.
Schwarz, U.J., 1978, Astron. Astrophys. (in press).
Stanier, H.M., 1950, Nature, 165, 354.
Steel, W.H., 1967, Interferometry, London: Cambridge Univ. Press.

DISCUSSION

Comment L.R.D.'ADDARIO
(1) "Aperture synthesis" is a misnomer, in that the synthesis telescopes in operation today have no equivalent aperture. A better term would be "Fourier synthesis". (Ryle's original use of the term "aperture synthesis" was in a different context.)
(2) The direct Fourier transform should not be abbreviated "DFT", since the latter has long been used in the literature to mean discrete Fourier transform.
(3) I think it is more useful to consider the fundamental equation of synthesis as two dimensional, because (a) the sky is two-dimensional, and (b) measurements in a plane are sufficient. The problem is that we find it impractical to make all our measurements in one plane.
Reply E.B. FOMALONT
I agree with your first two points. I see no way of avoiding an integration in the w-direction for a non-planar sampling of the spatial coherence function in order to obtain an explicit solution for the brightness distribution. For this reason I would consider the 3-D expression more fundamental.

Comment T.W. COLE
The van Cittert-Zernike theorem contains severe restrictions to plane apertures and narrow bandwidths. The fundamental theory can be extended to include the real situation of finite fields of view and finite bandwidths. This is very important now that techniques of point by point Fourier transformation are becoming faster. With such methods compensation is possible for aberrations such as the radial beam smearing. Would you agree that there is still a lot to be done in developing the fundamental theory?

Reply E.B. FOMALONT
I would disagree that there is much to be done in developing the fundamental theory, the van Cittert-Zernike theorem for use in radio astronomy. For example, I believe the radial smearing caused by large bandwidths is unavoidable. I do agree that the smearing is correctable in a variety of ways which I would consider as a deconvolution process.

THE TWO DIMENSIONAL REPRESENTATION OF THREE DIMENSIONAL INTERFEROMETER MEASUREMENTS

R.H. Frater
School of Electrical Engineering
University of Sydney

SUMMARY

A convolution technique for the reduction of three dimensional interferometer measurements to two dimensions is described. With the addition of relatively simple hardware to a general purpose computer the technique allows fast, efficient processing of three dimensional data.

1. INTRODUCTION

In the reduction of rotational synthesis data, measurements are often projected onto a plane normal to the direction of the source, then Fourier transformed for the final map. If the measurements are themselves on a plane having the same origin as the projected plane, a known distortion of the map co-ordinates results. When the measurements are not on such a plane, dispersion effects result. The production of a correct map in this situation requires the use of a transform from a three dimensional measurement set to a two dimensional map plane (Brouw, 1975, Clark,1975, Frater, 1978)

To compute the required transform, a set of two dimensional transforms may be carried out on the three dimensional data. A final stage involving multiplication and summation operations leads to the required map. Alternately, this final stage can be replaced by convolution and summation operations before the transforms. The field size limits in radio interferometry are such that the required convolution process can be greatly simplified resulting in an efficient approach to the processing of three dimensional data.

2. MEASUREMENTS IN (u,v,w) SPACE

If a set of measurements is made of a spatial autocorrelation distribution P(u,v,w) in a co-ordinate system (u,v,w) with the w co-ordinate directed to the centre of the field to be mapped (see Fig.1),

C. van Schooneveld (ed.), Image Formation from Coherence Functions in Astronomy, 19-26.

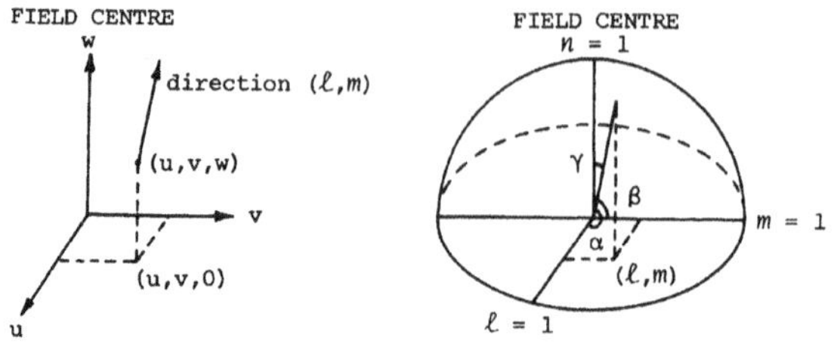

Figure 1. The co-ordinate system used for measurements made in (u,v,w) space

an estimate of the sky brightness distribution $B(\ell,m)$ can be obtained from the relationship

$$B(\ell,m) = \iiint P(u,v,w)g(u,v,w)\exp -2\pi i(u\ell+vm+w(1-\ell^2-m^2)^{\frac{1}{2}})\, dudvdw \quad (1)$$

where ℓ,m are direction cosines specifying a point in the map (ℓ=0, m=0 at the map centre) and $g(u,v,w)$ is a graded sampling function.

The term $-2\pi(u\ell+vm+w(1-\ell^2-m^2)^{\frac{1}{2}})$ may be identified as a phase. The w component may be expanded to give

$$w(1-\ell^2-m^2)^{\frac{1}{2}} = w - w(\ell^2+m^2)/2 - w(\ell^2+m^2)^2/8 - \ldots \quad (2)$$

The error resulting if only the first two terms are retained is small in practical cases. For example if $w=10^5$ wavelengths an error of 1° occurs at the corner of a 1.8° field while for $w=10^4$ wavelengths the error is 1° at the corner of a 3.1° field.

The first w in equation 2 can be represented as a phase correction to the data included in $g(u,v,w)$ so that equation 1 can be rewritten

$$B(\ell,m) = \iiint P(u,v,w)g(u,v,w)\exp -2\pi i(u\ell+vm-w(\ell^2+m^2)/2)\, dudvdw \quad (3)$$

To clarify operations in u and v for a fixed w, write

$$P_w(u,v)g_w(u,v) = P(u,v,w)g(u,v,w) \quad (4)$$

and for a fixed w using F to denote Fourier transformation, write

$$\exp 2\pi iw(\ell^2+m^2)/2 = F^{-1}\{C_w(u)C_w(v)\} \quad (5)$$

Now using ** to denote two dimensional convolution, the convolution theorem yields

$$B(\ell,m) = \iiint \{P_w(u,v)g_w(u,v)\}**\{C_w(u)C_w(v)\}\exp -2\pi i(u\ell+vm)\, dudvdw \quad (6)$$

$$= \iint P_c(u,v)\exp -2\pi i(u\ell+vm)\, dudv \quad (7)$$

The required convolution will be recognised as successive one dimensional convolutions in u and v with functions of the form

$$C_w(u) = F(\exp 2\pi i w \ell^2/2) \tag{8}$$

Equation 6 can perhaps be interpreted best by considering discrete values of w. For each w there are convolving functions $C_w(u)$ and $C_w(v)$ which are convolved with the correlation function for that w. The resultant values for each (u,v) are summed in w completing the outer integral and yielding a two dimensional correlation function $P_c(u,v)$ which represents the original three dimensional data set. $B(\ell,m)$ is now obtained by a two dimensional transform. The grading may, in fact, be conveniently applied at this stage.

At first sight the above procedure would seem to offer little when compared to a conventional three dimensional transform. Substantial advantage arises, however, by suitable choice of the convolving functions $C_w(u)$ and $C_w(v)$ and the implementation chosen for the process.

3. THE CONVOLVING FUNCTION

In equation 6, no bounds were set for ℓ and m and the convolving function is indeterminate. In the practical situation, data in the (u,v,w) domain must be convolved onto a three dimensional grid. The grid spacing determines the map dimensions. If co-ordinates u,v and w are expressed in wavelengths and ℓ and m in radians, the map width $(2L)$ will be the inverse of the grid spacing ΔF. The resultant transform will be replicated at intervals $(2L)$.

It is seen from equation 8 that for maps of equal dimensions in ℓ and m the convolving functions $C_w(u)$ and $C_w(v)$ are identical. Further consideration will therefore be restricted to $C_w(u)$. The function

$$\phi(\ell) = w\ell^2/2 \tag{9}$$

can be approximated in the range $-L$ to L by the Fourier series.

$$H(\ell) = \Phi/3 + \sum_{n=1}^{\infty} ((-1)^n 4\Phi/(\pi n)^2)\cos(n\pi\ell/L) \tag{10}$$

where $\Phi = wL^2/2$

The convolving function $C_w(u)$ can now be written

$$C_w(u) = F\{\exp(2\pi i H(\ell))\} \tag{11}$$

$$= F\{\exp(2\pi i\Phi/3) \times \prod_{n=o}^{\infty} \exp((2\pi i(-1)^n 4\Phi/(n\pi)^2)\cos(n\pi\ell/L)) \tag{12}$$

$$= F\{\exp(2\pi i\Phi/3)\} * F\{\exp((-2\pi i 4\Phi/\pi^2)\cos(\pi\ell/L))\}$$

$$* F\{\exp((2\pi i 4\Phi/4\pi^2)\cos(2\pi\ell/L))\} * \ldots \tag{13}$$

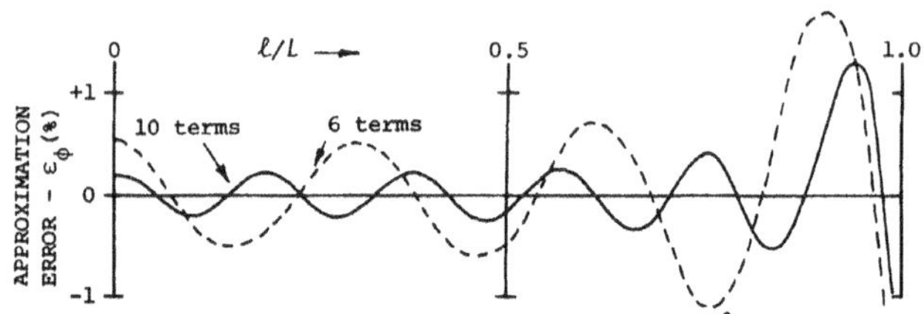

Figure 2. The error in the approximation of $w\ell^2/2$ by Fourier series with (a) 6 terms and (b) 10 terms.

Each of the transforms in equation 13 gives rise to a spectrum whose harmonic components have amplitude $J_n(4\Phi/\pi n^2)$ where J_n are Bessel functions of the first kind.

For small values of Φ (and hence small values for the arguments of the Bessel functions) the total spectrum from equation 13 approximates the sum of spectra resulting from the individual components since for small x

$$J_n(x) \approx (x/2)^n/n! \tag{14}$$

and only $J_1(x)$ need be considered. The spectrum thus falls of in amplitude as n^2.

The question now arises as to how many terms should be retained in order to have a manageable convolving function.

It is seen that the first order effect of restricting the number of components in the convolving function to 2n-1 is to restrict the number of terms in equation 10 to n. This gives rise to an error that can be calculated as

$$\varepsilon_\phi = 100\{H_n(\ell)-w\ell^2/2\}/(wL^2/2) \tag{15}$$

where subscript n denotes the number of terms retained in $H(\ell)$. This error is plotted in figure 2 for 10 terms and for 6 terms.

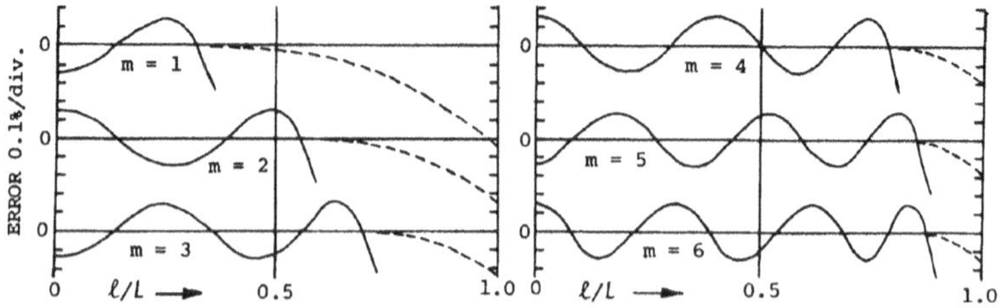

Figure 3. The error in the approximation for $w\ell^2/2$ as a percentage of the field edge phase Φ. The dotted curves are reduced in scale by 100.

m	a_0	a_1	a_2	a_3	a_4	a_5	a_6
1	0.2221	0.2236					
2	0.2845	0.3160	0.0330				
3	0.3085	0.3580	0.0600	0.0120			
4	0.3194	0.3782	0.0761	0.0227	0.0067		
5	0.3248	0.3884	0.0852	0.0300	0.0117	0.0046	
6	0.3274	0.3936	0.0900	0.0342	0.0152	0.0071	0.0036

Table 1. Coefficients for the approximation of $w\ell^2/2$ for orders 1 to 6 with 0.15% ripple.

While the Fourier series gives a least squares fit to the phase in equation 11 it is by no means optimum for present needs. It is possible to optimise series having a limited number of terms to meet a prescribed error over the central part of the range in ℓ at the expense of the extremes in ℓ. The result of such an optimisation with an error limit of 0.15% is shown in Fig. 3 for phase approximation functions of order 1 to 6. These functions are described for order m by the equation

$$H_m(\ell) = \sum_{n=0}^{m} (-1)^n a_n \cos(n\pi\ell/L) \qquad (16)$$

normalised for unity phase at $\ell=L$. Coefficients for orders 1 to 6 are given in table 1.

The functions may be optimised at various levels of maximum error. However, it must be considered that these errors are not random and limits must be determined with this in mind.

The convolving function can now be calculated from the equation

$$C_w(u) = F\{\exp 2\pi i \Phi H_m(\ell)\} \qquad (17)$$

The spectrum of $C_w(u)$ will have (2m+1) terms for small Φ but other terms will become significant for larger Φ. If the convolving function is truncated to (2n+1) terms for all values of Φ then this will

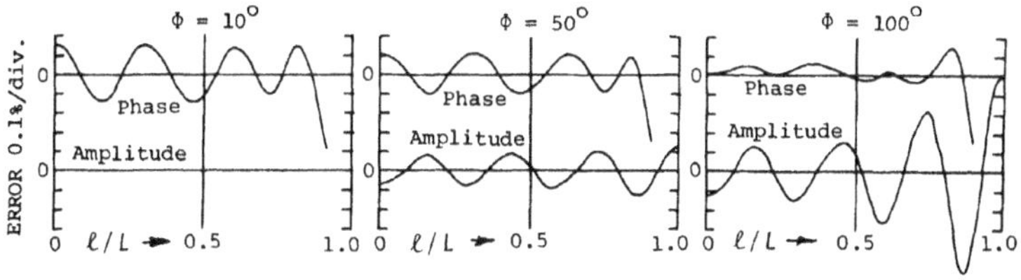

Figure 4. The effect of truncating the 6th order convolving function to 13 points. As Φ increases, the phase error reduces to a minimum at $\Phi=100^{\circ}$ while the amplitude error increases. The truncation error calculated in equation 18 corresponds to the region where ℓ is close to L.

give rise to phase and amplitude errors in the final map. Fig. 4 shows the total error including the truncation error produced using the convolution approach when compared to the exact computation, for a sixth order approximation function and for three values of Φ. For small Φ the error is in fact the error of the approximation function itself. At $\Phi=50^o$ the percentage phase and amplitude errors are equal while at 100°, the phase error has almost been nulled while the maximum amplitude error has risen to 0.5%. (0.3% within the range of interest). Similar behaviour occurs for other orders. The maximum truncation error ε_T is given approximately for values of n up to 6 by

$$\varepsilon_T(\%) = \{45.9/n + 0.85n - 7.05\}\Phi^2/10^5 \qquad (18)$$

In the range of interest, truncation principally affects the amplitude.

The final choice of convolving function will depend on the area of the map over which high accuracy is required and the value of Φ, which itself depends on the type of implementation chosen.

4. IMPLEMENTATION

In order to make use of either the convolution approach or a three dimensional transform, the measured data must be convolved onto a three dimensional "grid". The number of samples required in the w dimension depends on Φ and the extent of the convolving function used in the w dimension. For a two point function (corresponding to linear interpolation) the spacing in w giving 1° maximum phase error would be that corresponding to a difference of 35° in Φ for the two values of w.

There are two ways of utilizing the convolution approach. The first is to calculate a separate convolving function for each value of w, perform a two dimensional convolution with the appropriate function on each w "layer", then combine all data onto the w=0 plane. A two dimensional transform to obtain the map completes the process. This last transform requires only Real output. The second approach involves calculation of the appropriate convolving function for the grid spacing in w. Denoting the w=0 plane as w_o, the next as w_1 etc., this convolving function, which will be referred to as the Δw function, is that required for the w_1 plane. If the same convolution is performed twice, this corresponds to the required convolution for the w_2 plane.

To use this approach, the data from the w_n plane is first convolved with the Δw function. The data from the w_{n-1} plane is added and the convolution repeated. This process is repeated until the addition of the w_o data when again a two dimensional transform is used to produce the final map. The approach has the advantage that the truncation errors are very low. However, attention must be paid to rounding errors in the successive convolutions if a large number of w planes is involved. The best utilization of the Hermitian property of the measurements will be achieved by minimising the number of w layers.

Otherwise, special attention will be needed to "wrap around" the convolution in an appropriate way.

To obtain maximum benefit from the convolution approach some special purpose hardware is required on a general purpose computer. The function of this hardware can be simply described. It accepts, under d.m.a. transfer, a row or column of data from the computer, convolves it with the appropriate function and returns it under d.m.a. to the computer. It can be shown that the effective real multiply time required for the convolver to handle the data as fast as it can be accessed is $t_c/(4(2m-1))$ where t_c is the "read plus write" time for the computer. If t_c is 1.5 usec the multiply time is 29nsec. This time can be effectively achieved in hardware using eight separate 16 bit 240nsec multipliers.

It can be seen that for the purpose of comparison with the three dimensional transform approach the convolution can be completed within the access times required to send data to FFT hardware, resulting in a time saving equal to the transform time.

5. CONCLUSION

A new approach to the processing of three dimensional data has been discussed in relation to rotational synthesis applications although the principles involved may be rather more widely applied where three dimensional transforms are used.

The approach will allow high speed processing with a saving both in time and fast memory requirements. The availability of a two dimensional data set from original three dimensional data may be of value in optical processing and indeed the processes leading to the two dimensional data set may themselves be implemented in optics.

ACKNOWLEDGEMENTS

The author wishes to thank many people in the School of Electrical Engineering for useful discussions and indeed arguments in the course of the preparation of this paper. Thanks go in particular to I.S. Docherty for detailed discussions of transform theory, D.J. Skellern, J.G. Lucas, W.N. Christiansen, R.F. Mullaly and I.G. Jones for their many useful comments and criticisms.

REFERENCES

Brouw, W.N.: 1975, "Aperture Synthesis" in Methods in Computational Physics, 14, 131.
Clark, B.G.: 1975, VLA Computer Memorandum, No.127, NRAO, U.S.A.
Frater, R.H.: 1978, Astronomy and Astrophysics. In print.

DISCUSSION.

Comment L.R. D'ADDARIO
As this is an alternative to the three-dimensional transform, I would like to comment on each method. 1) The three-dimensional integral over the visibility measurements written by several of today's speakers cannot represent correctly the brightness distribution, if only because the physical units are incorrect. This is a consequence of the fact that the sky is inherently two-dimensional. Nevertheless, the map obtained by summing over all measurements with the three-term exponential kernel has properties much better than those obtained with the two-term exponential (ignoring the w-term). 2) With respect to your method, it is clear that measurements made everywhere in a w = const. $\neq 0$ planes can be "reduced" to the $w = 0$ plane by convolution. But what happens if the w = const. slice has measurements at only a few values of (u,v), as actually happens with rotation synthesis?
Reply R.H. FRATER
The method makes no assumption about the location of measurements in (u, v, w) space. Each measurement made for $w \neq 0$ is represented in the $w = 0$ plane by a set of "measurements" which produce the same fringe as the original measurement over most of the map area.

Comment T.W. COLE
My background makes it easier to visualize the method as a Fresnel transform or as propagation of spherical waves from one plane to another. But is it not true that the sampled measurements rather than continuous measurements still give you a map with beamshape varying over the field?
Reply R.H. FRATER
Since the measurement set in the $w = 0$ plane accurately represents the original set over most of the map the original property of a varying beamshape is retained. The new measurement set is not the set that would be obtained by measuring originally in the $w = 0$ plane.

MEASUREMENT ERRORS IN ROTATIONAL SYNTHESIS AND THEIR EFFECTS IN THE MAP PLANE

(Invited paper)

J.P. Hamaker
Netherlands Foundation for Radio Astronomy
Dwingeloo, the Netherlands

SUMMARY

The effect of errors in an aperture synthesis map is intimately related to the geometry of the synthesis instrument and the properties of the two-dimensional Fourier transformation. This paper concentrates on E-W rotational synthesis, the simple geometry of which allows a systematic treatment. After some preliminary remarks about error statistics and tolerances, the relevant properties of the Fourier transformation in polar coordinates are first reviewed. A classification of error types as "global", "circular", "radial" or "local" is introduced, based on the geometry of the area in the aperture over which they are coherent. A few special errors not fitting into this classification are also discussed. The characteristics of the corresponding patterns in the synthesized beam are described and illustrated with computer simulations and real-life examples.

1. PROLOGUE

1.1. Introduction

In the most general context of this Colloquium, very little can be said about errors that is worth writing about. Although most of them spring from sources common to all observing systems, the distribution of the aperture points they affect is entirely determined by the array configuration and, consequently, so are the error patterns in the map. I shall therefore narrow the scope of my contribution to East-West rotational synthesis, using the Westerbork Telescope in its original 10 x 2 configuration (Baars et al. 1973; Högbom and Brouw 1974) as a representative example.

Furthermore, I shall confine myself to continuum observations. Multichannel observations, either line or polarisation, do not differ fundamentally; in practice many errors track between channels and cancel in the subtractions that are an inherent step in the reduction (Weiler, 1973; Rots, 1974; for an example see fig. 3.7. below). The difficulties of these observations are associated mainly with the properties of the

C. van Schooneveld (ed.), Image Formation from Coherence Functions in Astronomy, 27–45.

antenna elements (e.g. primary beam polarisation, standing waves) or with fundamental difficulties in synthesis (the missing short spacings), both of which are outside the scope of this paper.

Even with all these restrictions, classifying and describing the errors is by no means simple. One fundamental reason is that the two-dimensional Fourier and convolution integrals in polar coordinates are not separable; a direct consequence is that there is no such thing as a scaling theorem for the angular coordinate. These complications make it extremely difficult to visualise the workings of the Fourier transformation. Moreover, the effect of the convolution of error beams and source brightnesses is not always easy to predict, as illustrated by fig. 1.1.

All these difficulties together probably account for the virtual non existence of litterature to be reviewed on the subject. My presentation will be mostly based upon our experience with the Westerbork Telescope. Before starting, a few general remarks are in order.

1.2. The meaninglessness of statistics

In many fields of science, errors can be treated meaningfully in terms of the theory of stationary random processes. The assumption that this is also true in aperture synthesis is implicit in many discussions and statements about map errors. In fact, however, the non-stationarity of the important errors is perhaps their most prominent general property.

Since most errors are related, directly or indirectly, to weather influences, they are highly nonstationary in time. Sensitivity of the equipment to these influences differing from antenna to antenna, there is spatial nonstationarity along the array. As a result of this and of the relatively small number of independent error parameters, a few errors usually dominate the scene in a map; the conspicuous patterns they produce are again anything but stationary. It is meaningless to describe them in terms of the familiar averages and variances of random process theory.

The concept of *dynamic range* is sometimes introduced and vaguely defined as the ratio of peak to weakest detectable brightness or flux in a map. A proper definition must state *what* one wants to detect *where* in the map (and perhaps also what constitutes a "detection").

The only possibly sensible use of statistical averages is in a comparison of large sets of maps, e.g. those made by different instruments or by one instrument at different frequencies, as was done by Hinder and Ryle (1971) in predicting the performance of synthesis arrays as a function of their size.

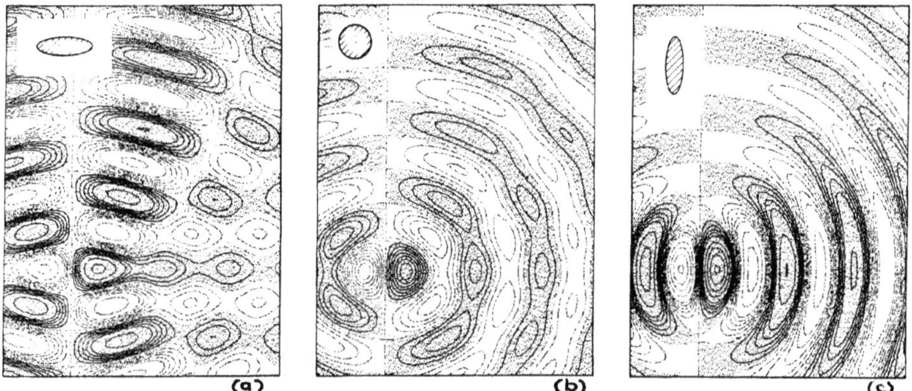

Fig. 1.1. The effect of convolving an error beam with a source distribution. Center: The error beam for an imaginary error $\Delta\varepsilon = 1$ in one interferometer; the circle indicates the half-power width of the corresponding interferometer beam J_0 $(2\pi Rr)$ (cf. sec. 2.2.). Contour interval is 1%; positive areas are shown shaded. Left and right: The result of a convolution of this error beam with elliptical gaussian sources of half-power sizes as shown.

1.3. Tolerances for synthesis telescopes

The definition of tolerances is in a sense the inverse of error analysis, and therefore equally difficult for a synthesis instrument. What is specified usually reflects what is considered feasible, and is at best supported by some superficial computer simulations.

Stability goals in the order of 1% for gain and 1 mm for phase are typical and have proved to be adequate for important astronomical work in the lower GHz range. As observing frequencies are pushed up, more stringent phase specifications must be met in order to retain a good dynamic range. At the lower frequencies the same may become necessary for more subtle investigations in the wake of the earlier great discoveries.

The general properties of the source population must also be considered. At high frequencies sources are weaker with respect to receiver noise and relatively large beam errors may be tolerable. At low frequencies the field of view contains many more sources and observations become confusion-limited. For satisfactory "cleaning" a very tight control of error sidelobes is required.

1.4. Approach and layout of this paper

In the simple geometry of rotational synthesis, a reasonable approach is to consider the radial and azimuthal coherence of the various errors and the associated error beam patterns. (Note that I prefer the use of "coherence" rather than "correlation", which has a strong statistical connotation). Analytical expressions will be derived only where they are simple enough to be readily visualized. Where this is not the case, computer-simulated error pattern displays will be included.

Chapter 2 reviews the relevant properties of the two-dimensional Fourier transform. Chapter 3 discusses the various errors on the basis of a spatial coherence classification. Because of space and time limitations, the originally planned discussion of the more conventional processing methods of error control will be submitted elsewhere (Hamaker, 1979).

Terminology: The term *telescope* will be reserved for a synthesis *array* as a whole, the individual elements of which are the *antennas*. Both the separation vector of an interferometer and its length will be referred to as the *baseline*. The domain of the measured *complex visibilities* is the *aperture* (plane). The *map* is the image that is obtained by a two-dimensional Fourier transformation of the visibilities.

2. THE TWO-DIMENSIONAL FOURIER TRANSFORM

For the basic theory of Fourier transformations several excellent texts exist (e.g. Bracewell). I review here some of the specific aspects relevant to the present application.

2.1. Fundamental properties and assumptions

Symmetries. Brightness B being real, the visibility V is hermitian: $V(u,v) = V^*(-u,-v)$. An important consequence is that only one of each pair of points (u,v), (-u,-v) need be sampled; in rotational synthesis, a full observation lasts 12 rather than 24 hours. Re V, when transformed back, yields the *even* or *symmetrical* part of B; Im V reproduces the *odd* or *antisymmetrical* part.

"Energy". In the context of this paper this term may give rise to confusion. Physically, it is closely related to *flux* and *brightness*, whereas a common interpretation of certain theorems in Fourier theory relates it to the squares of these latter quantities. I shall use the term in its Fourier sense, retaining the quotes to remind you of the non-physical meaning to be attached to it.

The synthesized beam. With the standard assumption of *shift invariance*, *stationarity* or *isoplanasy*, the telescope's effect in the output map is described by a convolution with a *synthesized beam*, which ideally is the transform of a *taper* function W. Errors modify the latter by multiplying it with an *error function* E. For the illustrations in this article, the standard Westerbork taper (a Gaussian with a value of 0.25 at the aperture edge) will be assumed.

A few errors do not fit in the framework thus provided. They will be dealt with separately in sec. 3.5.

The error beam. The error function E should ideally be identically equal to 1. It is convenient to consider the *deviations*, that is, we write

$$E = 1 + \Delta g + i\Delta\varepsilon \qquad (2.1.1.)$$

and define the *error beam* S_e as the transform of $W.(\Delta g + i\Delta\varepsilon)$. In a first-order approximation (Δg, $\Delta\varepsilon << 1$) we may identify Δg with *gain* and $\Delta\varepsilon$ with *phase* errors. The small-error approximation Δg, $\Delta\varepsilon << 1$ will be used in most of this paper, although the displays of beam error patterns will be normalized to unity errors.

It is convenient to consider E as continuous, so that S_e has no grating responses. One must remember, however, that such responses are generated through the convolution with the error-free beam. For those error patterns which are confined to a small area around the beam centre, the effects may be more conspicuous in the grating rings than in the central response.

2.2. The transform in polar coordinates

Using polar coordinates, the general Fourier transform relation between a map or beam pattern $P(R,\Phi)$ and its aperture counterpart $Q(r,\phi)$ reads

$$P(R,\Phi) = \int_{-\pi}^{\pi} d\phi \int_{0}^{\infty} rdr \quad Q(r,\phi) \exp\,(2\pi i rR \cos(\Phi-\phi)) \qquad (2.2.1.)$$

(The projection factor sin δ is not introduced here since it tends to obscure the inherent circular symmetries; it is better applied as a coordinate correction in the final result).

For the synthesized beam, the angular dependences disappear. The way the taper $W(r)$ transforms into the beam $S(R)$ can be visualised in two ways. Integrating over ϕ one gets

$$S(R) = 2\pi \int_{0}^{\infty} rdr\; W(r)\; J_o\,(2\pi rR), \qquad (2.2.2.)$$

showing S as a sum over all baselines of a series of circularly symmetric patterns with cross sections described by the zero-order Bessel function J_o. On the other hand, substitution of the one-dimensional Fourier transform

$$H(X) = \int_{-\infty}^{\infty} |r|\,dr\; W(r)\; \exp 2\pi i r X \tag{2.2.3.}$$

into (2.2.1.) yields

$$S(R) = \int_{-\pi/2}^{\pi/2} d\phi\; H(R\cos\phi). \tag{2.2.4.}$$

Thus S can also be viewed as the sum over all position angles of "*corrugations*" with cross sections H(X) (cf. Bracewell and Thompson 1973). These two quite different visualisations are relevant to the understanding of the effect of "circular" and "radial" errors, respectively (see Ch. 3 below).

A more general relation which is sometimes useful is

$$\int_{-\pi}^{\pi} d\phi\; \cos n(\phi-\alpha)\; \exp\,(2\pi i r R \cos(\Phi-\phi)) =$$

$$= 2\pi i^n\; J_n(2\pi\; rR)\; \cos n(\Phi-\alpha). \tag{2.2.5.}$$

For discrete radial sampling the pattern H(X) in (2.2.3.) becomes periodic, the repetitions giving rise to grating rings (Bracewell et al. 1973). The angular sampling is normally chosen dense enough to render its effects negligible.

2.3. Basic qualitative relationships between patterns in the two domains

Most of the relationships to be discussed here can be viewed as manifestations of the *scaling theorem* of Fourier theory. Qualitatively, this theorem states that the Fourier transform converts spatially extended patterns into narrow features and vice versa, while conserving the total "energy" in them. The precise meaning of "extended" and "narrow" is defined by the transformation formulae. I consider some examples.

The first is provided by fig. 2.1., which illustrates the familiar fact that a radial pattern transforms into another perpendicularly directed radial pattern. The same can be derived analytically from 2.2.3. and 2.2.4.

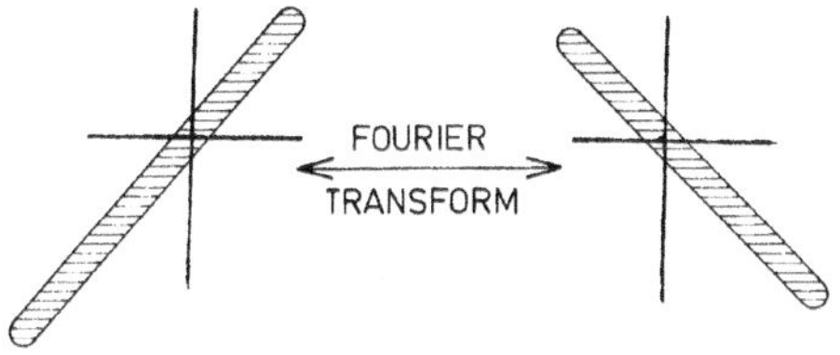

Fig. 2.1. See text

For stationary random noise, the power distribution in the beam is the transform of the autocorrelation function of the tapered error function E.W. and vice versa. If there is little correlation between error samples at different points, the autocorrelation function is concentrated in a narrow area and the beam noise is spread out widely. Also since the error "energy" is spread over the whole plane, the noise values in the beam are uncorrelated between points.

Upon the introduction of correlation between error samples the autocorrelation function widens, and consequently beam power concentrates toward the origin, while remaining uncorrelated between points. On the other hand, as one reduces the extent of the error distribution, the beam noise, while retaining the same extent, becomes more and more correlated between points. One manifestation of this is, that no noise "blobs" of dimensions smaller than the main response may occur anywhere in the synthesized beam, - a fact which is well-known in the analogous situation for a map.

Equation (2.2.2.) does not allow a simple interpretation in terms of the scaling theorem. It does indicate that circularly symmetrical features transform into patterns with similar symmetry, and is in this sense a counterpart to the radial pattern example above. More generally, (2.2.5.) indicates that angular modulation is conserved in the transformation. This also follows from elementary symmetry considerations.

3. ERRORS AND ERROR PATTERNS IN ROTATIONAL SYNTHESIS

In this chapter I consider the various types of errors in terms of the perturbations $\Delta g + i\Delta\varepsilon$ and the corresponding error beams S_e (sec. 2.1.).

3.1. General classification

The errors may be grouped in various ways, e.g. according to their origin, magnitude, etc. Guided by the developments of the previous chapter, I choose to classify them on the basis of their coherence properties in the visibility plane (c.f. fig. 3.1.).

Global errors: Errors which coherently affect the whole aperture or a large part of it. Example: Clock errors.

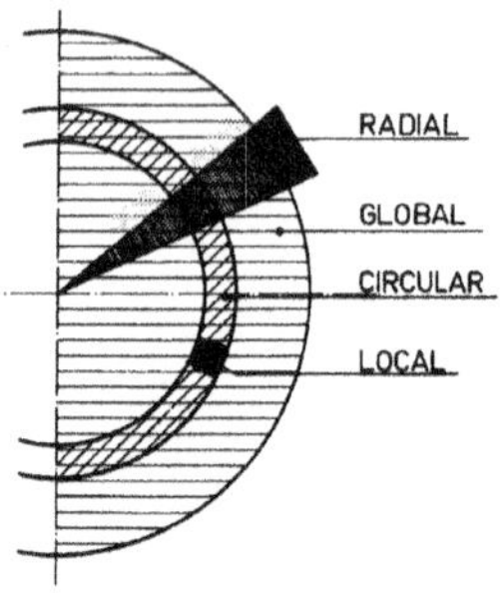

Fig. 3.1. Classification of error patterns in the aperture plane.

Radial errors: Errors coherent over a sector of the aperture; i.e. affecting many interferometers simultaneously over a limited period of time. Example: large clouds.

Circular errors: Errors coherent over large parts of circles in the aperture, i.e. affecting single interferometers over a sizable fraction of a full 12-hour period. Examples: Antenna position errors, electronic drifts.

Small-scale errors: Errors which are both short-lived and spatially confined, and therefore are incoherent in all directions in the aperture. Example: Small clouds.

3.2. Global errors

Coordinates. The orientation with respect to the source and the scale of the telescope baseline are affected by a considerable number of effects that are accurately predictable (precession, nutation, aberration, equation of equinoxes), measurable (observatory clock, polar motion, baseline in wavelengths), or (probably) small enough to be ignored at present (Earth tides).

I consider here two errors that are relatively difficult to calibrate accurately enough, and as a result frequently manifest themselves in Westerbork maps.

Clock errors result in map offsets in right ascension. When two or more observations with unequal offsets are combined, the most conspicuous effect is the incomplete cancellation of grating rings. The distortion of the main synthesized response becomes apparent when one tries to remove a strong point source from the combined map.

Tilts of the telescope as a whole with respect to the equatorial plane produce phase errors proportional to interferometer baseline. The map error consists of a slight source position shift and an antisymmetric East-West tail pair, see fig. 3.2.

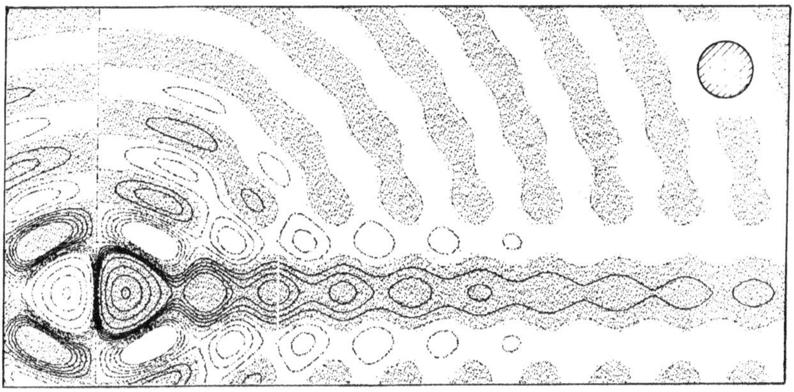

Fig. 3.2. The error pattern generated in a Westerbork beam by a baseline tilt out of the equatorial plane; the circle indicates the half-power size of the undisturbed beam. The pattern corresponds to errors $\Delta\varepsilon = r/r_{max}$ in the aperture. Contour levels are 0, $\pm$ 0.01, ..., $\pm$ 0.05, $\pm$ 0.10, ..., $\pm$ 0.25; positive areas are shown shaded.

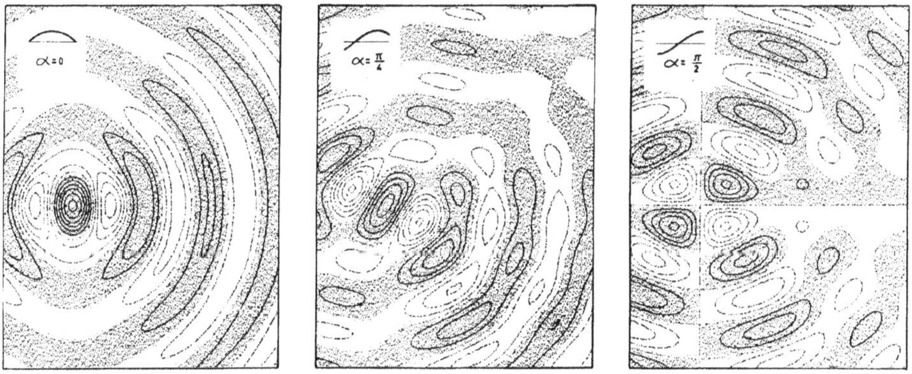

Fig. 3.3. The error pattern for sinusoidal gain errors $\Delta g = \cos(\phi + \alpha)$ in a single interferometer. Contour levels are the same as in fig. 1.1.

Horizontally stratified atmosphere. The fundamental processes in the atmosphere are absorption, thermal radiation, refraction and Faraday rotation. For an atmosphere without horizontal structure, they affect the whole aperture coherently.

Absorption and radiation are proportional to path length and therefore vary with diurnal motion of the source. Since in most practical detectors output voltage is inversely proportional to some power of system temperature, the effect of radiation reinforces that of absorption; it becomes more pronounced as electronic receiver temperatures are reduced. Through variations of the effective gain of the whole telescope, these effects may again give rise to residual 12-hour grating rings, this time with even symmetry.

The magnitudes are dependent on the type of air mass through which one observes. At Westerbork for $\lambda = 6$ cm, zenith radiation temperature may change by some 5K and absorption by a few percent upon the passage of a weather front (J.D. Bregman, private communication). These numbers (and those for small-scale effects to be cited below) would certainly justify the effort of a systematic investigation. A treasure of relevant information should be available in the telecommunications litterature.

Refraction produces a differential effect due to the fact that for any position outside the meridian the two elements of an interferometer observe the source at unequal elevations. The effect can be quite adequately described in terms of a homogeneous slab troposphere with appropriate fixed parameters (Ryle and Elsmore 1973). At frequencies below 1 GHz ionospheric refraction must also be considered (Hinder and Ryle 1971).

Faraday rotation in the ionosphere is mentioned for completeness. It produces a rotation of observed polarisations but no beam distortions. For more information see Strom (1972).

3.3. Errors in individual interferometers - circular errors

A large number of error sources are associated with individual antenna elements. We first consider their effect in a single interferometer, deferring a discussion of correlations between interferometers to the end of the section. According to sec. 3.1., we need only consider time scales comparable to a full 12-hour observing period.

Deviations relative to the assumed geometry (position, axis orientation, inter-axis separation, gravitational bending) affect both phase and, in a shift-variant way through pointing errors, gain. The variations have a fundamental 2π period in hour angle.

Most other effects are associated, in one way or another, with the state of the ambient atmosphere. Thermal expansion of the antenna structure again affects the geometry, but in a not so neatly periodic way. Electronic circuits are sensitive primarily to temperature, whereas the transmission of local oscillator signals is also influenced by pressure and humidity. Much can be done to suppress these influences, but the drifts with the longest time scales are particularly difficult to control. In the following I consider examples of error patterns to be expected as a result of such drifts.

A gain offset Δg in an interferometer with baseline r is represented in the aperture plane by

$$E(r,\phi) = \Delta g \,.\, \delta(r) \qquad (3.3.1.)$$

and hence, according to (2.2.2.), results in an error beam

$$S_e(R) = \Delta g \,.\, 2\pi r\, W(r)\, J_o(2\pi rR). \qquad (3.3.2.)$$

24-hour sinusoidal phase error. Similarly, an error $i.\cos(\phi+\alpha)$ in an interferometer with baseline r produces an error beam (c.f. 2.2.5.)

$$S_e(R,\Phi) = \Delta\varepsilon \,.\, 2\pi r\, W(r)\, J_1(2\pi rR)\, \cos(\Phi+\alpha). \qquad (3.3.3.)$$

Note that the relation between the aperture and beam patterns is rotation-invariant, i.e. rotation of the error pattern in the "observed half" of the aperture imparts the same rotation to the beam.

Phase offset. In the above two examples, the required hermiticity of the errors is provided by elementary analytical expressions. For a phase offset, however, the aperture error becomes

$$E(r,\phi) = i\Delta\varepsilon \,.\, \mathrm{sgn}\,(\cos\phi). \qquad (3.3.4.)$$

The function sgn (cos ϕ) is a square wave. Thus E may be expanded in an infinite Fourier series, each term of which can be transformed according to (2.2.5.). Alternatively, the transform can be shown to be related to $J_o(2\pi rR)$ through a horizontal line-by-line Hilbert transformation (Bracewell). Neither representation seems to be particularly useful.

The numerically evaluated beam is shown in fig. 1.1.(b). It exhibits the circular patterns one would have expected, but the structure within these is at first sight surprising.. At closer inspection, one finds the maxima and minima to align, forming horizontal ridges and throughs of wave patterns which are associated with the pair of singularities at $(r,\pm\pi/2)$ in the aperture.

24-hour sinusoidal gain drift. The same difficulties are encountered in a treatment of a gain error $\Delta g.\cos(\phi+\alpha)$, which may serve as an appro-

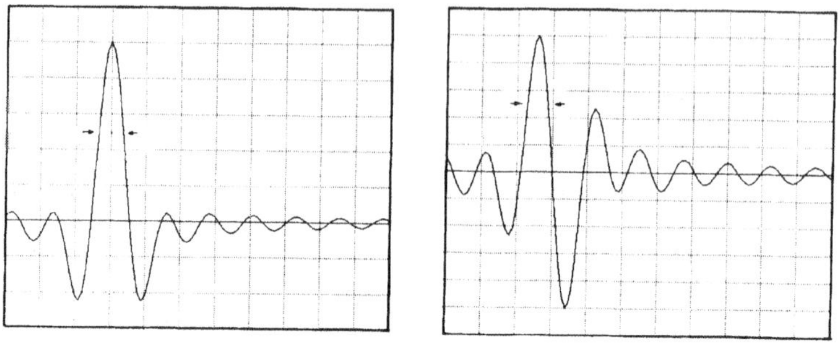

Fig. 3.4. The profiles H and H_e (see text) for the standard Westerbork taper. The width of the circular synthesized beam is indicated by arrows.

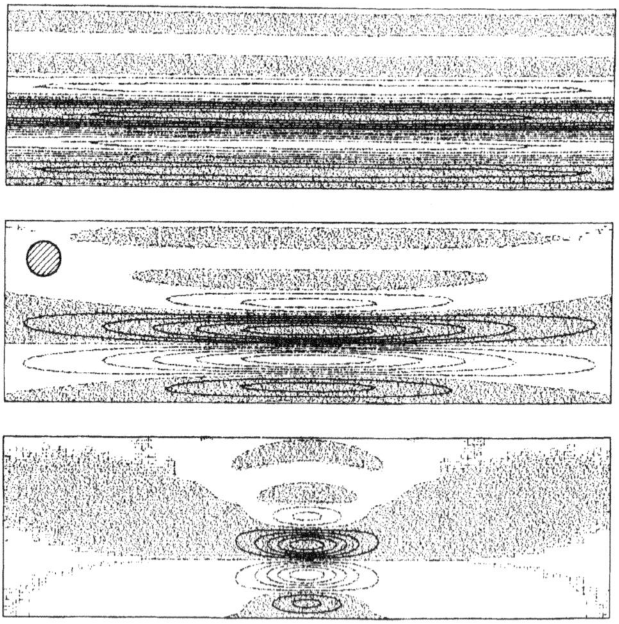

a

b

c

Fig. 3.5. Error beams corresponding to shifts in the apparent source position over prolonged periods of time. The shifts are assumed to produce phase errors $\Delta\varepsilon = (r/r_{max})\exp(-\phi^2/\phi_o^2)$. From top to bottom the time span ϕ_o increases and, consequently, so does the maximum of the error beam. (a): ϕ_o = 6 min, contour interval Δc= .00125 x the peak value of the unperturbed beam: (b) ϕ_o = 22 min, Δc= .005; (c) ϕ_o = 90 min, Δc = 0.02.

ximation for drifts with a linear or quadratic time dependence. In the aperture

$$E(r,\phi) = \Delta g \cdot \cos(\phi+\alpha) \cdot \mathrm{sgn}(\cos\phi). \qquad (3.3.5.)$$

There is no rotation-invariance as for a sinusoidal phase error and the beam patterns are quite different for different values of α, as shown in fig. 3.3. Horizontal wave patterns are discernible similar to that of fig. 1.1.(b), except for $\alpha = 0$, where the discontinuity in the visibility is of a higher order.

Higher harmonics. According to sec. 2.2. analytical transformations are possible for gain errors with frequencies of 2n/24h and phase errors at (2n+1)/24h. These transforms may be occasionally helpful but no new insights are involved.

Radial correlations. If an error occurs in a single antenna, it affects all the interferometers of which it is a member, thus introducing a correlation between errors at different radii in the aperture. The nature of these correlations depends entirely on the layout of the array.

Quite generally, the beams shown above have a wavy nature along map radii, the quasiperiod of the waves being inversely proportional to baseline. For an error occurring in two baselines simultaneously, two identical patterns with different quasiperiods are superimposed in the map, and we may expect beats to occur.

When a long-lived error occurs coherently in a considerable fraction of all interferometers, it is more properly classified as a global one. An example is provided by the secondary grating rings that frequently occur in Westerbork maps halfway between the radii of the standard 12-hour rings. They are associated with calibration errors in one of the movable antennas, which appear coherently in 10 simultaneous interferometers. Examples of a related error will be shown in sec. 3.5. The incomplete cancellation of regular grating rings (sec. 3.2.) can also be interpreted from this viewpoint. Since at the ring positions many interferometer responses add coherently, a very small error suffices to bring about a visible effect.

3.4. Atmospheric inhomogeneities - small scale and radial errors

Atmospheric inhomogeneities give rise to a distortion of the transmitted radiation wavefronts in addition to the effects already discussed in sec. 3.2. Such inhomogeneities occur both in the troposphere (primarily in water vapour concentration) and in the ionosphere (electron density). According to their size relative to the observing instrument they may be classified as "large" or "small".

Small-scale irregularities. The effects associated with small irregularities are by definition confined to only a part of the telescope. Moreover, since irregularities are always associated with motion, they donot persist for long. The corresponding error function then consists of noise with an autocorrelation function that is narrow in both dimensions and thus, according to sec. 2.3, results in an extended low-level noise component in the error beam. Because of its noiselike character, the effect is difficult to assess in actual observations. In comparison with the large-scale errors presently to be discussed, the small-scale ones appear to be unimportant (Hamaker, 1978).

Large-scale phase irregularities. The existence of large-scale irregularities in ionospheric electron density has long been known. Recent work (Hardy 1974; Hargrave and Shaw 1978; Hamaker 1978) has revealed the frequent occurrence of refractive disturbances in the troposphere, with scale sizes spanning a range from about one to many tens of kilometers. According to the rate at which they move and/or change, they may persist for times ranging from minutes to hours or even the better part of a full half-day (Ryle and Elsmore 1973).

Being larger by definition than the telescope, these errors classify as "radial". Their primary effect is to shift the apparent instantaneous position of the source by an unpredictable amount, and the corresponding corrugation in (2.2.4.) undergoes a similar shift. The cross-section of the resulting error pattern is, in first order, given by

$$H_e(X) = \Delta X \, . \, dH/dX \qquad (3.4.1.)$$

where H is the cross-section profile defined by (2.2.3.) and ΔX is the instantaneous position shift perpendicular to the corrugations, i.e. parallel to the telescope baseline. The cross-sectionsfor the Westerbork taper are shown in fig. 3.4. The main response is seen to consist of a positive-negative parallel ridge pair.

If the position shift persists for some time, these dual ridges are superimposed over the corresponding range of hour angles. Far out in the map, the result will tend to zero, positive and negative contributions neutralising each other. In the middle, all negative contributions coincide on one side of the beam centre, all positive ones on the other, so that constructive interference occurs. One thus finds a concentration and buildup of error power near the origin which becomes more pronounced as the error persists over a larger hour angle range; this is as it should be according to sec. 2.3. A computed example is shown in fig. 3.5., a practical example from Westerbork in fig. 3.6.

Absorption and radiation. The general remarks made in sec. 3.2. on our ignorance about these processes apply here as well. Small-scale effects noted at Westerbork are mostly associated with rain clouds, which, at 5 GHz in the zenith, produce antenna temperature variations of

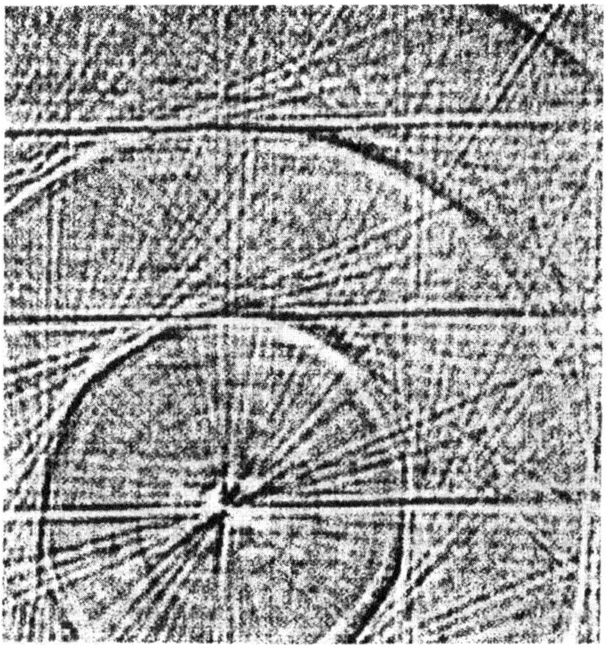

Fig. 3.6. The error beam for a 1x12 hr map of 3C48 made at Westerbork; large scale atmospheric refraction effects dominate the scene. Note the variety of radial extents corresponding to different time scales and the repetition at grating intervals. Near the position where the theoretical response was subtracted, error structures with even symmetry are seen which must correspond to gain errors in the observations.

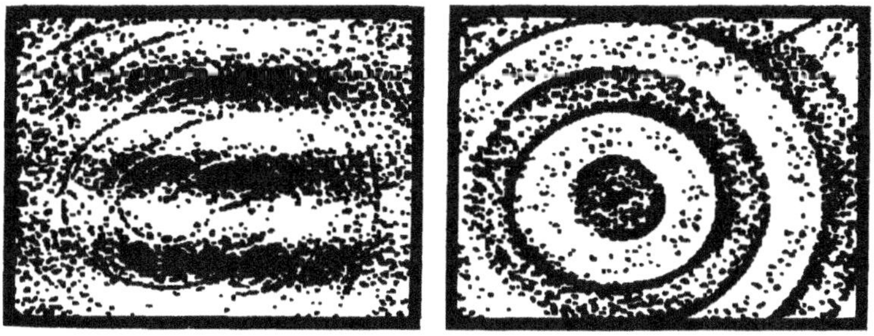

Fig. 3.7. (a) One-bit representation of a map of a single channel in a line observation at λ=21 cm made at Westerbork. The vertical-running wave pattern is due to cross-talk on the shortest baseline. (b) The same map with the continuum radiation (estimated from other channels) subtracted, demonstrating the ease with which many instrumental effects may be removed in a multichannel observation (courtesy R. Sancisi).

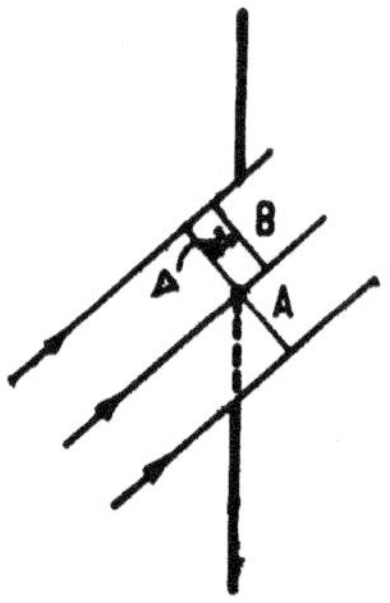

Fig. 3.8. Qualitative demonstration of the effect of shadowing for an off-axis source. For the unobstructed aperture, A is the phase centre. Under partial blocking (dashed line), this center shifts to B, receding over a distance Δ relative to the incident wavefront.

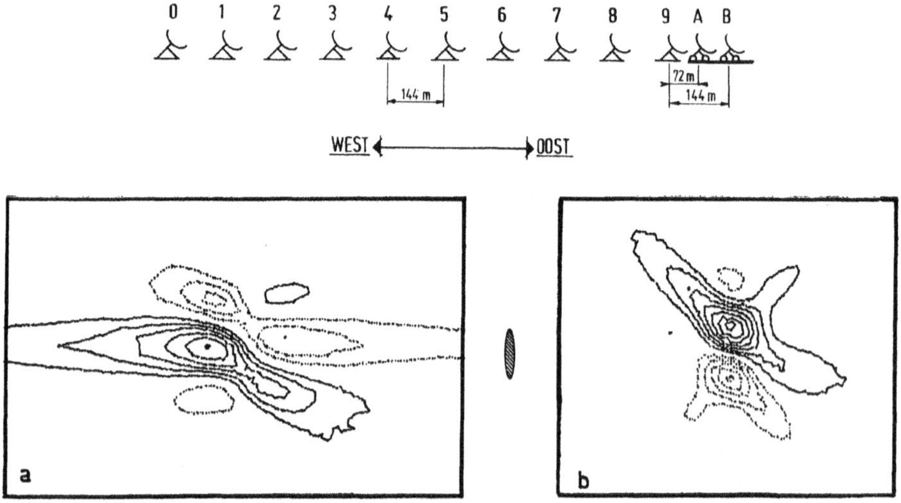

Fig. 3.9. Error beams caused by shadowing in a 12 hour observation with the Westerbork Telescope in the configuration shown. Declination is 10°. The source is at the 3 -dB point of the field of view (a) toward the right, contour interval 0.25% of synthesized beam maximum; (b) toward the top, contour interval 0.5%. The ellipse shows the half-power size of the synthesized beam. (Courtesy J.D. O'Sullivan)

the order of 5 to 10 K and signal power variations of several percent. Snow clouds are particularly heavy absorbers: Signal losses of 30% and antenna temperature rises of 100 K are typical, the phase at the same time remaining perfectly undisturbed (J.D. Bregman, private communication).

Fig. 3.10. Shadowing effects in a map of the Virgo cluster ($\delta=12^{o}$) at λ= 49 cm. The field centre is near the top right corner. The map is made from two 12-hour observations, in which the distance between antennas 9 and A (cf. fig. 3.9.) was 36 and 72 metres, respectively. The strong grating rings thus correspond to a 36 m radial sampling interval of the aperture. (Courtesy C.G. Kotanyi.)

3.5. Mutual interference between antennas

It is unavoidable to put some antennas in close proximity to each other in order to sample the aperture plane as close as possible to the origin. Some peculiar effects result which are of considerable practical importance. Owing to their special nature, they can only be discussed outside the framework used so far.

Cross-talk. For source directions close to the interferometer baseline vector, receiver noise radiated by an antenna nearer to the source may be picked up by the next antenna behind it. The picked-up signal has just the right amount of delay to produce an output when the signals from the two antennas are correlated. Since the effect only occurs for short baselines and hour angles close to $\pm \pi/2$, the corresponding map pattern consists of long "plane" waves "running" verically accross the map. Fig. 3.7 gives an example.

Shadowing. Under conditions similar to those for cross-talk, part of the aperture of one or more antennas becomes blocked or "shadowed" by another antenna in the line of sight. The loss of signal that results can easily be calculated; in a first-order approximation its shift-variance may be neglected and appropriate gain corrections applied.

It was recognized only recently that this is not the whole story. (J.D. O'Sullivan, unpublished). Indeed, for sources away from the field centre, the effective phase centre of the shadowed antenna shifts, as shown qualitatively by fig. 3.8. This shift is strongly dependent on source position in the observed field, and consequently, so are the error beams. Two calculated examples are shown in fig. 3.9. It is seen that shadowing is apt to produce artefacts that look just like the source tails which have aroused so much interest with astrophysicists.

An example from Westerbork is shown in fig. 3.10. It is seen that in this 2x12 hour observation the mock tail repeats at the AB grating interval. This is a clear sign that something is wrong; we owe its occurrence to the regularity of the layout of our array, combined with the fact that the shadowing environments for antennas A and B are different. For less regular arrays such as the VLA, such conspicuous warning signs may very well fail to occur.

Being strongly shift-variant, this effect is not easily corrected for. The procedure commonly used here is to simply delete all aperture points affected and restore the resulting map to a decent appearance through CLEANing. The CLEANed map might serve as a starting point for more refined iterative correcting procedures to be developed in the future.

ACKNOWLEDGEMENTS

I am indebted to W.N. Brouw, R.M. Forster, E. Raimond and G. Robertson for useful comments on earlier drafts of the manuscript, and in particular to the editor of this volume for very thorough screening of the final version. Miss W. Smit and Mrs. J.H. Millenaar were responsible for careful typework; the figures were prepared by R. van Dalen and H.W.H. Meijer. The Westerbork Synthesis Radio Telescope is operated by the Netherlands Foundation for Radio Astronomy with financial support from the Netherlands Organization for the Advancement of Pure Research.

REFERENCES

Baars, J.W.M., van der Brugge, J.F., Casse, J.L., Hamaker, J.P., Sondaar, L.H., Visser, J.J. and Wellington, K.J., Proc. Inst. Elec. Electron. Engrs. 61, 1258.
Bracewell, R.N., "The Fourier Transform and its Applications", New York: Mc Graw-Hill., 1965.
Bracewell, R.N. and Thompson, A.R., 1974, Astrophys. J. 182, 77.
Hamaker, J.P., 1978, to be published in Radio Science 13, Sept.-Oct. 1978.
Hamaker, J.P., 1979, to be submitted to Astron. Astrophys.
Hardy, R.N., 1974, Nature 249, 431.
Hargrave, P.J. and Shaw, L.J., 1978, Monthly Not. Roy. Astr. Soc. 182, 233.
Hinder, Richard and Ryle, Martin, 1971, Monthly Not. Roy. Astr. Soc. 154, 229.
Högbom, J.A. and Brouw, W.N., Astron. Astrophys. 33, 289.
Rots, A.H., 1974, Thesis, University of Groningen, Netherlands.
Ryle, M. and Elsmore, B., 1973, Monthly Not. Roy. Astr. Soc. 164, 223.
Schwarz, U.J., 1978, this volume.
Strom, R.G., 1972, Neth. Found. Radio Astr., Internal Techn. Rep. ITR-99.
Thompson, A.R. and Bracewell, R.N., 1974, Astron. J. 79, 11.
Weiler, K.W., 1973, Astron. Astrophys. 26, 403.

DISCUSSION (see also discussion after paper by Wilkinson-Readhead)

Comment J.R. FORSTER
Were the errors in visibility used to produce the error beams associated with antenna shadowing symmetrical in hour angle?
Reply J.P. HAMAKER
The shadowing effect itself is symmetrical in a sense, that is, the shadowing antenna at negative hour angles becomes the shadowed one at the other side of the meridian. However, since the antennas occupy different positions in the array (e.g. antennas 9 and A at Westerbork) the result in the aperture is by no means symmetrical, and this is reflected in the absence of any symmetry in the error beams.

Comment R.D. EKERS
Even if you have an error which is symmetrical the source may have a complicated asymmetric shape which when convolved with the error lobe will give an asymmetric result. In general it can be misleading to rely heavily on the expected shapes of error lobes for a point source if you have a field with extended objects.

"KNEADING": THE ADJUSTMENT OF INSTRUMENTAL PHASE AND GAIN PARAMETERS TO SUPPRESS ERROR PATTERNS IN A SYNTHESIS MAP.

J.P. Hamaker
Netherlands Foundation for Radio Astronomy
Dwingeloo, the Netherlands

SUMMARY

Often, particular unnatural-looking patterns appear in a synthesis map. An experienced observer can recognize the type of error that produced them and may even be able to apply the corrections necessary to get a better-looking map. A systematic approach is described in this paper: A model of the errors most likely to occur is fitted to the patterns in the map, and the fitted parameters are used to apply corrections. The presence of a point source that can serve as a reference is required. This requirement limits the applicability of the method, but is met by quite a few objects of interest. Where interference from other sources would be too strong, their influence may be reduced by CLEANing the map prior to model fitting. One example of the results for atmospheric position scintillation and one for baseline calibration errors will be shown.

1. INTRODUCTION

In this short contribution I shall report some work on adaptive error removal which I have been doing over the past few years. In doing so, I cannot avoid returning to some of the discussion we had on the application of closure phase techniques in connected-interferometer systems. I hope to do so in a way that may clarify the situation, although I have not been able in this short time to put every detail in its correct place.

Adaptive correction of observational data is a procedure familiar to most experimentalists. One usually has some a priori notions about the possible outcomes of an experiment, as well as some knowledge about the nature of artefacts that can be introduced through the measuring procedure. This information may be used to isolate the hard facts from the artefacts as well as possible. In this paper, I shall present examples of a systematic application of this concept to sky maps obtained with the Synthesis Radio Telescope at Westerbork.

C. van Schooneveld (ed.), Image Formation from Coherence Functions in Astronomy, 47-53.

The nature of some of the simple and most common error patterns to be found in these maps has been described in my previous contribution (Hamaker, this volume). Such errors as 12-hour grating rings and AB-rings are easily identified, and instrumentel parameters specified in the data reduction process may be adjusted so as to make them disappear. This is what one might call the manual version of the procedure which has come to be known in our Institute as "kneading". Some observers have developed considerable skill at it, even for less easily identifiable errors such as deviations in a single antenna (R.G.Strom, private communication).

Computerised kneading is a logical next step. It allows much more complex situationsto be handled at the expense of much less labour. On the other hand there is the obvious danger that the computer will carry the process too far; it may mistake parts of the true source for errors and, in "correcting" for them, mutilate the source and introduce new spurious structures. Close guidance by a human operator must be considered an indispensible ingredient to a successful kneading programme. Furthermore, the error model to be used should be designed with the smallest number of degrees of freedom that is adequate to describe the important errors.

2. FIRST EXAMPLE: LARGE-SCALE ATMOSPHERIC ERRORS

My first example concerns large-scale atmospheric errors. The uncorrected map, fig. 1a, clearly shows the characteristic streaks discussed earlier (Hamaker, this volume). As I described, these result from an apparent random wandering of the source field during the observation. To get rid of them one must somehow align the individual corrugations (Hamaker, loc. cit.) to their correct positions.

It should be noted at this point that the map (including the strong point source) is directly derived from the measured data. It satisfies whatever closure phase conditions one might derive from the same data and, since these closure phases contain no position information, will continue to do so regardless of any shifting around of the contributing corrugations. Other information is needed for making the alignments.

In the present case we may use the strong unresolved source as a reference, aligning its responses in all corrugations to coincide in a single point. This procedure is highly successful, as demonstrated by the result in fig. 1b. It should be emphasized however, that reference

---

Fig. 1. (opposite page). a (top): Map of 3C236 made from four 12-hour observations. To better display the weak features, a point source has been subtracted at the centre, whose strength would correspond to about twice the height of the picture. The residual map clearly shows the atmospheric streaks associated with this source superimposed on the extended source structure. b (bottom): The same map after application of the "kneading" procedure described in Chapter 2.

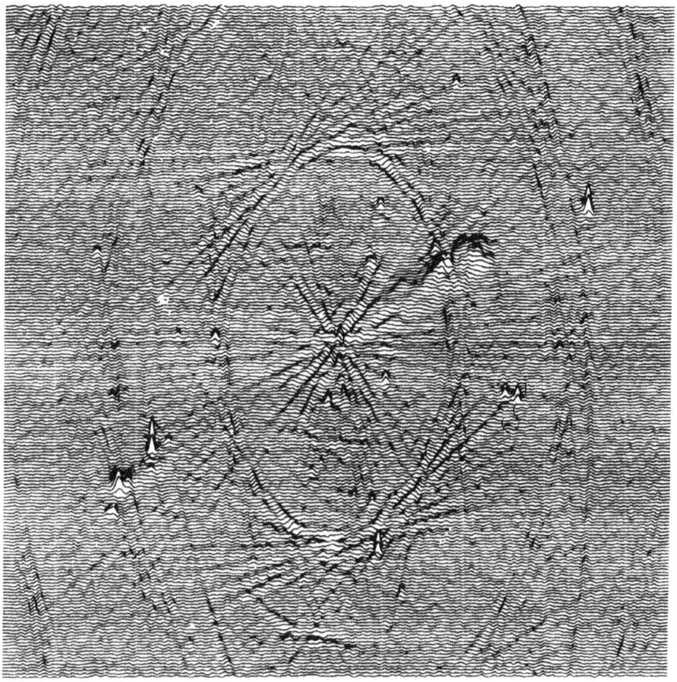

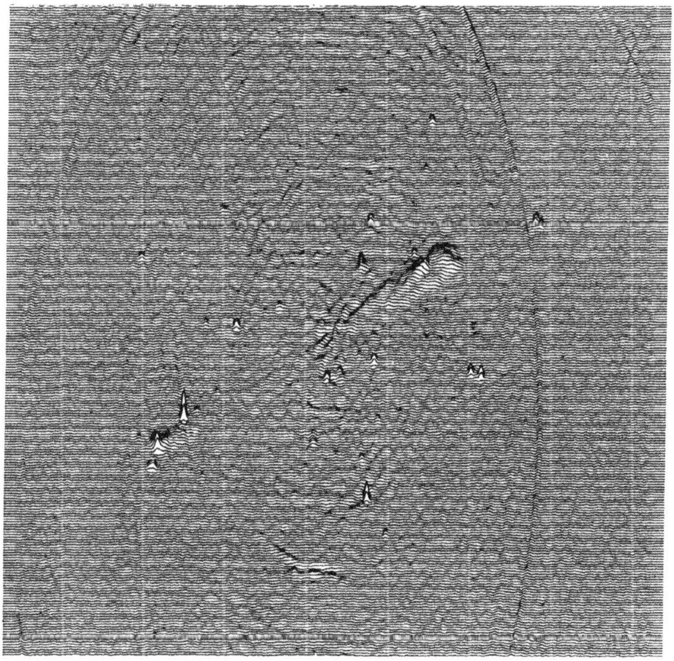

sources as outstanding as this one within an observed field are rare. Where they are either weaker or partly resolved it remains to be seen how much one can achieve.

Particularly in these more complicated situations, methods of the type described by Wilkinson (this volume) to handle closure phase data could be valuable. One might, for instance, from the "dirty" map derive a window defining the extent of the source and then use a Wilkinson-type iterative procedure to eliminate all structures outside and ensure positivity inside. Such possibilities are certainly worth further investigation.

3. SECOND EXAMPLE: INSTRUMENTAL DRIFTS

Fig. 2a shows a map in which the effects of the known deficiencies of the synthesized beam have been removed by CLEANing and restoring with a beam having no sidelobes (Schwarz, this volume). The presence of residual circular structures reveals the occurrence of instrumental drifts. The patterns around the central source stand out quite clearly: one should be able to analyse them in terms of contributions from individual interferometers and apply appropriate corrections.

To do this, an error model containing the major effects of simple offsets and drifts (Hamaker, this volume) is constructed and its parameter values determined by a least-squares fit. With these values, the model is then added as a correction to the theoretical synthesized beam. The corrected beam, representing a better approximation to the true instrumental profile, should give better results when used instead of the theoretical beam in CLEANing.

Contrary to the previous example, the error pattern is weak compared to the brightness of the extended source component. Precautions must therefore be taken to prevent contamination of the fitted error parameters by this component. This can be attempted in either of two ways. One may remove most of the effect of the source by CLEANing with the theoretical beam before proceeding with the model fit; or the region of the contaminating source may be excluded from the data base for the fit. The best results are to be expected when the two methods are applied together.

Fig. 2. (opposite page). a (top): The central part of a map of 3C66 made from two 12-hour observations. The largest ring is a residual 12-hour grating response. Halfway out an AB-ring can be recognized. The remaining quasi-circular structures are not readily identified. b (bottom): The same map after application of the "kneading" procedure described in Chapter 3. Three features formerly masked by the instrumenten errors now stand out clearly: There is no bridge between the two source components; the extended source can be traced to its very ends; and the point source is recognized to be extended.

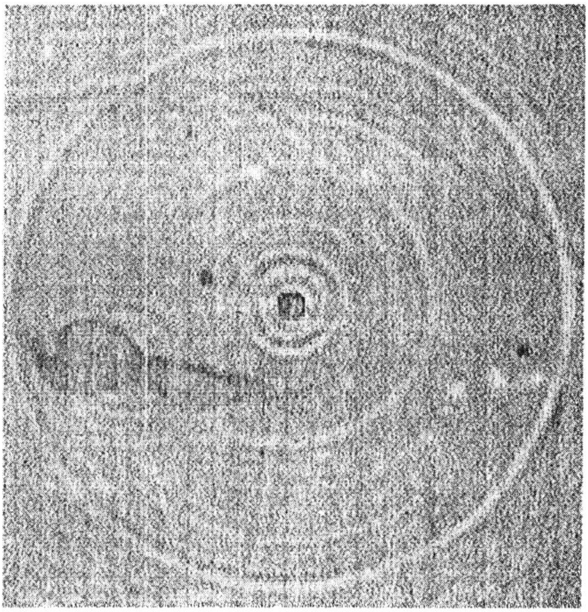

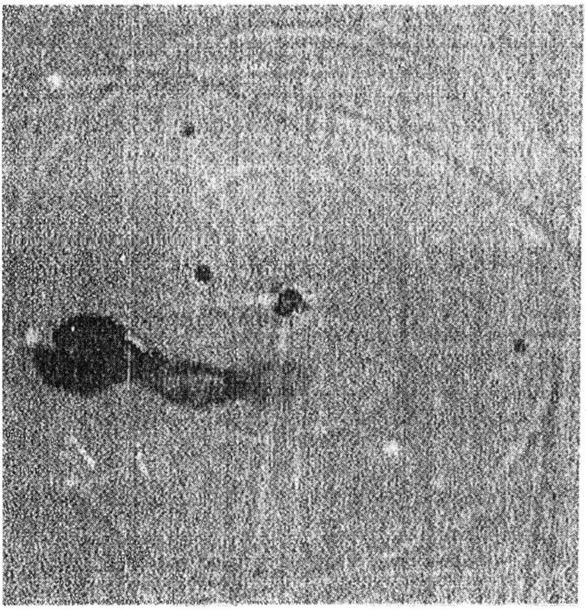

For the present example the end result is shown in fig. 2b. Indeed the rings are all gone except for some weak "high-order" parts, which were deliberately not included in the model in order to minimize the risk of spurious source pickup. For the peak in the extended source, the residue of the 12-hour ring has changed in appearance but has not completely vanished. I suspect that some weakly shift-variant error must be responsible (e.g. antenna pointing).

The figure shows the central source to be slightly extended in position angle 45°. If the program had been left unguided, it would have proceeded on the assumption of a point source and produced an incorrect estimate of the error beam. The occurrence of this sort of situation not expected by the computer must continually be anticipated.

In cases where the source and error distribution are difficult to disentangle, iterating the process is conceivable. One then arrives at a procedure again closely resembling Wilkinson's (this volume) although differing in detail due to the different nature of the missing information.

4. CONCLUDING REMARKS

The examples shown are the only two in which the kneading technique has been applied so far. They demonstrate the potential of the method but give no clue as to the range of situations in which it may be successfully applied. The impact of various sources of trouble will have to be assessed in further practical tests and simulations. Owing to the large programming efforts involved, progress is expected to be slow.

Due to time and space limitations, this presentation has by necessity been brief and sketchy. More elaborate accounts will be submitted elsewhere (Astronomy and Astrophysics, 1979?).

ACKNOWLEDGEMENTS

I am indebted to R.G. Strom for prompting the study of 3C236 and sorting out the problems of producing a fine map from my somewhat messy computer output. The study of 3C66 could not have been done without the hard work of A.M.J. Vossen, who has spent a year translating theoretical concepts into a program package that worked in spite of frightening hardware limitations. The Westerbork Synthesis Radio Telescope is operated by the Netherlands Foundation for Radio Astronomy with the financial support of the Netherlands Organization for the Advancement of Pure Research (Z.W.O.).

DISCUSSION

Comment E.B. FOMALONT
After removing the extended structure we are left with a dirty beam associated with the point-source which applies to the observation coverage and imperfections. Instead of analyzing the error beam for the phase imperfections, why not use this dirty beam directly to clean the source?
Reply J.P. HAMAKER
One problem is that you need an antenna pattern of linear size twice that of the map, so you must somehow entrapolate. I could not think of a better way to do this than by model fitting.
Reply R.J. ALLEN
In relation to this question I wish to make the following comment. The procedure of using the actual response of a point source on the map as the "dirty beam" in a clean operation has indeed been tried in a number of cases. It is unsatisfactory in that the noise inevitably present on the map seriously limits the level to which the clean operation can be carried. Since astronomers are almost always interested in looking at the faintest extended features which may be present on the map, the procedure outlined above is not useful. The method used by Hamaker employs as much of the physics as is known about the actual instrumental instabilities in order to construct a model of them, thereby rejecting the effects of noise to a high degree.

Comment D. SRAMEK
It was shown earlier by Dr. Baldwin that self calibrating on a point source in the field can lead to a false symmetric image on the map. How is this avoided with your technique?
Reply J.P. HAMAKER
I do use part of the phase information rather than completely disposing with it as Baldwin does. What I do is to fit a linear slope to the array of observed phases for each hour angle and retain the residual phases. The underlying assumption (which I did not have time to discuss) is that the slope represents the large-scale atmosphere, whereas the residuals carry information related to the source structure.

A PROPOSAL TO OBTAIN SHORT SPACING INFORMATION BY INTERPOLATION

A.H. Rots
Netherlands Foundation for Radio Astronomy, Dwingeloo, the Netherlands

1. INTRODUCTION

For a short description of the problem posed by missing information at short interferometer spacings in aperture synthesis we would like to refer to the introduction of the previous paper by Ekers and Rots. The solution that is suggested in this paper is not a true interpolation but somewhat intermediate between interpolation and Clean.

If we confine ourselves for a moment to one dimension, it can be understood quite easily that sampling at baselines which are multiples of 36 m is adequate for sources that are relatively small. For several reasons, however, one may choose to observe with an increment of 18 m so that one is left with the problem of the missing 18 m spacing (see also previous paper). Since the 36 m increment fulfils the sampling criterium, however, one can interpolate this 18 m visibility, provided one has a sample at 0 m. The latter can usually be obtained in a variety of ways, so we shall assume in this paper that it is available.

The consideration outlined in the preceding paragraph is the basis of the method, but in a two-dimensional case we would not like to do a straight interpolation. As a priori knowledge of the source we shall use spatial confinement and positivity.

2. THE METHOD

We first assume that the source as it appears in the map obtained from the available spacings (n x 18 m, n = 0,2,3,...,N) is a reasonable approximation of the true brightness distribution. The part of the map in which we believe the source to be confined is cut out and all negative intensities in that area are set to zero.

This first approximation is then Fourier Transformed (by direct transformation) onto an 18 m spacing track in the (u,v)-plane. This visibility function is transformed back and added to the map, with the proper

C. van Schooneveld (ed.), Image Formation from Coherence Functions in Astronomy, 55-58.

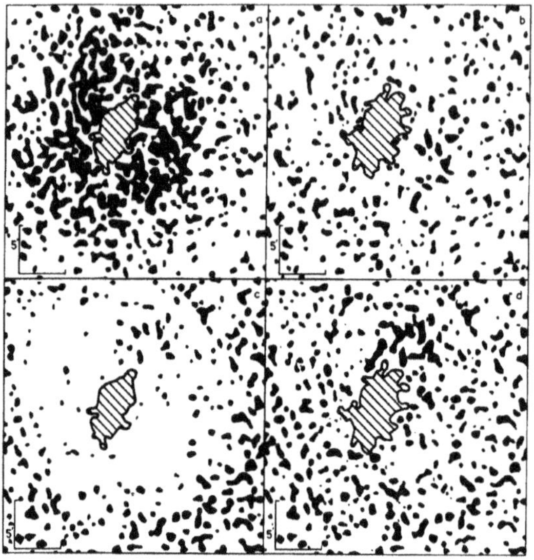

Fig. 1a-d. The effect of adding short baseline information obtained by one iteration of the procedure described in section 2. This example is taken from WSRT observations of M81 in the 21-cm line of neutral hydrogen; the figure shows a single channel map at half resolution (50 arcsec). Panels (a) and (b) display all areas in black where the brightness temperature $T_b < -\sigma$; (c) and (d) show $T_b > +\sigma$, where σ is the r.m.s. noise. The main body of the source has been hatched in all panels. The straight WSRT map is shown in (a) and (c); (b) and (d) display the same map after adding 0 m and 18 m spacing information.

weight. One can actually view this process as an interpolation where the interpolating function in the (u,v)-plane is optimized for the source under consideration: it is the Fourier Transformation of the source shape.

We have used the procedure described so far for the neutral hydrogen observations of M81 (A.H.Rots, 1974, Dissertation, University of Groningen; A.H.Rots and W.W.Shane, 1975, Astron. Astrophys. 45, 25). An example is shown in Fig. 1.

At that time Ekers already suggested to iterate on the process. One takes the same area of the map again, sets the negative intensities to zero, repeats the two transformations, and adds the fresh part of the 18 m baseline contribution into the map.

Recently we have done some simulations with model sources in such an iterative scheme. At 1.4 GHz the primary beam of the WSRT has a FWHP of 37 arcmin; the radius of the first point source grating response is 40 arcmin. We have tried circular pillbox sources with diameters of 20 and 30 arcmin, a bar of 10 x 35 arcmin, and circular Gaussians with FWHP's of 10 and 20 arcmin. One cannot extend the sources beyond the grating ring radius since the grating response of one side is then affecting the other side of the source. All results were satisfactory. The next section is a compilation of comments and conclusions based on these experiments.

3. COMMENTS AND CONCLUSIONS

(i) Apparently not both dimensions of the source have to be smaller than a certain size in order to fulfil the sampling criterium; it can be quite large in one direction. What seems to be important is the area over which the source extends: the 20' pillbox and the 10' x 35' bar gave similar results notwithstanding the length of the bar; their areas are very close. This obviously is an advantage over the one-dimensional case.

(ii) Rather precise knowledge (probably 5-10%) of the zero spacing flux is very important; not only for its own sake, but also because it appears to be useful to monitor the integral over the map and stop the iteration process when it equals the zero spacing flux. There is, however, a connection with the next point.

(iii) A good estimate of the area in which the source is confined is of equal importance. Under certain conditions a wrongly defined area may lead to spurious results. On the other hand, a good estimate may relax the requirements for the accuracy of the zero spacing flux. For "small" sources ($\lesssim$ 400 square armin at 1.4 GHz) where the exact shape is known the iteration process will converge to the correct value even without knowing the zero spacing flux; one can obtain that flux at the same time.

(iv) For obtaining the 18 m visibility function a smoothed version of the map can be used. Notwithstanding the use of direct Fourier Transforms the CPU time per iteration can easily be less than 1 sec. and all data can be kept in core. Hence the process is suitable for interactive use; this should actually be encouraged.

(v) It should be ackowledged that other techniques than direct Fourier Transforms can be used, like convolution with the zero order Bessel function corresponding to 18 m, or Clean. Application of the latter may very well face serious problems in the presence of noise.

(vi) The advantage of the method described in this paper is that one does not need separate observations requiring special treatment to obtain the 18 m spacing visibilities. It is reduced to a rather straightforward data processing problem.

(vii) We have to make the proviso, however, that the process really works. The method should now be tested on real data, and extensively be compared with other schemes. To us it seems that it will probably work quite satisfactorily in many cases.

DISCUSSION

Comment J.E. BALDWIN
Do you believe that your procedure is logically different from CLEAN?
Reply A.H. ROTS
To a certain extent the procedures are logically similar, except that CLEAN does not use the positivity constraint.

Comment J.P. HAMAKER
I think what your scheme does is to develop a solution that is compatible with your data and equal to zero outside your window. Isn't that correct?
Reply A.H. ROTS
No, not in the sense that I am forcing the resultant map to zero outside the window.
Comment J.P. HAMAKER
Well, by zeroing the area outside your window in every iteration, you force your solution into that direction.
Reply A.H. ROTS
Yes, in so far that there is not supposed to be any contribution to the 18 m spacing visibilities from outside the window.

Comment C. van SCHOONEVELD
Did you consider to apply your method to the missing long spacings, rather than to the short spacings?
Reply A.H. ROTS
No. I would like to refer to the work that Trevor Cole has done in that field.
Reply T.W. COLE
I cannot now remember the actual reference in A&A but yes, I did a matrix inversion to exploit redundancy and get missing spacings. It works for missing innerspacings but the linear algebraic computer implementation could be much, much better.

Comment U.J. SCHWARZ
The noise inherent to the observation or interpolation plays an important role. Short baselines are more difficult to measure accurately, due to all sorts of non-random effects (confusion, interference), than long base-lines. If one makes the optimistic assumption, that the short spacings can at best be measured as accurately as the long spacings, is then an interpolation scheme competing with observations (that means, are systematic effects smaller than the random errors)? I made an analysis of this problem and the answer is yes, except in the case of a strong complex source outside the extended region of interest.

THE SHORT SPACINGS PROBLEM IN CORRELATION INTERFEROMETERS AND A PROCEDURE FOR SOLVING IT

E. Bajaja
Leiden University, Dept. of Astronomy
Leiden, the Netherlands

ABSTRACT

The effects of the missing short spacings in correlation interferometers measurements on the synthesized antenna pattern and on the maps produced, are analysed. A method for correcting those effects, based on the obtention of the missing components from single dish observations, is described. The procedure has been developed for the WSRT observations of extended objects and it has been applied on several cases. Some results are shown for M31 and NGC4258.

C. van Schooneveld (ed.), Image Formation from Coherence Functions in Astronomy, 59.

SHORT SPACING SYNTHESIS FROM A PRIMARY BEAM SCANNED INTERFEROMETER

R.D. Ekers[1] and A.H. Rots[2]
[1]Kapteyn Astronomical Institute, University of Groningen, Groningen, the Netherlands.
[2]Netherlands Foundation for Radio Astronomy, Dwingeloo, the Netherlands.

1. INTRODUCTION

Aperture synthesis instruments providing a generally highly uniform sampling of the visibility function often leave an unsampled hole near the origin of the (u,v)-plane. One of the common configurations of the Westerbork Synthesis Radio Telescope (WSRT) provides sampling in concentric rings with radii of n x 18 m. However, the samples with n=0 and n=1 cannot be obtained directly with the instrument because of rather obvious physical constraints. This missing short spacing information causes a central depression of the synthesized beam pattern, resulting in large negative, bowl shaped areas around (especially) extended sources. Very large sources will only partially be present in the maps. Annoying as this can be for continuum observations, in the case of spectral line data it causes serious distortions of the individual line profiles.

In this paper we shall use the following nomenclature. Coordinates on the sky are l and m; in the aperture plane u and v. Capital letters will indicate functions in the plane of the sky, lower case letters their Fourier Transformations in the (u,v)-plane. B is the brightness distribution on the sky, A the antenna pattern of the individual 25 m elements of the array (the so-called primary beam), and S is a bed-of-nails sampling function on the sky; δ signifies the usual delta function. Multiplications are represented by a period (.), convolutions by an asterisk ($\ast$).

If t(u,v) is the sampling of the entire array, then we have observed:

$$b^1(u,v) = (b(u,v) \ast a(u,v)) \,.\, t(u,v) \qquad (1^a)$$

resulting in a map:

$$B^1(l,m) = (B(l,m) \,.\, A(l,m)) \ast T(l,m) \qquad (1^b)$$

The information we would like to add to the map is:

C. van Schooneveld (ed.), Image Formation from Coherence Functions in Astronomy, 61-63.

$$b^{1}(0,0) = (b(u,v) * a(u,v)) \cdot \delta(u,v) = \iint B(l,m) \cdot A(l,m)\, dl\, dm \quad (2)$$

for the missing zero spacing flux, and:

$$b^{1}(18m) = (b(u,v) * a(u,v)) \cdot t_{18}(u,v) \quad (3)$$

where t_{18} represents the sampling along a hypothetical 18 m spacing track.

2. A SINGLE TELESCOPE

Just one observation with one telescope, at the proper position, will only provide the zero spacing flux as expressed in Eq. 2. It has been recognized, however, for many years that one may obtain more information about the visibility function by scanning the whole map area in the sky with a single telescope. In that way one can theoretically derive the visibility function in a circle with a radius equal to the diameter of the dish, provided one knows the antenna pattern A_s of this telescope. The result of the scanning is:

$$B^{11}(l,m) = (B(l,m) * A_s(l,m)) \cdot S_s(l,m) \quad (4^a)$$

which transforms into (neglecting S_s):

$$b_s^{11}(u,v) = b(u,v) \cdot a_s(u,v) \quad (4^b)$$

It is quite easy to see that one can obtain the desired functions in Eq. 2 and 3 from these data.

Usually one requires a dish with a diameter of at least 60 m to fill in the hole in WSRT observations.

3. SHORT BASELINE INTERFEROMETER

Single observations with a short baseline interferometer (e.g. 36 m) only give visibility data along the track of the interferometer in the (u,v)-plane. In analogy with the single dish case it is possible, however, to get information about the visibility function in a broad annulus around that track by scanning an area in the sky.

Assuming that the present orientation of the interferometer corresponds to $u = u_o$, $v = v_o$; leaving the delay center at $l = o$, $m = o$; and pointing the telescopes into the direction $l = l^1$, $m = m^1$, the output of the interferometer is:

$$\iint B(l,m) \cdot A(l-l^1, m-m^1) \cdot e^{2\pi i(u_o l+v_o m)}\, dl\, dm =$$

$$= \{b(u,v) * (a(u,v) \cdot e^{-2\pi i(ul^1+vm^1)})\} \cdot \delta(u-u_o, v-v_o) =$$

$$= \iint b(u_o-u,v_o-v) \;.\; a\,(u,v) \;.\; e^{-2\pi i(ul^1+vm^1)} du\; dv =$$

$$= \{(B(-l,-m) \;.\; e^{-2\pi i(u_o l+v_o m)}) \star A(l,m)\} \;.\; \delta(l-l',m-m') \qquad (5)$$

If one observes this quantity on a grid $S(l^1,m^1)$ and performs the Fourier Transformation over l^1 and m^1, one obtains:

$$b^{111}(u,v) = b\;(u_o-u,v_o-v) \;.\; a(u,v) \qquad (6)$$

4. RETRIEVING SHORT SPACING INFORMATION

Scanning the sky with the 36 m baseline interferometer and recording the cross correlation as well as the total power of (one of) the telescopes yields $b_s^{11}= b^{11}$(since a_s=a) from Eq. 4^b and b^{111} from Eq. 6. The telescopes are 25 m in diameter and hence b^{11} extends to a radius of 25 m, while b^{111} is defined in an annulus between radius 36$\pm$25m. Although in practice the overlap between b^{11} and b^{111} will be smaller, it should nevertheless be possible to reconstruct b(u,v) in the area of interest such that b^1(o) and b^1(18m) can be constructed.

If one may assume that

$$a(u,v) = p(u) \;.\; p(v) \qquad (7)$$

-as would be the case if we can approximate a(u,v) by a circular gaussian-, one only needs scanning in one dimension. Let us for simplicity assume that v_o=o. Then m^1=o and one has to scan in l-direction. Fourier Transformation over l^1 yields (instead of Eq. 6):

$$b^{1111}(u) = p(u) \;.\; \int b(u_o-u,v) \;.\; p(v)\; dv \qquad (8)$$

which already contains half of the convolution needed in Eq. 3.

PART II

APERTURE SYNTHESIS WITH LIMITED OR NO PHASE INFORMATION

FUNDAMENTAL ASPECTS OF APERTURE SYNTHESIS WITH LIMITED OR NO PHASE INFORMATION

(Invited paper)

J.E. Baldwin and P.J. Warner
Cavendish Laboratory, Cambridge, England.

Much ingenuity has been devoted to the determination of mutual coherence functions at radio wavelengths. Measurements of the amplitude and phase of the interference fringes with uncertainties of only 1 percent are now attainable. Unfortunately the atmosphere intervenes between the measured quantities and the mutual coherence function itself and the phase uncertainties introduced by the ionospheric and tropospheric irregularities are often the factor limiting the quality of radioastronomical maps.

In this paper we describe methods for making maps from data with poorly determined or even unknown phases. We discuss in turn the effects of errors on the maps, methods of phase determination using reference sources, the properties of hybrid maps and their application to mapping of fields containing point sources, extended sources or a mixture of the two. We assume in all our discussions that the data are adequately sampled for the problem in hand and will leave the problems associated with incomplete sampling of the mutual coherence function to the cleaners and those able to maximise entropy.

1. THE EFFECTS OF ERRORS ON THE MAPS

We enquire first under what circumstances the phases limit the quality of a map. Suppose the 2-dimensional mutual coherence function comprises a set of amplitudes and phases $A(x,y), \phi(x,y)$ having uncertainties ΔA, $\Delta\phi$. Since the fourier transform is a linear process, the resulting images can be considered as the sum of the ideal map and those obtained by fourier transformation of $(\Delta A, \phi)$ and $(iA\Delta\phi, \phi)$. These two error maps have similar properties; the only outstanding difference arising from the fact that in most work $A(-x,-y)$ and $\phi(-x,-y)$ are not measured but derived assuming the mutual coherence function to be conjugate symmetric. Thus the error map associated with ΔA is symmetric whilst that associated with $\Delta\phi$ is antisymmetric. The rms sidelobe levels are proportional to ΔA and $\Delta\phi$. There is little published data on the amplitude stability of aperture synthesis observations but there can be little doubt that the variations are mainly

C. van Schooneveld (ed.), Image Formation from Coherence Functions in Astronomy, 67–82.

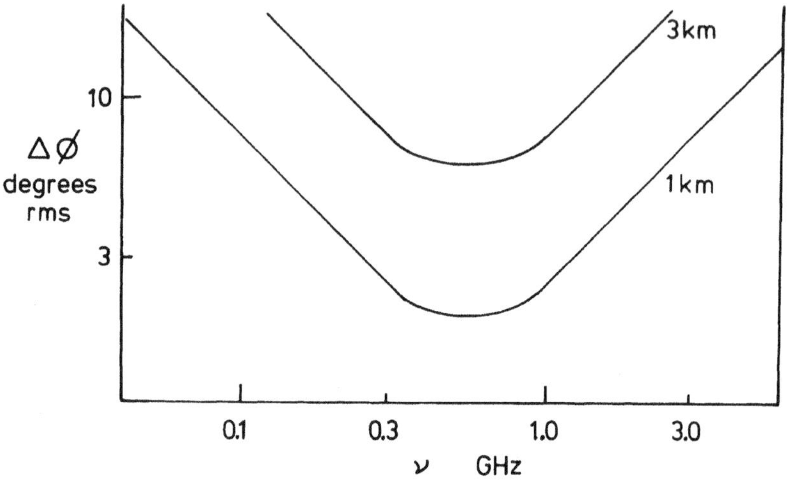

Fig 1. Typical rms phase variations due to ionospheric and tropospheric irregularities as a function of frequency and baseline.

instrumental in origin. Experience at Cambridge suggests that $\Delta A/A$ may be as small as 1 per cent under good conditions. The vagueness about ΔA merely reflects the fact that it is usually small compared with $\Delta\phi$. No authors have claimed phase uncertainties as small as 0.5 deg (i.e. 1 per cent of a radian). Even if instrumental sources of error are excluded the atmospheric contribution to ϕ is important for most synthesis telescopes. Fig 1 shows the rms phase deviations expected under average conditions as a function of frequency and baseline. Variations in atmospheric conditions may cause changes of at least a factor of 3 about these mean values. It is clear that for all existing aperture synthesis instruments the effects of phase deviations are likely to outweigh those of amplitude.

There is a wide range in the time scales and linear scales of the phase fluctuations. The most important ones are the long period large scale features, which produce an rms sidelobe level of roughly $0.2\ t^{\frac{1}{2}}(\mathrm{hrs})\Delta\phi_{max}(\mathrm{rad})$ over an area of sky of $\sim 16\ t^{-2}(\mathrm{hrs})$ beam areas. As an example the 5 km telescope operating at 15 GHz is affected most by fluctuations with time scales of roughly 30 mins giving rms phase fluctuations $\Delta\phi_{max}$ (at the largest baseline) of 30 deg. The resulting sidelobes are 6 per cent over an area of 64 beams.

2. PHASE DETERMINATION FROM REFERENCE SOURCES

Astronomers have devised many calibration techniques to overcome the desperation caused by poor phases. All rely on the idea of a reference source. In what follows we shall assume that everything possible by way of phase calibration using sources external to the field of view has been done, but that the resulting maps are still limited by unknown atmospheric or instrumental phase fluctuations.

There are a number of methods of improving maps which use reference sources *within* the area of the map. Any point source on the map may be used to give the interferometer phase for each spacing by 12 hour averaging, but signal-to-noise and confusion will limit accuracy. It may even be possible to use shorter term phase averages to correct the phase. At first sight this seems an ideal way of overcoming atmospheric phase problems. The source is exactly where it is needed. But the apparent fluctuations in the phase of the reference are partly due to the atmosphere and partly real, i.e. due to the interference between the reference and other sources in the field. Removal of the fluctuations by some type of running mean distorts the map. In general the map is accurate in the outer parts and progressively less so as the distance from the reference source is decreased.

The limit of this procedure is reached when the time scale for correcting the phase is extremely short. All variations of the phase are attributed to the atmosphere and the apparent phase of the reference source is set equal to zero. We have discussed this approximation elsewhere (Baldwin & Warner, 1976).

The reference source need not be a single object. Any distribution of sources whose relative positions and flux densities are known can be used to calculate reference amplitudes and phases as a function of HA and spacing which can then be used for calibrating the observation. An example of this technique applied to observations of the field of view of the 5C7 survey (Pearson & Kus, 1978) is shown in Fig 2. There are several sources having flux densities in the range 0.5 to 1 Jy and the comparison between reference phase and the observation is so good that the interferometer phase can be measured to an accuracy of 2 deg.

These techniques of internal calibration require the isolation of an observed phase and its correction to some reference level. Two separate effects can combine to make this impossible: the signal-to-noise may be inadequate to correct variations on short time scales and it may not be possible to construct a good enough phase reference. An obvious criticism of this method is that it implies prior knowledge of the distribution of sources in the field of view. How this difficulty is overcome is described below.

3. HYBRID MAPS AND THEIR APPLICATION

We have recently described an iterative technique derived from crystallographic procedures to make maps from amplitude data alone (Baldwin & Warner, 1978). The starting point is a map made using the square of the observed amplitudes, which is independent of the observed phases. It is the autocorrelation of the true distribution: we call it a *Map of* A^2. It is possible to analyse this map to obtain an

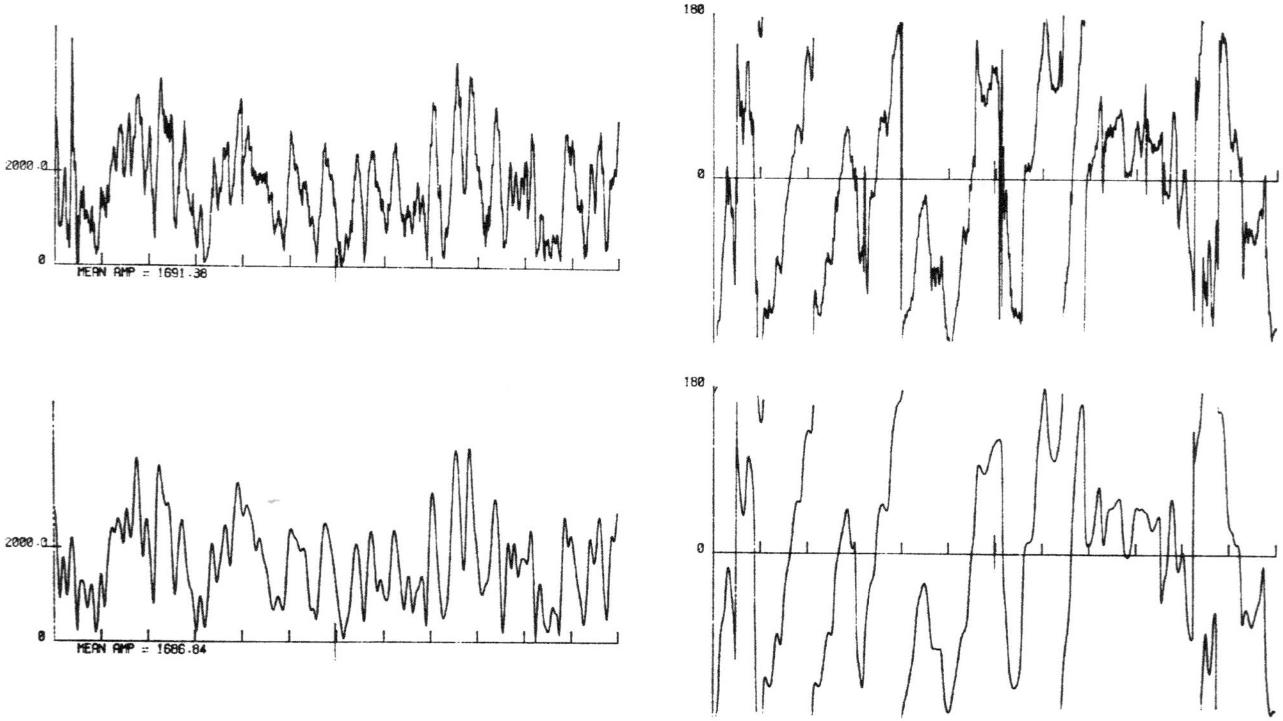

Fig 2. Calibration using field sources as an internal reference. Amplitudes and phases as a function of Hour Angle at a baseline of 718λ for the 5C7 survey. Upper curves; observations. Lower curves; 186 source model.

approximate model of the true brightness distribution. However, if observed phases are available, the map based on them can provide the starting model. This is used to calculate phases which are combined with the observed amplitudes in a *Hybrid map*. Strictly, hybrid maps are made using amplitudes which are appropriate to one brightness distribution with phases which are appropriate to a different one. They have been used by several authors (e.g. Fort & Yee, 1976; Baldwin & Warner, 1978; Readhead & Wilkinson, 1978). An understanding of their properties is important as a basis for assessing the progress of the iteration, its convergence and speed of convergence.

The simplest case is that mentioned above, the *phase zero map*. The amplitude is correct for the particular distribution of emission whilst the model for the phase is a point source at the centre of the map. Where there is a bright source in the field of view the map made from this combination has the following properties:

(1) The bright reference source is at the map centre.
(2) Every other source occurs at its correct relative position in the map but with half its true intensity.
(3) There is an identical spurious image for every source situated with the reference source as a centre of symmetry.
(4) There are fainter spurious peaks which correspond to the harmonics of, and cross modulation terms between, the various sources in the map. They occur at positions in simple geometrical relationships to the positions of the true sources.

(We shall show an example of the more complicated case where no source dominates later).

The analysis of such a map is straightforward only if the reference source is situated at the edge of the field of view so that the correct and spurious images can be distinguished easily. A case where this technique was applicable has been described by Riley & Pooley (1978) who used it for studies of 3C 123 at 15 GHz with the 5 km telescope. The structure of 3C 123 shows a large range in surface brightness and a bright, only slightly resolved feature, occurs at one side of the source. The original map using the measured phases is shown in Fig 3a whilst that obtained by setting the phases to zero is illustrated in Fig 3b. The improvement over the uncorrected map for the discrimination of faint features is a factor of about 3.

A more commonly occurring case is that of a hybrid map containing many point sources. If the relative positions and flux densities of some of the sources in the field of view are known then a hybrid map can be made with the observed amplitudes and the phases of a trial distribution composed of these point sources. The properties of this kind of hybrid map are:

(1) The sources in the trial occur in their correct relative positions and with nearly their correct intensities.
(2) Features in the true map but not in the trial appear in their correct positions but with half their true intensities.
(3) A large number of fainter spurious responses, both positive

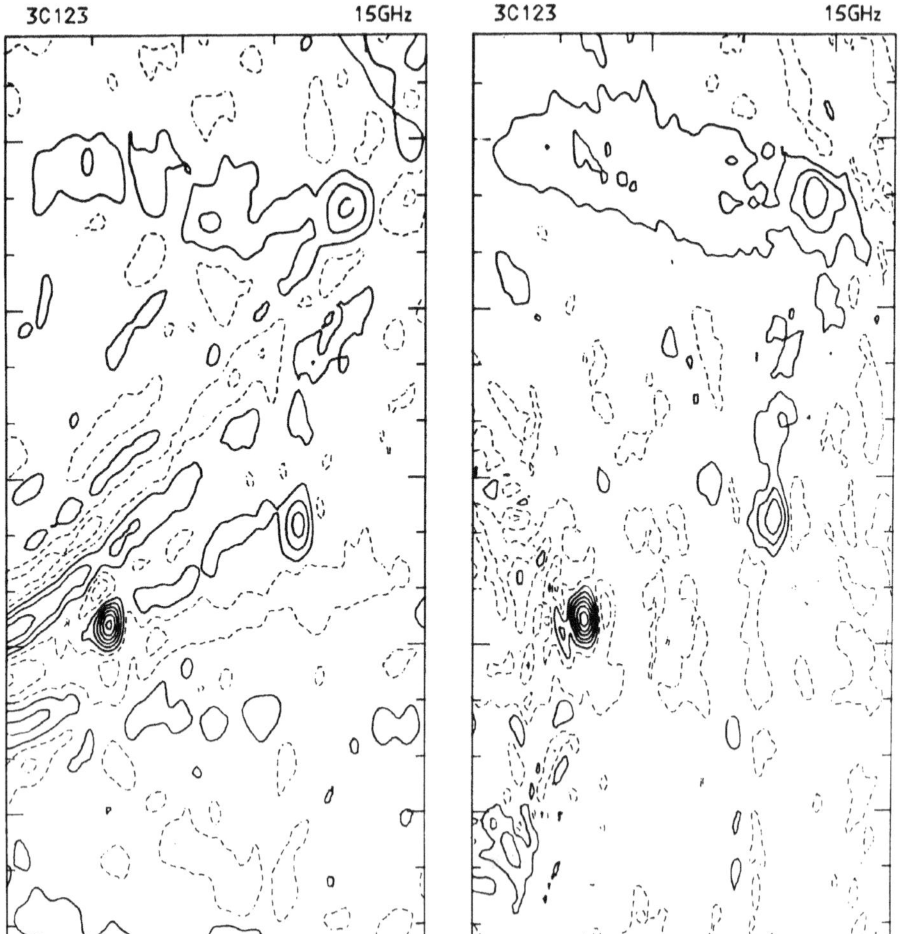

Fig 3. Improvement of sidelobe levels by using phase zero maps. Observations of 3C 123 at 15 GHz with a resolution of 0.65 arcsec. Left hand map; observed amplitudes and phases. Right hand map; observed amplitudes and phase zero. The bright reference source is the SE component which lies off the lower left hand edge of each picture. It would have 70 contours in Fig 3a and 140 in Fig 3b.

and negative occur at positions in a simple geometrical relation to the sources in the true map.

In practice it is often useful to plot the *difference map* i.e. hybrid map minus trial map.

It is clear that the best way of improving the trial is to add to it those features in (2) but at twice their apparent intensity. It is this property of hybrid maps which is the basis for the iterative improvement of the trial map. The problem is how to distinguish the real features from the spurious peaks. The brightest feature in the hybrid map will almost always correspond to a real source; the spurious peaks are fainter by a factor $a_m a_n/\Sigma a^2$ where a_m is the flux density of a source in the trial map and the sum is over all sources in the trial map. If the number of sources of comparable intensity in the trial map is large then the factor may be << 1, many new peaks can be included correctly at each iteration and convergence to the true map is extremely rapid.

We have described a test of this procedure using data from a 5C survey. The 5C7 survey (Pearson & Kus, 1978) at 408 MHz found 5 sources between 0.4 and 1 Jy and about a further 250 above the limiting flux density of 12 mJy. The noise level on such maps should be 0.5 mJy but Pearson & Kus found fluctuation with an rms of 2 mJy; they attributed this discrepancy to the effects of confusion and to calibration errors.

Using observed amplitudes and model phases the noise level is expected to be slightly more than $\sqrt{2}$ times worse than the ideal value, i.e. 0.8 mJy. On a difference map the effects of confusion are greatly reduced so the phaseless method offers the possibility of reaching fainter sources. Recent analysis shows that our earlier failure to reach the expected limiting sensitivity was mainly due to a lack of perseverence. Now after 6 iterations, 186 sources have been incorporated in the trial map down to a flux density of 14 mJy. The difference map shows about a further 100 sources down to 10 mJy and the rms noise level, excluding these sources, is 1.8 mJy or about 2.2 times that expected theoretically. The position and flux densities are in good agreement with those of Pearson & Kus. We do not know the reason for the remaining discrepancy in the noise level. A necessary precaution in reaching a result as good as this was the omission from the analysis of those samples in the aperture plane in which the amplitude fell below 2σ. Inclusion of such points both raises the apparent noise level on the difference map and was found to increase the flux densities of the sources in the trial map by about 4 per cent above their true values.

We conclude that this technique is successful for fields composed of point sources and that it may provide a useful extension of the dynamic range in maps which are limited by phase fluctuations in the data.

An important question concerning the method is how good must the starting model be? A simple estimate is that at least half the amplitude in any sample must be due to sources in the trial map. Under these circumstances the true phase must always lie within $\pi/2$ of the trial phase and the synthesis, though of poor quality, is certainly good enough to identify correctly the brightest peaks in the difference map. A series of model examples is shown in Fig 4. The true map contains 10 sources (in fact lying at the positions of the 10 brightest sources in the 5C7 survey). Difference maps corresponding to trial maps containing 1, 2, 3, and 4 sources respectively are illustrated in Figs 4a, 4b, 4c, 4d. The positions of the 10 sources have been marked by dots. All the other peaks are spurious images. In Fig 4d it is clear that all of the remaining true sources are brighter than the spurious images. In Fig 4b the brightest true source is only 1.7 times brighter than the largest spurious image and at most 2 sources could be added to the trial map at the next iteration. It is not disastrous to the method if a few spurious peaks are incorrectly assumed to be true sources, it only causes some oscillation in the convergence. Our experience in analysing the 5C7 data was that, at each iteration, the flux density of the faintest sources incorporated in the trial map could be reduced by a factor of about 2.

The quality of the initial trial map would be of no great importance if the analysis could be shown to converge from any starting point and that the result was unique. This last point is an important one. For any given distribution of amplitudes in the aperture plane there may be a large number of distributions of phase which would give acceptable maps. Provided that $\phi(x,y) = -\phi(-x,-y)$ the map will be real and the only criterion for choice then is that the intensity must be positive and that there may be some knowledge of the extent of the source. However, alternative solutions satisfying both criteria can sometimes be found in simple one dimensional cases.

Even if the positivity constraint is not satisfied strictly the region of negative intensity may be shallow and the map appear to be of acceptable quality. To illustrate how misleading an incorrect trial can be, a hybrid map made from a uniform amplitude A with trial phases appropriate to the same 10 sources used in Figs 4 is shown in Fig 5. Strong peaks occur at the positions of the 10 sources with flux densities in the ratio used in the trial. There are many negative sources on the map but the most intense are only about 10 per cent of the intensity of the main sources. This result bears no relationship to what all astronomers would agree is the correct solution to the uniform A problem, namely a single point source, and it illustrates how dangerous wrong initial models can be.

This ambiguity in the solution contrasts sharply with the apparent uniqueness of the result in the analysis of the 5C7 data. Could other solutions have been found for 5C7? Not if the basic assumption that the map is composed of point sources is adhered to. Each unresolved peak in the map of A^2 arises from the coincidence of two point sources in the autocorrelation process rather than from some ingenious distribution of positive and negative amplitudes which gives a value of the

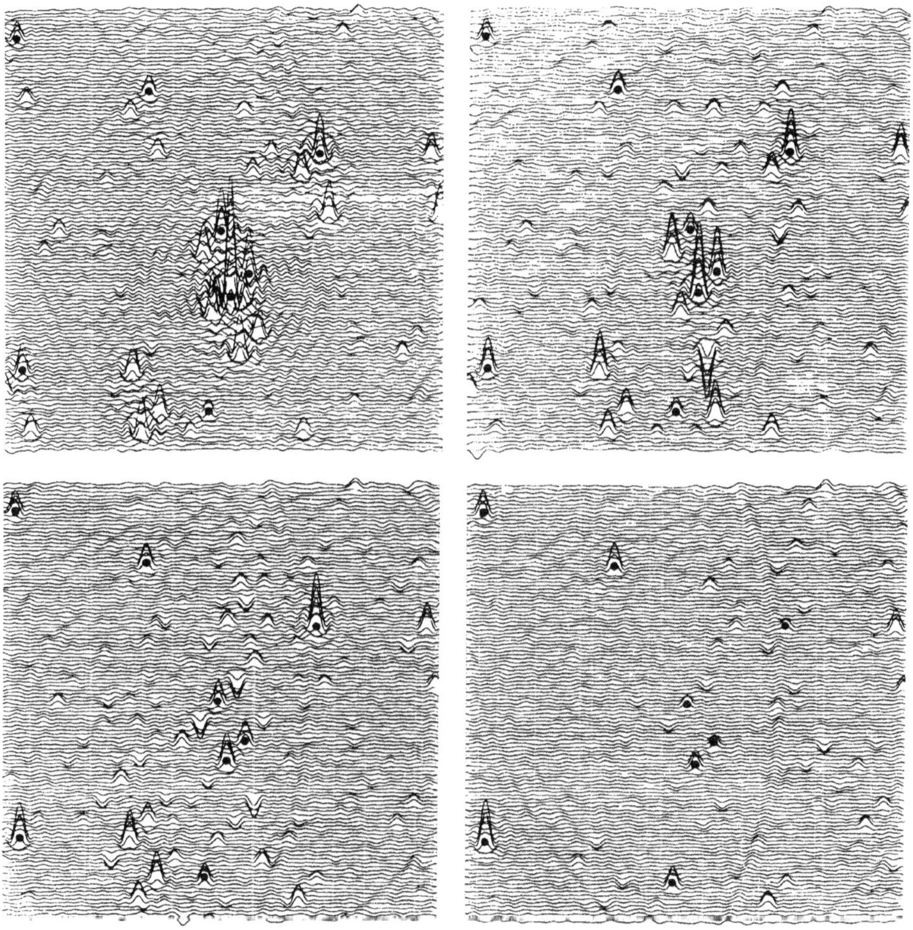

Fig 4. The effect of different trial distributions on the interpretation of hybrid maps. Difference maps made for a distribution of 10 point sources and trial distributions comprising successively 1, 2, 3 and 4 of the sources. Positions of 8 of the sources are marked by • , the remaining 2 lie outside the plotted area. The four sources of the trials are the central three and the upper right hand one. Notice that residual peaks at their positions gradually become smaller.

autocorrelation function at that point and zero at all surrounding points. The N(N-1) vectors in the A^2 map then uniquely define the arrangement of the N sources. Even if some of the faint vectors are undetected, there is usually no possibility for ambiguity in the arrangement of the bright sources.

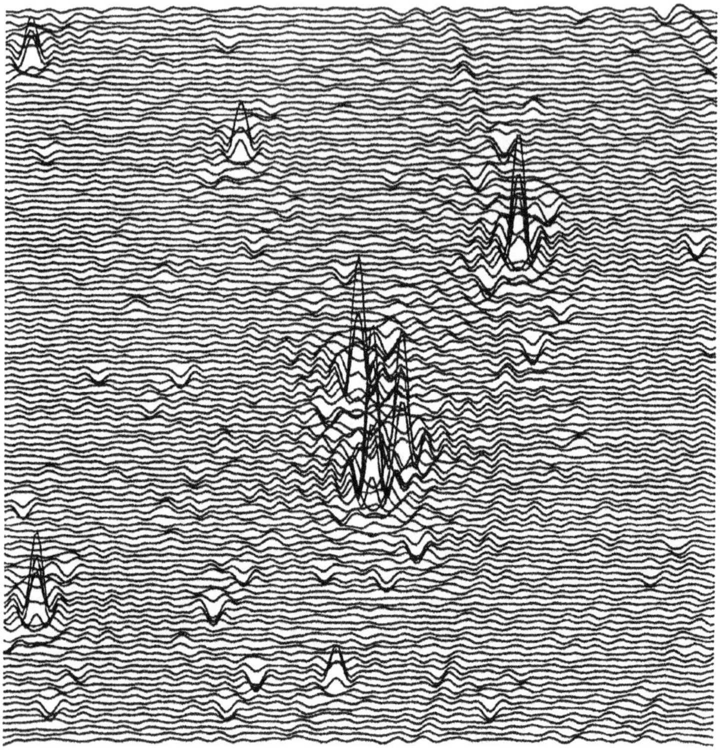

Fig 5. Hybrid map made from constant amplitude and phases appropriate to the distribution of the 10 sources in Fig 4.

4. EXTENDED SOURCES

Can this method be applied to extended sources? Where the extended structure accounts for less than about half the total flux density there seem to be no problems in its application as the discussion of 3C 123 demonstrates. But in cases where no trial map can be made from point sources the problem is much more difficult. An essential feature of the analysis of an A^2 map to give a trial distribution is the recognition of three vectors which together make up a closed triangle in the true distribution, a process which is simple only when there are discrete peaks in the A^2 map. For extended sources the A^2 map has a continuous distribution of brightness and there is no way of picking out particular sets of three vectors. The trial distribution is likely to be a much poorer first guess. Furthermore there can be substantial overlapping of correct and spurious images in the hybrid map which makes interpretation and advancement of the iteration difficult.

In this case progress can be made on the assumption that large areas of the map are zero unless there is definite evidence to the contrary. The limits of the area can be obtained from a study of the A^2 map. The idea of restricting the extent of a source to a 'window' is related to the use of a 'window' in the CLEAN procedure (Schwarz 1978) though in that case the phases (or beam shape) are known. In our case it is the phases which are unknown. This type of window has been used in the iterative procedure of Readhead & Wilkinson (1978) in which they also incorporate phase closure measurements. Successive iterations differ because the decomposition of the map into point sources by CLEAN is restricted to the region of the window. It has not been clear to us exactly what function the window serves in connection with the cleaning procedure or how important it is compared with the closure phase measurements in guiding successive iterations towards the correct distribution of brightness.

At first sight it is surprising that setting a region surrounding the source to zero can exert any useful influence. A study of the measurements necessary to distinguish an asymmetric pair of point sources from a symmetric triple source using amplitude data alone, helps to clarify the situation and is of interest in its own right for sources such as 3C 345 whose structure is a matter of argument. A triple source, comprising a central object of amplitude A and two outer sources of amplitude a/2 at separations of θ and $-\theta$ from the central source, gives an amplitude as a function of spacing d(wavelengths) of

$$A + a\cos(2\pi d\theta)$$

whereas a double source with amplitudes A and a and separation θ gives an amplitude

$$(A^2+a^2)^{\frac{1}{2}}\left\{1 + \frac{aA}{(A^2+a^2)}\cos(2\pi d\theta) - \frac{1}{2}\frac{(aA)^2}{(A^2+a^2)^2}\cos^2(2\pi d\theta) + \ldots\right\}$$

which contains higher harmonics of the terms $\cos(2\pi d\theta)$. It is these terms which distinguish symmetric and asymmetric models. They give rise to the sharp features in the amplitude - spacing curves noticeable where the amplitude falls towards zero. The presence of these high harmonics in the amplitude implies that the sampling of the aperture plane necessary to specify the amplitude completely, for a source whose over-all dimension is θ, is much finer than the normal information theory interval of $^1/_{2\theta}$. If the aperture plane is sampled at intervals of $^1/_{2\theta}$ it is impossible to distinguish the double and the triple models from amplitude measurements alone. If samples of A are taken at intervals of $^1/_{4\theta}$ then the fourier transform, taking an initial guess for the phase distribution of $\phi = 0$, correctly represents the triple source whereas the double source appears as a similar triple with extra responses at distances of 2θ, 3θ etc. from the central peak. It is at this point that knowledge of the maximum extent of the source becomes

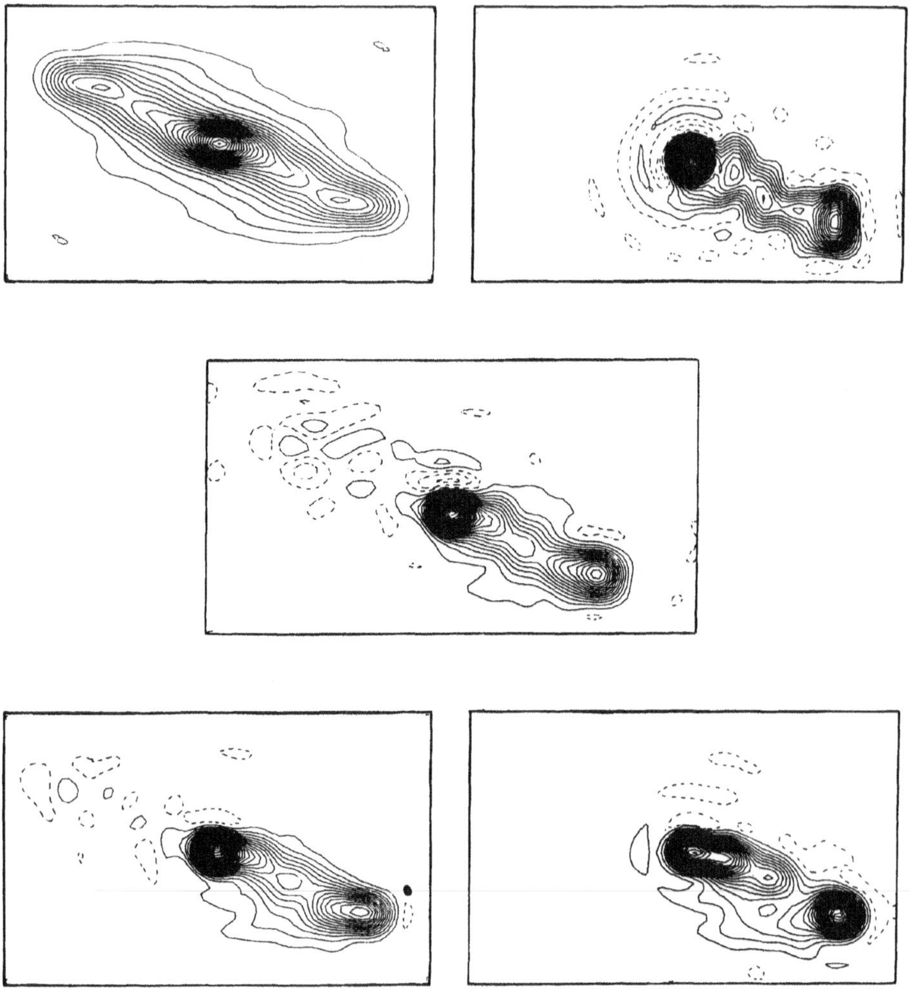

Fig 6. An attempt to derive the structure of 3C 382 from phaseless data using a window technique.
(a) Map of A^2, the autocorrelation of the true distribution.
(b) The initial trial map deduced from (a).
(c) The first hybrid map obtained from observed amplitudes and phases from (b).
(d) The hybrid map after 4 iterations. The window used to define the model for the phase calculation corresponds approximately to the first contour of the source.
(e) The map obtained from observed amplitudes and observed phases.

valuable and enables one to reject the multiple interpretation.

We must now enquire whether this information is powerful enough to provide an iterative scheme which will converge to the correct distribution of intensity. As an example we consider observations of 3C 382 made with the One Mile Telescope at 1407 MHz (S.F. Burch private communication). Fig 6a shows the map of A^2: the extent of the contours is a direct measure of the maximum vector separations which exist between regions of emission in the true map. An accurate trial distribution is hard to deduce from the map of A^2 but it is clear that it must have a predominant axis. The rather uniform intensity away from the peak at the origin and along the axis indicates that the true distribution must have peaks at each end which are joined by some form of bridge. There is no evidence whether the peaks are symmetric or not. An arbitrary ratio of 2:1 in their flux densities was adopted. This initial trial distribution is shown in Fig 6b. The hybrid map, made from the observed amplitudes and the trial phases is shown in Fig 6c. Comparison with the true map (Fig 6e) demonstrates both the improvement over the initial trial and the existence of spurious regions well away from the source. For the next iteration the main body of the source was represented by a set of point sources, following the practice with CLEAN. After 4 cycles the resultant map (Fig 6d) was changing very slowly. In some respects it closely resembles the true map: the shape of the bridge, the faint extensions to the SE and the sharp cut off to the N are all reproduced. The intensity ratio of the two main peaks and their shapes are less well reproduced: the valley running across the bridge does not appear.

Can we distinguish whether the result is an alternative solution to the correct one or whether the speed of convergence is so slow that very many iterations are needed to reach it? A simpler case than 3C 382 is more easily examined. Consider a source distribution composed of four point sources. Suppose that the trial solution exists comprising the three outer sources in their correct positions. The overall extent of the distribution is easily obtained by examination of the map of A^2. The hybrid map (Fig 7a) shows 6 positive peaks inside that overall extent (the 3 trial sources and the peaks A, B, C). These are returned as the next trial distribution for calculating the phases of the second hybrid map. The convergence towards the correct solution is much slower than that discussed for 5C7 since spurious peaks are incorporated into the trial in successive iterations. The normal factor of 2 applied to features in the difference map cannot be applied since the procedure then diverges in this case. The intensities of the correct peaks and the 2 spurious ones are shown as a function of the number of iterations in Fig 7b. Roughly 20 iterations are required to reach a good solution. We do not know how much improved our bybrid map of 3C 382 would be after so many iterations.

5. CONCLUSIONS

Maps from aperture synthesis measurements are usually limited by the errors in the phases. We have investigated methods for improving

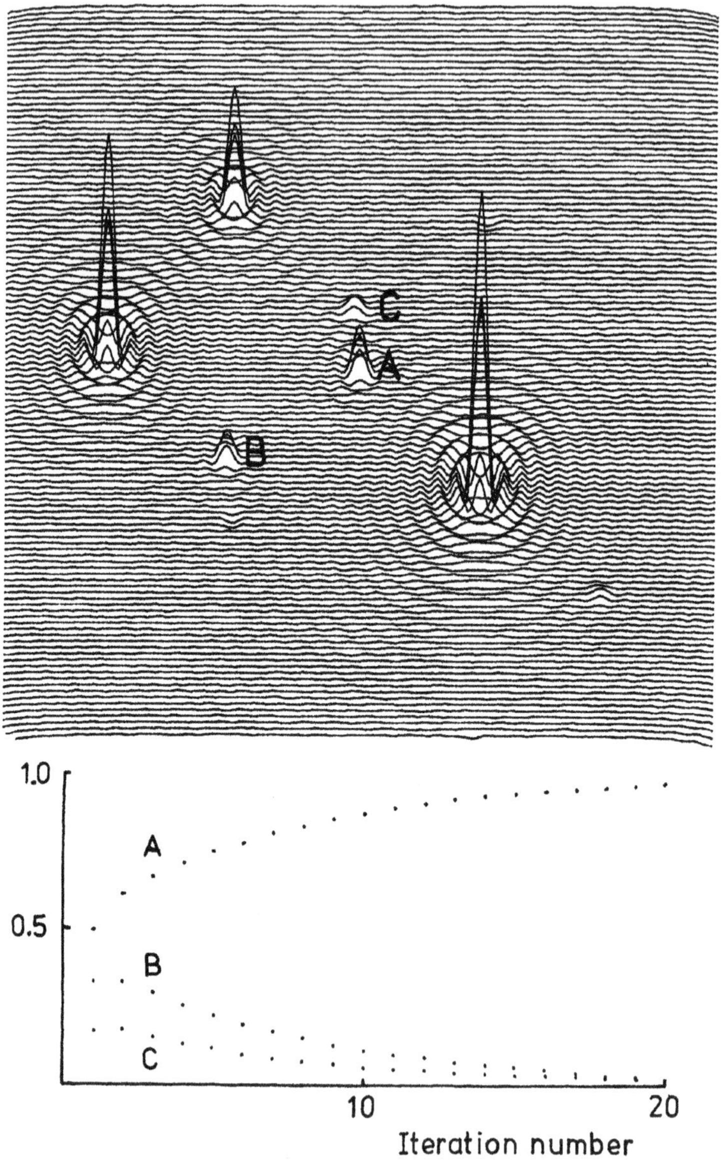

Fig 7. The use of a window technique on a field of 4 point sources.
(a) Hybrid map (1st iteration) made from amplitudes (3 bright sources and source A), together with phases (3 bright sources).
(b) Changes in the correct peak A and spurious peaks B and C in the hybrid map with successive iterations. The intensity scale is normalized to the correct value for peak A.

their quality by using the observed phases, if any, only to provide an initial model of the intensity distribution and thereafter relying solely on the amplitude measurements. The use of hybrid maps provides an easily understood basis for a scheme of iterative improvement and analysis of their properties leads us to believe:

(1) The reconstruction of fields of point sources can be effected to a high degree of accuracy even in the absence of phase information.
(2) The main features of the structure of extended sources can be derived provided that the overall extent of the source is defined and that the amplitudes in the aperture plane have been sampled adequately.

For extended sources the following questions need to be answered:

(1) What characterises a sufficiently good initial model?
(2) How many acceptable alternative solutions exist and how similar are they?
(3) Can the 'window' iteration be improved on?

Although the problems tackled here have arisen in a purely radio-astronomical context, it is evident that the results can be carried over to other fields where the phases are poor or unmeasured such as in speckle interferometry.

REFERENCES

Baldwin, J.E. & Warner, P.J. 1976. Mon. Not. R. astr. Soc., 175, pp 345-354.
Baldwin, J.E. & Warner, P.J. 1978. Mon. Not. R. astr. Soc., 182, pp 411-422.
Fort, D.N. & Yee, H.K.C. 1976. Astron. Astrophys., 50, pp 19-22.
Pearson, T.J. & Kus, A.J. 1978. Mon. Not. R. astr. Soc., 182, pp 273-274.
Readhead, A.C.S. & Wilkinson, P.N. 1978. Astrophys. J., 223, pp 25-36.
Riley, J.M. & Pooley, G.G. 1978. Mon. Not. R. astr. Soc., 183, pp 245-256.
Schwarz, U.J. 1978. Astron. Astrophys., 65, pp 345-356.

DISCUSSION (see also discussion after paper by Wilkinson - Readhead.)

Comment R.H. HARTEN
If you choose a "wrong" component early in the iteration scheme, i.e. a wrong source position, will the method converge and is it self correcting or does it require a critical eye to spot it?
Reply J.E. BALDWIN
If an incorrect component is put into the trial at an early stage it

can cause serious problems, as was shown by the constant amplitude case I described. At a later stage an incorrect component merely causes some oscillations which slow down, but do not prevent, convergence.

Comment J.J. WITTELS
1) It is necessary to be very careful in choosing the box in which the source is considered to be confined, especially early in the modelling, and especially if an FFT algorithm is used.
2) Computer size constraints create, or can create, a difficulty in choosing the correct grid spacing and a large enough box size to adequately model both the long- and short-baseline data simultaneously.
3) Is it not preferable to use closure phase with a connected interferometer rather than to throw out the noisy phase data completely? Have you tried it?
Reply J.E. BALDWIN
1,2) A good, i.e. conservative, method of defining the windows is to look at the A^2 map. That map gives a correct measure of the maximum vector separations in the source and hence of the area which wholly contains the source.
3) The main phase problems in connected interferometers are associated with uniform phase gradients covering the whole telescope array. The closure phases provide no information about the presence or absence of such gradients. We tend to think of this situation as one in which there is one unknown, not in terms of phase closure relations. Otherwise I agree with what you say. We did not try it in the cases I have just described.

Comment T.W. COLE
Did the talk misrepresent the relevance of "information theory"? It gives one a direct measure of redundancies, the need for more measurements if one throws out phases, and how windowing reduces the amount of independent information in the spatial frequency plane.
Reply J.E. BALDWIN
I hope it didn't! I agree that more amplitude information is clearly needed if we have no phase information, but it is much more than a factor of 2. For instance an equal double source has an amplitude which drops to zero at some spacings and the cusps there require indefinitely fine sampling in the aperture plane to specify then completely.

Comment R. GORDON
You noted that the error in phase can be estimated as a function of wavelength and spacing of antennae. It seems that this a priori information could be useful in constructing hybrid maps. During each iteration each phase could be constrained to remain within the error range of its measurement.
Reply J.E. BALDWIN
My impression is that if the observed phases are sufficiently good to provide a map which is the basis of the initial trial, then subsequent interations are unlikely to depart significantly from those input phases. The point is that the amplitude information is a much less powerful constraint than the phases. But I agree that your suggestion might help if only to a small extent.

MAP MAKING WITH NON-PHASE STABLE INTERFEROMETERS

P. N. Wilkinson
University of Manchester, Jodrell Bank, U.K.
A. C. S. Readhead
California Institute of Technology, Pasadena, U.S.A.

1. INTRODUCTION AND BACKGROUND

The ability to make reliable maps of radio sources at any available frequency and on all feasible angular scales is obviously a fundamental goal of observational radio astronomy. Unfortunately, until very recently, our map-making capability has been extremely restricted, limited in fact to those physical baselines and frequencies where phase-stable interferometers could be operated. However it is now becoming increasingly clear that phase stability is not an essential prerequisite for reliable mapping. The basic point is that whereas in principle there are an infinite number of brightness distributions which could give rise to the observed amplitudes of a set of Fourier components whose phases are unknown, in practice the *a priori* information that the required distribution is real and positive severely constrains the range of possibilities. Both Bates and his co-workers (e.g. Bates and Napier 1974) and Ross *et al*. (1978) have pointed out that the brightness distributions we are seeking are examples of 'entire' functions. In particular it is known that the real and imaginary parts of such functions are not independent (see references in Ross *et al*.) and that by studying the positions of their complex zeroes one may well be able to deduce the brightness distribution from knowing only the modulus of its Fourier transform. Unfortunately such a rigorous approach appears to be rather difficult to implement in practice. However the results of Napier and Bates (1974) have confirmed that in two dimensions reliable structure determinations can be made without phase information.

Based on this knowledge one can have some confidence in the various pragmatic approaches to the problem which have been developed. Of course the best known method of analysing fringe amplitude data is that of model fitting. In our opinion this method has obtained an unduly bad reputation for producing ambiguous results when in fact the ambiguity really arises because of the paucity of the observational data. When more extensive data have been collected, for example by Purcell on 3C147 (see Wilkinson *et al*.1977) a careful model fitting analysis assuming Gaussian components produced what has subsequently

C. van Schooneveld (ed.), Image Formation from Coherence Functions in Astronomy, 83-91.

been confirmed to be a good representation of the source structure. In fact the basic precepts of model fitting are sensible. If, for example, gaussian components are used they are: (i) assume that the sky is basically smooth and non-negative (i.e. blank) (ii) assume that the source is basically smooth (Gaussians are infinitely differentiable) (iii) use the minimum number of components to fit the data within the noise (i.e. extract the least amount of information while remaining consistent with the data). These are just the ideas underlying the maximum entropy method! The major problem with model-fitting - like all non-linear approaches - is that it can take extravagant amounts of computer time to arrive at an acceptable solution for all but the simplest sources. However for simple sources it remains a convenient way of parametrising the visibility data. Baldwin and Warner (1978) have demonstrated another approach which seems to work well for complex sources - at least if they contain several well separated compact regions or one dominant compact feature. This method may run into difficulties if the emission is smoothly distributed however and Baldwin and Warner give little indication of the computing effort required to reach an acceptable solution. The method applied by Conway and Stannard (1975) to map the 3C273 jet also depends critically on the presence of a strong compact feature in the source.

In many cases of practical interest some phase information is always available no matter how badly disturbed are the individual phases in an interferometer network. The 'closure' phase (Jennison 1958; Rogers et al. 1974) is a linear combination of observed phases around closed loops of baselines and contains no systematic errors. The only restriction its use implies is that the fringe amplitude must always be above the noise on all baselines in the loop. For N telescopes there are at most (N-1)(N-2)/2 independent closure phase relationships; one may in fact include all baselines with fewer than this but this is not desirable. Each closure phase tells us as much about the shape of the source as does each fringe amplitude and thus it is clearly logical to include this phase information in the reconstruction. By this means one expects to reduce the residual uncertainties in amplitude only reconstructions. Chief among these uncertainties is the $\pm 180^{\circ}$ position angle ambiguity and this is resolved by the closure phase. Still however the absolute position of the source on the sky is lost. Three methods of using this information have been published (Fort and Yee 1976; Wittels et al. 1977; Readhead and Wilkinson 1978) and the results confirm that with the extra, and extremely strong, constraints on the visibility phase imposed by the closure phase one may anticipate being able to reconstruct arbitrary brightness distributions with high reliability.

A way of obtaining relative position information and of ensuring that at least the relative phases of the Fourier components are correct is to calibrate the interferometer phase with a nearby point source (e.g. Peckham 1973). Clearly then a map can be made using conventional techniques for the analysis of synthesis data. The success of this method depends entirely on the presence of a sufficiently strong, compact, reference object close enough to the source of interest and this angular distance is a function of interferometer baseline and observing frequency. Phase-referencing has been successfully employed at 408 and 1660 MHz on the MK IA-MK III and MK IA-Defford baselines (24 km and 127 km) using reference sources of ≳100 mJy. If the switching period is ≲10 mins relative phase deviations ≲0.2 rad can be achieved on arbitrary sources. At 408 MHz the reference source is usually in the same beam whereas at 1660 MHz angular throws of up to $\sim 2^{\circ}$ have been used. At lower and higher frequencies, especially on longer baselines, little work has yet been done but it is to be expected in the near future. However if phase-referencing does prove to be difficult, at metre and short centimetre wavelengths on VLBI baselines, the closure phase will always be available and we shall now discuss our method of using it.

2. A PRACTICAL RECONSTRUCTION METHOD USING CLOSURE PHASE

Readhead and Wilkinson (1978) have demonstrated a linear approach to the use of closure phase which can best be understood with an example. From a four telescope six baseline interferometer network one can obtain three independent closure phase relations, C, Thus:

$$C_{123} = \phi_1 + \phi_2 - \phi_3$$

$$C_{245} = \phi_2 + \phi_4 - \phi_5$$

$$C_{346} = \phi_3 + \phi_4 - \phi_6$$

where the ϕ_i are the six unknown visibility phases. If we have *a priori* knowledge which enables us to establish three of the phases independently then the other three can be calculated from these closure phase relations. This *a priori* information exists in the form of the fringe amplitudes for, as we have discussed above, much can be said about the visibility phases from the amplitudes alone. In practice simple models have proved quite adequate for providing such starting phases. The flow diagram for the iterative procedure we have adopted is shown in Fig. 1. Note that after the initial use of the closure phases the method uses exactly the same software as is contained in any standard CLEAN (Högbom 1974) package.

Fig. 2 shows the method at work on a simple source and enables us to understand why convergence to a stable solution takes place. The source in the top left hand corner is a point double with 2:1 flux ratio as would be mapped with a four telescope, phase-stable, network using CLEAN. Below it we show the first iteration obtained using the three closure phases and assuming that three of the visibility phases are zero. Obviously as all the phases are incorrect the resulting map contains spurious features and the essence of the method lies in its use of CLEAN to remove these errors. The fundamental point is that if we subtracted sources over a large area of sky we should, on Fourier

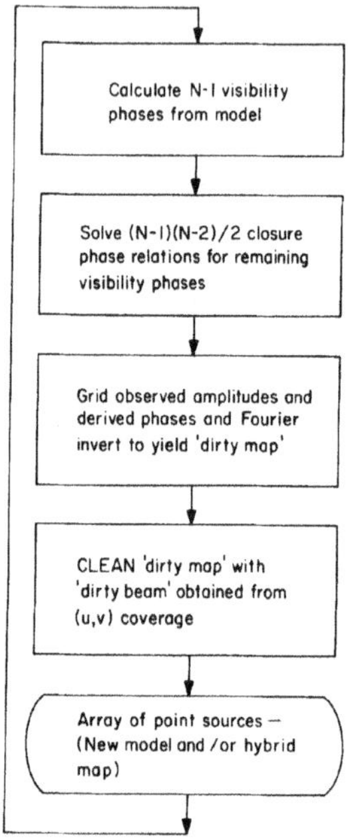

Fig. 1. Flow diagram for a linear reconstruction method using closure phase

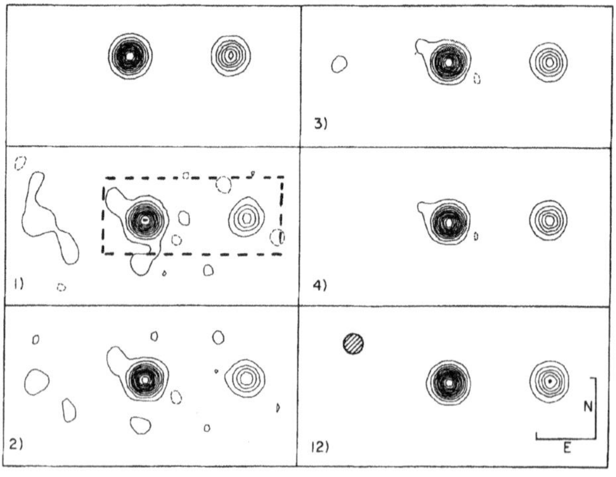

Fig. 2. Illustration of method on 2:1 point double source

transformation, reproduce the input phases exactly since CLEAN is a linear process; in this case no convergence would occur. However in line with the original philosophy behind CLEAN we assume that most of the sky is empty and restrict our search to a 'window' around where we suspect that the source lies. We thereby systematically reject spurious components and the map therefore tends towards the true distribution as the iterations proceed. In Fig. 2 the CLEAN window is the whole area of each map and the iteration number is indicated in the bottom left hand corner. The effect of the window size on the rejection and hence the speed of convergence can be judged by noting that when we only subtracted sources within the area inside the dashed line convergence was twice as fast. More visibility data is obviously helpful and with a five telescope network and the smaller window convergence was three times as fast.

Thus for a simple source no initial subjective judgement at all is required to arrive at the correct structure. For more complicated sources it is sensible to try and start with a plausible model to reduce the number of iterations required to arrive at the correct solution. In order to convince ourselves of the efficacy of our approach on non-trivial sources we performed 'blind' tests on simulated amplitude and closure phase data to reproduce as closely as possible the real situation. We regard such tests as vital for any brightness reconstruction method where the observer can interact to aid convergence. The results of two such tests are shown in Figs. 3 and 4. In each case a) is the test source (not convolved with the restoring beam) b) is the CLEAN solution from full amplitude and phase data from a four station network and c) is the solution obtained from amplitude and closure phase. In Fig. 4(d) we show the marked improvement obtained by adding a fifth station to the array.

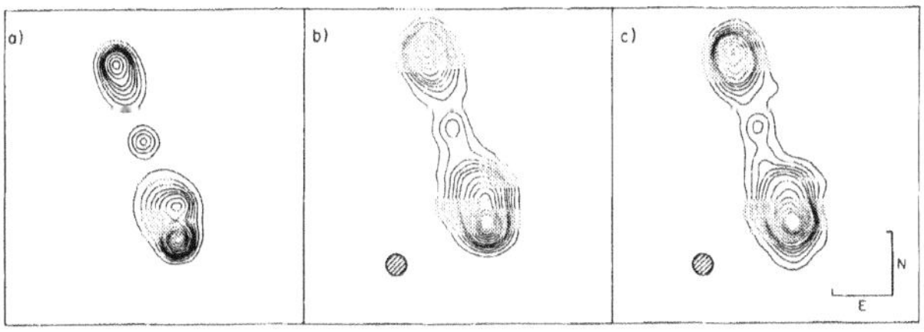

Fig. 3. 'Blind' test of method

To help the method to converge in these more complicated examples we have used three other ways of reducing the spurious features: (i) return only positive point sources for use in the initial phase calculation (ii) in the first few iterations, when many weak spurious features exist, do not clean very deeply - increase the depth as the

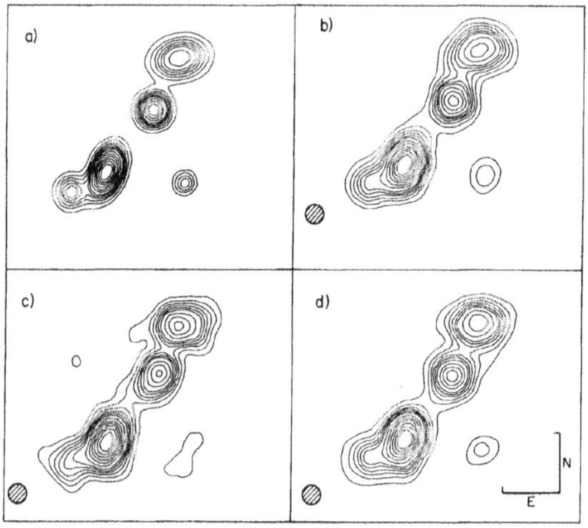

Fig. 4. 'Blind' test of method

iterations proceed (iii) starting from a stable map which is suspected to be close to the true distribution derive other maps by changing the N-1 baselines on which the phase is calculated in the initial step. If the first map is the true distribution then these subsequent maps will all be the same to within the noise. If it contains minor errors then the other maps will be different and the error regions tend to move to new locations in each map. Thus a better map for use as the input to the next iteration can be obtained by adding these trial maps together, for the correct areas in them add coherently and are thereby enhanced with respect to the spurious ones.

It is important to give some idea of the speed of this process. In the examples shown in Figs. 3 and 4 (array size about 30 x 30) each iteration (one job) took typically 1 minute CPU time on an IBM 370/158; this involved cleaning and returning $\lesssim$200 point sources per iteration. On average we allowed ourselves $\sim$20 jobs, trying various different approaches, before comparing our preferred solution with the test source.

A particularly neat application of this method arises when the source contains a bright compact region which can be used as a phase reference. In this case convergence is extremely rapid (e.g. Wilkinson et al. 1977). Fig. 5 shows a test on simulated four station data on Purcell's model of 3C147. Here again a) is the model b) is the CLEAN solution and c) is the solution using closure phase. Two things are noteworthy: first the good agreement between the full and the closure phase solutions and second the fact that one can clearly reproduce most of the features in such an extended source out to $\sim$20 beam diameters. This test gives us confidence that our actual maps of 3C147 (Wilkinson

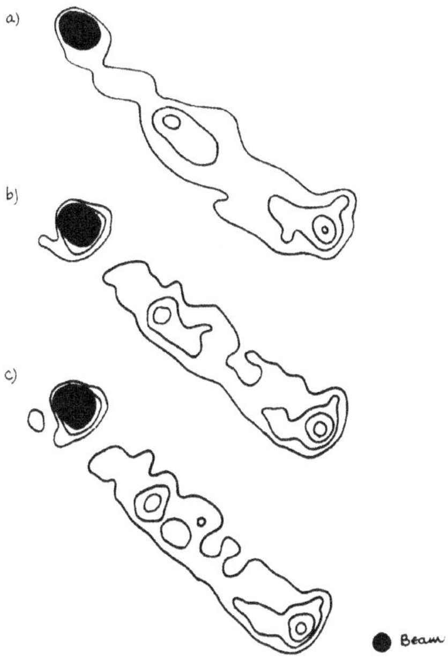

Fig. 5. Test on Purcell's model of 3C147

et al.1977) accurately represent the compact features in the source.

3. CONCLUSIONS

The obvious way to achieve a degree of phase stability in an interferometer network is to employ phase-referencing techniques. Not only is positional information collected but the ability to make long coherent integrations means that weak sources can be studied. However even if this is not possible it is clear that one can still obtain data which are sufficient for determining the structures of the stronger sources. The closure phase is particularly easy and error-free data to collect and indeed there may well be cases where individual relative phases can only be determined to say $\pm$ 30° r.m.s. while the more restricted but more accurate closure phase (say $\pm$ 5° r.m.s. on a strong source) can lead to a more reliable map.

ACKNOWLEDGEMENTS

We thank M.H. Cohen for his enthusiastic support and D.H. Rogstad for many discussions regarding his CLEAN package. PNW gratefully acknowledges the receipt of the Royal Society Weis Fellowship. ACSR acknowledges support via NSF grant no. AST 77-00247.

REFERENCES

Baldwin, J. E. and Warner, P. J.: 1978, Mon.Not.R.astr.Soc., 182, 411.
Fort, D. N. and Yee, H. K. C.: 1976, Astron.Astrophys., 50, 19.
Högbom, J. A.: 1974, Astron.Astrophys.Suppl., 15, 417.
Jennison, R.C.: 1958, Mon.Roy.astr.Soc., 118, 276.
Napier, P.J. and Bates, R. H. T.: 1974, Astron.Astrophys.Suppl., 15,427.
Peckham, R. J.: 1973, Mon.Not.Roy.astr.Soc., 165 , 25.
Readhead, A. C. S. and Wilkinson, P. N.: 1978, Astrophys.J., 223, 25.
Rogers, A. E. E., et al.: 1974, Astrophys.J., 193, 293.
Ross, G., et al.: 1978, Proc.Roy.Soc., 360, 25.
Wilkinson, P.N. et al.: 1977, Nature, 269, 764.
Wittels, J. J., et al.: 1976, Astron.J., 81, 933.

DISCUSSION

Comment T.A. CLARK
I concur with Wilkinson's comments that model-fitting has been given "bad press". When the model fitting is approached with statistical procedures, either as described by Cotton or below, they can be considered quite similar to maximum likelihood estimation.
An additional comment on phase closure analysis. A slightly different approach taken by L.K. Hutton and me was to form a two dimensional "bed-of-nails" delta-function representation of the source. By direct Fourier Transform, the entire source was simultaneously adjusted in a weighted non-linear least squares analysis to minimize the differences between the actual amplitude and closure phase data and that predicted by the model. An efficient gradient-search least-squares algorithm made this quite computationally efficient. Initial constraints consisted of a maximum window size determined by the shortest baselines and/or a priori source knowledge, source grid spacing determined by maximum resolution, and a positive flux constraint on each grid point. A starting point that was found necessary was to begin with one point non zero to "nail down" the final map. Generally the final map "grew" with the strongest component at the position of this "nail". The final delta-function map was convolved with a gaussian beam similar to the point source response determined from the U-V coverage.

After the presentation by Wilkinson and Readhead a general discussion developed on the subjects of closure phase and positivity. Several of the previous speakers answered questions and comments.

Comment G. POOLEY
Several people have mentioned using a constraint that the data are positive; in measuring polarization data you cannot use this constraint.

Question B.C. CLARK
In the case where one is approaching the signal-to-noise limit, one presumably wishes to prevent, or at least discourage, the phases measured on the baselines where the amplitudes are low, and signal-to-noise ratios bad, from entering into the closure phases, and hence affecting the solved phases on other baselines. Is there a systematic procedure for handling this?
Reply J.E. BALDWIN
Our analysis of 5C7 does not bear on the closure phase problem but we did indeed find it was essential to remove amplitude data where it fell below 2σ. Otherwise, this noisy data combined with the phase of the trial conspired to put extra flux density into the trial source in the next hybrid map. The 2σ limit was an arbitrary decision and there may be some more proper procedure.

Comment by D.B. SHAFFER to J.P. HAMAKER
If you are willing to give up all positional information, phase closure will give you the correct shape map (but not its position).
Reply J.P. HAMAKER
Let me answer your remark in terms of an array which produces not only closure phase observations but also contains enough redundancy to solve for all instrumental phase errors. What we can obtain then are one-dimensional strip scans of perfect shape but unknown position, as you suggest. To align these in a two-dimensional map, however, one does need the relative position of these scans, if one wants to avoid the radial streaks which I showed before. These relative positions are not provided by the closure phase or redundancy information and one must make an educated guess of one sort of another. Part of my short paper (J.P. Hamaker: "Kneading", this volume) has to do with precisely this point. I shall try to think your remark over and presently make some additional comments.

Comment R. SRAMEK
When discussing closure phase, it is best to think of the phase of a single baseline interferometer to consist of 1) a baseline dependent part ϕ_{12} resulting from e.g. source structure, correlator problems, etc. and 2) an antenna dependent part $\phi_1-\phi_2$ which is due to e.g. front-end problems, the atmosphere, etc. The closure phase is independent of the antenna dependent phase, no matter what the antenna geometry, VLBI or connected antennas, or the origin of ϕ_1 and ϕ_2.

IMAGE RECONSTRUCTION FROM VLBI AMPLITUDE AND PHASE CLOSURE DATA

William D. Cotton and Jill J. Wittels
Massachusetts Institute of Technology
Cambridge, Ma, U.S.A.

ABSTRACT

A two step procedure is developed for the determination of the angular structure of the compact components of radio sources from observations made with very long baseline interferometry (VLBI). The first step is to fit a model composed of circularly symmetric Gaussian components to the observations. Then this, or any other, model and closure phase information is used to predict visibility phases which, when paired with the corresponding visibility amplitude observation, are combined in a direct synthesis; the resulting map is deconvolved from the point source response by means of the "CLEAN" procedure. If necessary, the "cleaned" components can be used as a starting model to predict visibility phases and the process iterated. The initial model in many cases may be as simple as a point source. Convergence of this process appears to be quite rapid, requiring only a few iterations when the source is simple or the starting model is good. These two steps for source mapping can be used either together or separately. Examples are given using data generated from test models.

C. van Schooneveld (ed.), Image Formation from Coherence Functions in Astronomy, 93.

ON THE DANGER OF APPLYING STATISTICAL RECONSTRUCTION METHODS IN THE CASE OF MISSING PHASE INFORMATION

J.C. Dainty, M.A. Fiddy and A.H. Greenaway[+]
Department of Physics, Queen Elizabeth College,
Campden Hill Road, London W8 7AH, U.K.

[+]Département d'Astrophysique, Université de Nice,
Parc Valrose, 06034 Nice Cedex, FRANCE.

I. INTRODUCTION

The use of entropy as a basis for object/image reconstruction procedures is not new[1], but with the appearance of new, faster algorithms[2] the actual use of these algorithms for the reconstruction of objects from 'real' data is likely to increase.

The purpose of this contribution is not to discourage such applications, but to illustrate that, under certain circumstances, there is a need for caution in interpreting the results obtained from such algorithms. Specifically, we shall show that the application of statistical methods to problems of object reconstruction, in situations where only the modulus of the object Fourier transform is known, could lead to wholly false conclusions. Indeed, we shall primarily be concerned here with situations for which *there is no 'correct' solution.* In such situations it is pointless to speak of 'safe' object reconstruction algorithms. The important point here is that the user of a statistically based 'object reconstruction algorithm' may be totally ignorant of whether or not he is working in this régime.

The problem thus formulated, i.e. the reconstruction of an object from its autocorrelation, is the so-called phase problem, which occurs in fields as diverse as electron microscopy, radio and optical astronomy and X-ray diffraction. It has long been recognised that, in general, the solution to such problems cannot be unique. In all cases it is necessary to provide additional information, such as the phase of the object Fourier transform, in order to resolve the inherent ambiguities associated with this problem.

The key point to this paper, is that to speak of 'safe' reconstruction procedures, one must show that the problem posed has a unique solution. If one can show that this is indeed the case, any algorithm which yields a solution consistent with the defined problem will give the correct solution. In this sense all such algorithms may be regarded as 'safe'.

C. van Schooneveld (ed.), Image Formation from Coherence Functions in Astronomy, 95-101.

The choice of algorithm will then be dictated by the speed and ease with which it may be implemented, and its stability against the effects of noise and data truncation. It is in this latter sense that maximum entropy algorithms may be regarded as 'safe'.

By application of the theory of entire functions we shall demonstrate the fundamental ambiguity associated with phase problems and illustrate this with some specific examples. We shall also give consideration to the conditions under which the degree of ambiguity may be reduced, or where a unique solution may be obtained.

Initially, we shall consider the situation where the recorded modulus data is perfect and complete. The situation where the data is noisy and/or where some phase information is available, will be considered later.

2. THEORY

It is not our intention here to give a rigourous treatment of the phase problem in terms of entire function theory, but we shall briefly indicate the results of this analysis[3].

Let the object field to be reconstructed be represented by $f(\xi)$ and its Fourier transform by $F(x)$, then

$$F(x) = \int_a^b d\xi \; f(\xi) \; \exp(-i\xi x) \tag{1}$$

where the limits of integration (a,b) will always be finite, either because of the finite field of view of the instrument or due to the finite extent of the object. Attention will be confined to this one-dimensional case, since this simplifies the analysis and illustrates all the features we wish to discuss.

The use of equation (1) to describe the system, imposes some severe and far reaching restrictions on the behaviour of $F(x)$. Specifically, any $F(x)$ which is a solution to an equation such as (1), must be the limit on the real(x-) axis of an analytic function which is of order unity and of finite type. Thus $F(z)$, which may be generated by replacing x with $z = x + iy$ in equation (1), is analytic for all finite values of z and may be specified, to within trivial ambiguities, from a knowledge of the set of points at which it is identically zero. In this respect $F(z)$ is like a polynomial whose roots (i.e. zeros) must occur at isolated points. For an $F(z)$ generated from equation (1) there will always be a denumerable infinity of such roots. These zeros are distributed so that in any finite region of the z-plane there will only be a finite number of roots, their density being proportional to $(b-a)$[4]. Furthermore, all such zeros must lie close to the real axis and will tend to lie along the x-axis at the Nyquist sampling rate as $x \to \pm\infty$. Thus with any finite section of the x-axis we may associate a finite number of zeros. It has been noted[5] that this is the same as the number of degrees of freedom of the system.

Formally, we may express the relationship between F(x) and the roots of F(z) by

$$F(x) = F(0) \exp(-ix\{\frac{b-a}{2}\}) \prod_{j=1}^{\infty} (1-\frac{x}{z_j}) \tag{2}$$

where z_j, j=1,2,... are the roots of F(z) arranged such that $|z_{j+1}| \geq |z_j|$. In writing equation (2) it has been assumed that $F(0) \neq 0$; this is always true in astronomy where objects are real and positive; the condition can be relaxed if required.

When the phase of F(x) has been lost, the observable magnitude is the scattered intensity, which is proportional to

$$I(x) = |F(x)|^2 = F(x)\, F^{\star}(x) \tag{3}$$

Now visual inspection of (2) shows that $F^{\star}(x)$ will have its roots at $z_j^{\star}$, j=1,2,... . Thus I(z), the continuation of I(x) into the z-plane has zeros at z_j and $z_j^{\star}$ and so has twice as many zeros as F(x). Thus, when attempting to reconstruct F(x) , or equivalently f(ξ), from I(x) roots of F(x) which occur on the x-axis may be identified with certainty but for the complex roots, i.e. those not on the real axis, it is impossible to decide whether F(z) has a root above or below the x-axis. If there are M complex zeros, the repetition of this twofold ambiguity for each such zero given rise to 2^M possible alternative solutions for F(x), all of which correspond to the same intensity, I(x), and all of which have inverse Fourier transforms which are confined to $\xi\varepsilon(a,b)$. *On the basis of measurements of* I(x) *alone, there is no possible way to distinguish between these alternatives.* It has been known for some time that, with perfect data and given the constraints imposed by equation (1), this is the *only* source of ambiguity.[6]

In some situations it is possible to resolve or reduce this ambiguity by making f(ξ) satisfy certain *a priori* conditions. For example, if f(ξ) is forced to take a known value over some ξ-range within the interval (a,b)[7], or if it is known *a priori* that all zeros lie in one half of the complex plane[3], a unique solution may be ensured. A special case of all zeros being in one half plane occurs when they are all real. The existence of a unique solution in this case has been exploited to determine the structure of nerve myelin from X-ray data[8]. If f(ξ) is known to be real, the zeros of F(z) must satisfy certain symmetry properties which reduce the ambiguity by a factor of $2^{M/2}$.

Given that difficulties of this kind exist with perfect intensity data it is essential that any reconstruction algorithm which uses intensity data is capable of resolving this implicit ambiguity. It is not clear that statistically based 'object reconstruction' algorithms, such as maximum entropy, can take any account of such ambiguities, the importance of which are illustrated in the following section.

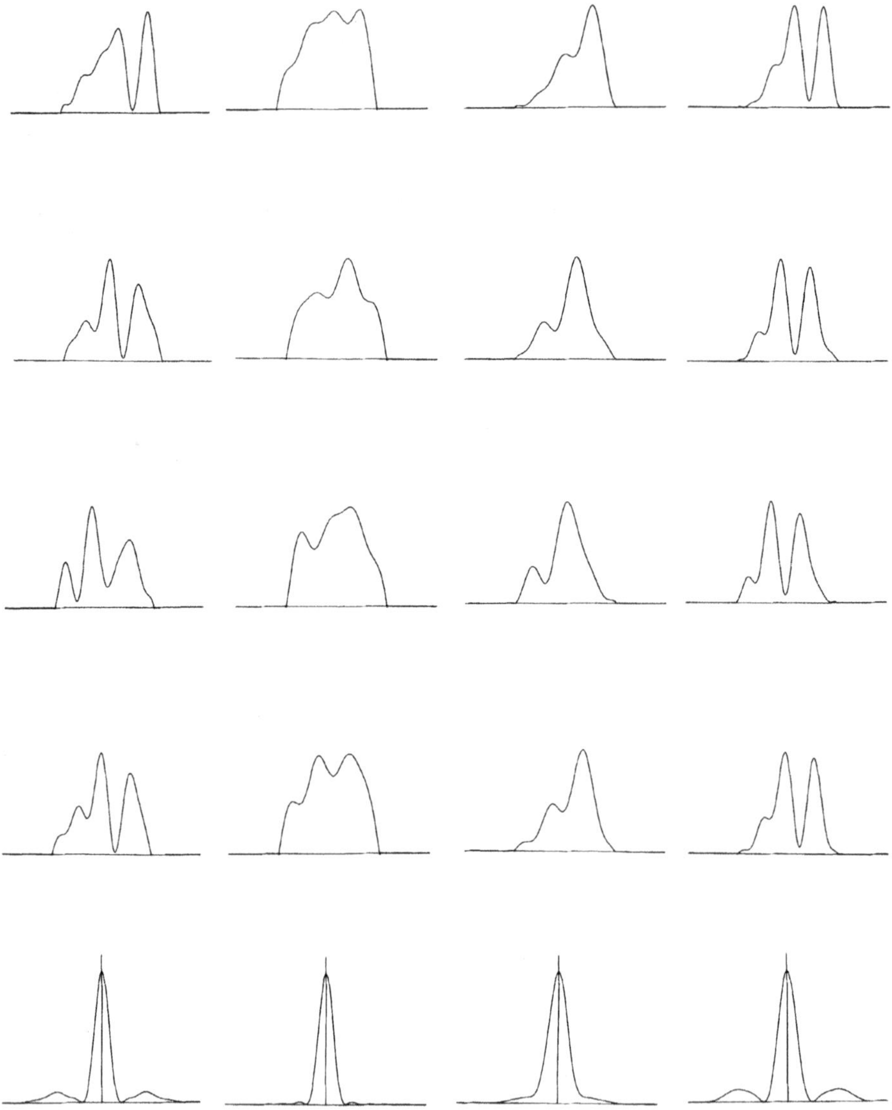

Figure 1 Four examples of ambiguous object reconstruction. The upper four graphs in each column show four different objects having the same I(x), which is shown in the lower graph. For discussion see the text.

3. EXAMPLES OF AMBIGUOUS RECONSTRUCTIONS

In figure 1 we show four examples of ambiguity in object reconstruction from measurements of I(x) alone. In each example we have constructed an F(z) which has just six complex zeros, the remainder being real. Each example therefore has 2^6 possible solutions for $f(\xi)$, all wholly consistent with equation (1). If we specify that $f(\xi)$ must be real, then there must be a symmetry of the zeros of F(z) about the y-axis. Enforcing this symmetry reduces the ambiguity to 2^3 equally acceptable real solutions for each example. Now, reflecting any given distribution of zeros about the x-axis is equivalent to replacing $f(\xi)$ by $f(-\xi)$. This difference we shall neglect, so reducing the ambiguity by another factor of two. For each example this leaves four equally valid, real and positive solutions for $f(\xi)$. These are shown in figure 1.

Each example occupies one column of the figure, the upper four graphs of each column being the four possible object reconstructions which may be made from the observed intensity shown in the lower graph of each column. In each case the top graph of each column corresponds to the minimal or Hilbert phase solution[3], with all zeros in one half plane.

Note that for none of these examples may one speak of a 'correct solution', in each case the four alternatives represent equally valid object reconstructions. There can be no question of *any* algorithm providing the 'correct solution' unless other information is available. The examples in figure 1 show that this statement is non-trivial.

In the first example, shown in the left hand column, one cannot even say with certainty how many peaks there are in the object distribution. In the second example it is clear that the object consists of a single peak of reasonably well defined width, but what can one say with certainty about the structure superposed on that peak? The ambiguities in the third and fourth examples are less severe but even here the width, relative position and height of each peak is somewhat uncertain.

4. INCOMPLETE AND NOISY DATA, WITH AND WITHOUT PHASES

So far it has been assumed that no information about the phase of F(x) is available. Clearly, if both the modulus and phase of F(x) are available and the data is perfect and complete, the ambiguity is resolved, thus the zero positions are specified exactly.

If we incorporate the effects of noise and data truncation into our previous description, it is evident that, even when phase information is available, we can only obtain *estimates* for the positions of the zeros of F(z). If no phase information is available and I(x) is contaminated with errors, not only is the exact position of each zero uncertain but the fundamental ambiguity, i.e. determining in which half plane each zero belongs, must also be resolved. Since errors are inherent in measured data we must resign ourselves to defining regions in which the zeros must be located, rather than exact zero positions. More exactly, we should define a probability density function for the position of each zero. *In the absence of phase information this p.d.f. will be symmetric with respect to the real axis.*

When some phase information is available we may expect the p.d.f.s to exhibit a maximum to one side of the real axis. Under these circumstances some algorithm for selecting an, in some sense, optimum set of zero positions consistent with the p.d.f.'s would be useful. Such a procedure, based on the theory of entire functions, has already been proposed and used to reduce radio astronomical data.[9] This method relies on the experimental observation that when both the phase and modulus are measured experimentally the uncertainty in the phase frequently exceeds that in the modulus. One may therefore use I(x) to determine the zero co-ordinates, with the phase used merely to decide in which half plane each zero belongs. The resulting reconstructions may be considerably more accurate than those made using the raw data. The maximum entropy algorithm may well fulfil the same rôle without the need to invoke the theory of entire functions explicitly.

5. CONCLUSIONS

We have demonstrated the well known result that in the absence of additional information it is, generally, impossible to restore an object unambiguously from measurements of the modulus of its Fourier transform. We have shown with counter examples that positivity of the solution does not constitute the additional information needed to resolve this ambiguity. Nor do statistically based algorithms, e.g. maximum entropy, provide such information. Thus they offer no way out of the dilemma and they cannot be regarded as giving a 'safe', i.e. unique, restoration of the objects from the modulus of their Fourier transforms. Arguments that errors in real data may negate these conclusions are equivalent to suggesting that ignorance provides a suitable basis for inference.

If phase information is available, the maximum entropy algorithm may represent a useful approach to the problem of selecting an optimum set of zero positions consistent with imperfect data. In the absence of phase information, statistical algorithms should provide an optimized version of one of the 2^M possible object reconstructions (perhaps the Hilbert solution). They will not, in general, provide any information about the other possible solutions. This 'lost' information could be retrieved by using the optimized $f(\xi)$ to calculate the roots of $F(z)$ and proceeding as above. Note, however, that in the examples discussed here M has been deliberately kept small. In 'real' experiments it may be much larger with a consequent increase in the number and range of solutions.

ACKNOWLEDGEMENTS

This work has been partially supported by the S.R.C. in the U.K. and by the C.N.R.S. in France under the European Scientific Exchange Scheme.

REFERENCES

1 Frieden B.R. : 1972,*J.Opt.Soc.Am.*,62,pp511-518
2 Gull S.F. and Daniell G.J. : 1978,*Nature*,272,pp686-690
3 Burge R.E., Fiddy M.A., Greenaway A.H. and Ross G. : 1976, *Proc.Roy.Soc.*,A350,pp191-212.
4 Titchmarsh E.C. : 1926,*Proc.Lond.Math.Soc.*,25,pp283-302.
5 Burge R.E., Fiddy M.A., Greenaway A.H. and Ross G. : 1976 *Recent Advances In Optical Physics*,(ICO-10, Prague),pp 687-694
6 Walther A. : 1963,*Optica Acta*,10,pp41-49.
7 Greenaway A.H. : 1977,*Optics Lett.*,pp10-12.
8 Burge R.E., Fiddy M.A. and Wheeler M.W.L. : submitted to *J.Mol.Biol.*
9 Bates R.H.T., and Napier J. : 1972,*Mon.Not.R.astr.Soc.*,158,pp405-424.

DISCUSSION

Comment J.E. BALDWIN
In the complex plane analysis it is easy, given one solution, to produce any other by transferring zeros across the real axis. Starting with a solution comprising point sources separated by regions of zero intensity, I believe that all the other solutions have strong negative features, so that the solution for a point source case is almost certainly unique.

For extended objects it may be that regions of zero intensity inside the area within which the source lies, even though we do not know where they are, may be sufficient to make the solution unique. Have you investigated this possibility or think it likely to be useful?

Reply A.H. GREENAWAY
Although I have not experimented with distributions comprised of isolated point sources I am inclined to agree with you that, if we ignore 180° ambiguities, the solution is probably rendered unique by application of the positivity constraint.

With regard to extended objects, the exact knowledge of any small internal region of the object is certainly sufficient to allow a unique reconstruction from the modulus data (see ref. 7), even in the absence of a positivity constraint. Your suggestion that for objects with internal gaps the positivity constraint alone may be sufficient is very interesting. One may certainly quote counter examples for which this information does not lead to a unique solution - for example an object consisting of two identical regions separated by a blank space would not give a unique solution if the regions in question individually give rise to ambiguities of the type shown in figure 1. If the two regions are dissimilar, but have one or more complex zeros in common one would also expect ambiguities which would not be resolved by these constraints. However, these are rather trivial examples and it may well be that the use of positivity with an internal gap in the object will, in many cases, lead to a unique solution.

REDUCTION OF PHASE NOISE IN INTERFEROMETRY WITH STATE-SPACE ANALYSIS

Ali Okatan and J. P. Basart
Electrical Engineering Department and
Engineering Research Institute
Iowa State University
Ames, Iowa 50011 U.S.A.

1. INTRODUCTION

In radio mapping, one of the problems encountered is the random bias in the visibility estimate. The bias can be divided into two parts: (a) the positive bias due to the common sky background seen by all elements of the interferometer, and (b) the negative bias due to phase noise present in the system. The first kind of bias can be easily removed by subtracting the correlation between the signals at two interferometer sites when the source is not in the antenna beams from that measured with the source in the antenna beams. This bias will therefore not be considered here. In contrast, the second kind of bias is more difficult to remove. When the signal-to-noise ratio of the interferometer system is high, incoherent averaging techniques can be utilized in the fringe frequency or in the time domain (Clark et al., 1969; Moran, 1973).

Incoherent averaging in the time domain essentially consists of measuring the correlation between the two interferometer signals more than once with integration times much less than the coherence time of the system. By using these correlations and the probability density function of the measured correlation amplitudes, a maximum likelihood estimate of the correlation amplitude is found. The method over-estimates the correlation by a factor of $1 + 0.5\ \mathrm{SNR}^{-2}$, and it loses the phase information about the source.

In removing phase noise effects from the interferometer data, the nature of the phase noise process at a given frequency must be determined. It is well-known that at low frequencies (e.g., 26 MHz), the ionosphere is the major source of the phase noise. Consequently, the stability of the ionosphere mainly determines the success or failure of an interferometer experiment at such low frequencies.

The system noise of the interferometer and imperfect local oscillators are also responsible for the phase fluctuations. At low frequencies, the phase noise contributed by the local oscillators is

C. van Schooneveld (ed.), Image Formation from Coherence Functions in Astronomy, 103-110.

usually a linear drift at a random rate. In the next section, a method will be presented for reducing different phase noise components from the interferometer phase.

2. METHOD

The method is based on empirical phase noise modeling combined with state-space formulation and discrete Kalman filtering. In the following discussion, it will be assumed that the interferometer data have been reduced by using standard reduction programs to remove local oscillator offsets, position errors for the source and interferometer baseline, and fringes. It is further assumed that the integration time used in this reduction process is sufficiently small in order to avoid degradation of the complex fringe function resulting from phase instabilities. Let there be N measurements of the complex fringe function for a point in the u-v plane. Let $\gamma(u,v,mT)$ be the complex visibility function measured for the m-th integration interval, where $m = 1, 2, \ldots N$ is the time index; T is the lowest integration time in the first data reduction process (for example, the Mark I system uses a lowest integration time of 0.2 sec (Moran, 1976)). The total observation time, NT, must be small enough so that u and v do not change significantly. The function γ (u, v, mT) can be written as

$$\gamma(u,v,mT) = A(u,v,mT) \exp\,[j\phi(u,v,mT)] \tag{1}$$

where $A(u,v,mT)$ and $\phi(u,v,mT)$ denote the amplitude and phase, respectively. At the end of the first data reduction process, N pairs of $A(u,v,mT)$ and $\phi(u,v,mT)$ values are available, and it is desired to remove the noise in $\phi(u,v,mT)$. The measured phase function is decomposed as

$$\phi(mT) = \phi_{source}(mT) + \phi_{ion}(mT) + \phi_{osc}(mT) + \phi_{stat}(mT) \tag{2}$$

where we omitted u and v for brevity. The subscripts 'source', 'ion', 'ocs', and 'stat' are used to denote, respectively, the phase components due to the radio source observed; the differential phase noise component due to the propagation media; the noise term introduced by the local oscillators; and the phase noise term due to the system noise and finite integration time. (Let $\phi_{source}(t)$, $\phi_{ion}(t)$, $\phi_{osc}(t)$, and $\phi_{stat}(t)$ be the continuous forms of $\phi_{source}(mT)$, $\phi_{ion}(mT)$, $\phi_{osc}(mT)$, and $\phi_{stat}(mT)$, respectively.) Each term in (2) can be modeled in terms of its continuous form as a random process satisfying a certain stochastic differential equation. Each equation takes a form imposed by the physical process generating a particular phase noise. $\phi_{source}(t)$ is modeled as a random constant satisfying

$$\frac{d}{dt}\,\phi_{source}(t) = 0 \tag{3}$$

Since for a single point (u,v), the visibility phase is unique. $\phi_{ion}(t)$ is taken as a solution of the following equation

$$\frac{d}{dt}\begin{bmatrix}\phi_{ion}(t)\\ \phi_{aux\ 1}(t)\end{bmatrix} = \begin{bmatrix}0 & 1\\ -\alpha^2 & -2\beta\end{bmatrix}\begin{bmatrix}\phi_{ion}(t)\\ \phi_{aux\ 1}(t)\end{bmatrix} + \begin{bmatrix}1\\ \alpha - 2\beta\end{bmatrix} w(t) \qquad (4)$$

where $\phi_{aux\ 1}(t)$ is an auxiliary random variable; α and β are the parameters describing the processes $\phi_{ion}(t)$ and $\phi_{aux\ 1}(t)$; and $w(t)$ is a white noise process. Equation (4) describes a random periodic process with auto-correlation function (Fitzgerald, 1966).

$$R(\tau) = \varphi_o^{\ 2} \exp\ (-\beta|\tau|)\ \cos\ \omega|\tau| \qquad (5)$$

where $\varphi_o^{\ 2}$ is the variance of the process; ω is the angular fluctuation frequency; and τ is the time lag. This type of stochastic process for $\phi_{ion}(t)$ is suggested by the data shown in Figure 1, which is characteristic of almost all observations.

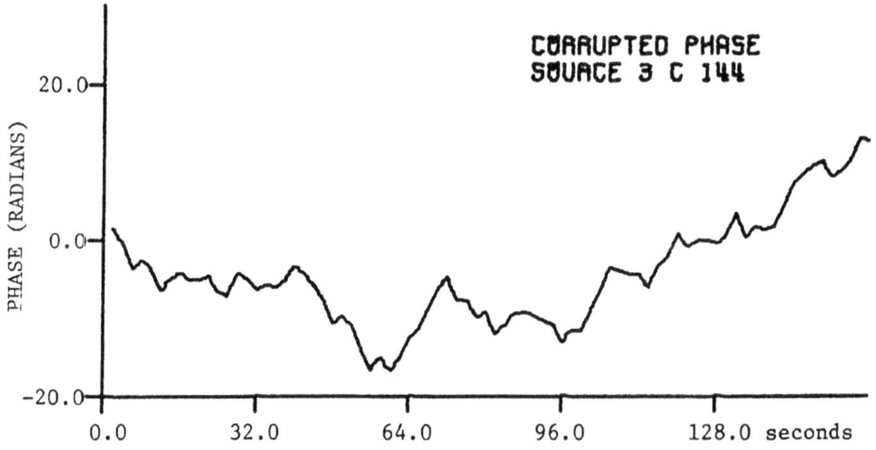

Figure 1. Phase data with oscillator drift and ionospheric perturbations. See also the note under "References".

$\phi_{osc}(T)$ was modeled as

$$\phi_{osc}(t) = ct \qquad (6)$$

where c is a random constant. This process satisfies the equation

$$\frac{d}{dt}\begin{bmatrix}\phi_{osc}(t)\\ \phi_{aux\ 2}(t)\end{bmatrix} = \begin{bmatrix}0 & 1\\ 0 & 0\end{bmatrix}\begin{bmatrix}\phi_{osc}(t)\\ \phi_{aux\ 2}(t)\end{bmatrix} \qquad (7)$$

where $\phi_{aux\ 2}(t)$ is the auxiliary variable whose initial value determines the value of the slope c of the process $\phi_{osc}(t)$. Equation (3),

(4), and (7) can be combined into the following matrix equation

$$\frac{d}{dt}\begin{bmatrix}\phi_{source}(t)\\ \phi_{ion}(t)\\ \phi_{aux\ 1}(t)\\ \phi_{osc}(t)\\ \phi_{aux\ 2}(t)\end{bmatrix} = \begin{bmatrix}0 & 0 & 0 & 0 & 0\\ 0 & 0 & 1 & 0 & 0\\ 0 & -\alpha^2 & -2\beta & 0 & 0\\ 0 & 0 & 0 & 0 & 1\\ 0 & 0 & 0 & 0 & 0\end{bmatrix}\begin{bmatrix}\phi_{source}(t)\\ \phi_{ion}(t)\\ \phi_{aux\ 1}(t)\\ \phi_{osc}(t)\\ \phi_{aux\ 2}(t)\end{bmatrix} + \begin{bmatrix}0\\ 1\\ \alpha-2\beta\\ 0\\ 0\end{bmatrix} w(t) \quad (8)$$

Equation (8) describes a 5-dimensional linear stochastic system whose state variables are $\phi_{source}(t)$, $\phi_{ion}(t)$, $\phi_{aux\ 1}(t)$, $\phi_{osc}(t)$, and $\phi_{aux\ 2}(t)$, the driving function being a vector white noise process. If α, β, and the initial values for the state variables are known, the state variables at any arbitrary time t can be generated by solving (8). In order to use these solutions for reducing phase noise, rewrite (2) in a continuous form:

$$\phi(t) = (1\ 1\ 0\ 1\ 0)\begin{bmatrix}\phi_{source}(t)\\ \phi_{ion}(t)\\ \phi_{aux\ 1}(t)\\ \phi_{osc}(t)\\ \phi_{aux\ 2}(t)\end{bmatrix} + \phi_{stat}(t) \quad (9)$$

By using (8), (9), and the recursive Kalman filter algorithm, estimates of $\phi_{source}(mT)$, $\phi_{ion}(mT)$ and $\phi_{osc}(mT)$ are obtained (Okatan, 1977) at discrete times mT. To obtain residual interferometer phase, $\phi_{res}(mT)$, these estimates are used in the following equation:

$$\phi_{res}(mT) = \phi(mT) - [\hat{\phi}_{source}(mT) + \hat{\phi}_{ion}(mT) + \hat{\phi}_{osc}(mT)], \quad m=1, 2, \ldots N \quad (10)$$

where the hat (^) stands for the predicted phase components. The quantity $\phi_{res}(mT)$ is now used with A(u,v,mT) to get a coherent average for the visibility by using the equation

$$\gamma_{coh}(u,v) = \frac{1}{N}\sum_{n=1}^{N} A(u,v,mT)\exp[j\phi_{res}(mT)]$$

In order to obtain α, β, and the initial values for the state variables, only a small portion of the phase data is used. The procedure is as follows:

(a) Remove the linear trend from the phase data by using the ordinary least squares method. The slope of the trend is the initial value for the state variable $\phi_{aux\ 2}(t)$. Set the initial value of $\phi_{osc}(t)$ to the value of the trend at the initial time.

(b) Fit the data from the preceding step to a second order autoregressive time series model having an autocorrelation function given by (5). From this process, β, φ_o^2, and ω are obtained. α is obtained from the relation

$$\alpha = (\beta^2 + \omega^2)^{1/2}$$

(c) Compute the variance of the residual process left over from the previous step to get an estimate for the variance of $\phi_{stat}(t)$. Set the initial value of $\phi_{ion}(t)$ to the initial value of the second order autoregressive process obtained at step (6) The initial value of $\phi_{aux\ 1}(t)$ is taken to be zero while the initial value of $\phi_{source}(t)$ is set to the best guess for a particular source.

In the Kalman filter algorithm, the uncertainties on the initial values of the state variables have to be supplied along with the variance of w(t). Once the initial parameters are obtained, they can be used for the other portions of the data. These parameters have to be updated as needed. Because the Kalman filter is recursive, the computer storage requirements are minimized. The computer time required for filtering phase noise is quite small and can be reduced further by decomposing the state matrix in (8). If we do not have a good guess for $\phi_{source}(t)$, it can be set to zero along with the uncertainty associated with it. Then, the Kalman filter assumes perfect information for this component and does not change it throughout the filtering process. Figure 2 shows a block diagram of the method.

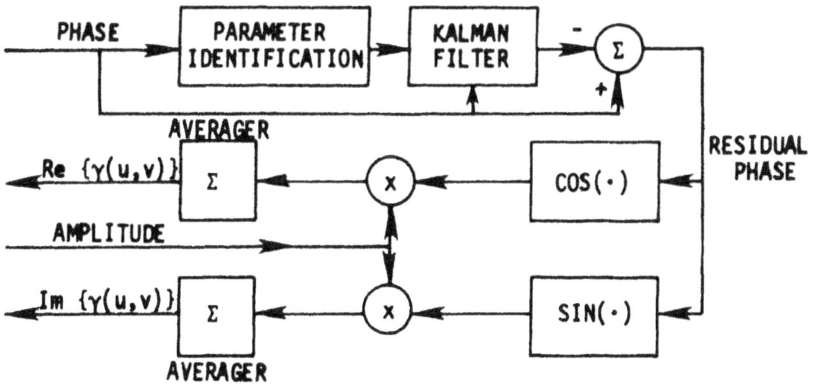

Fig. 2. Block diagram of the phase-noise reduction method.

3. RESULTS AND CONCLUSIONS

The method was applied to the data taken during VLBI experiments between Boulder, Colorado and Ames, Iowa in August, 1971. During these experiments, the ionospheric phase fluctuations were severe. The half-power Gaussian size of 3C 144 pulsar was determined as 2.20 ∓ 0.50 arc-sec before the technique was used. After using the method described in this paper, the size was determined as 1.34 ∓ 0.20 arc-sec, which is in close agreement with the recently published sizes of this source 1.3 (+0.23, -0.13) (Mutel et al., 1974). Figure 3 shows the residual interferometer phase obtained after the application of this method. The oscillation and linear trend are removed and the rms phase is small.

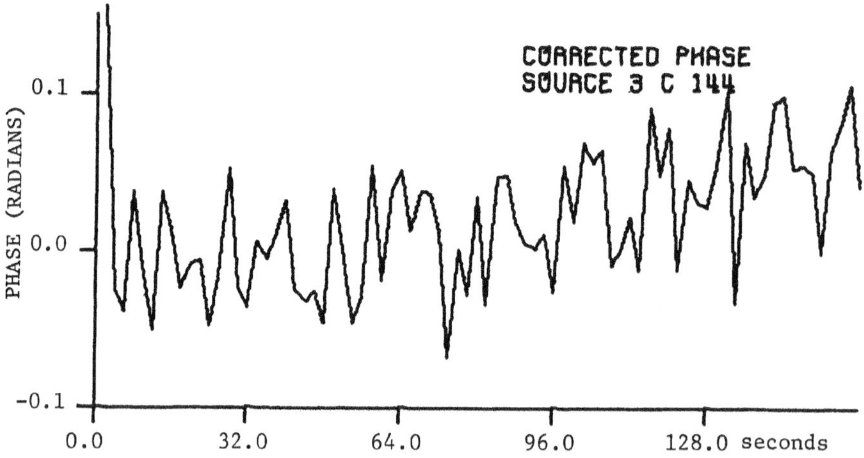

Figure 3. The phase data of fig. 1, corrected for oscillator drift and ionospheric perturbations.

4. RECOMMENDATIONS

A real time implementation of the method described here may be implemented in digital and analog interferometer systems. For the digital systems, high data rate can be a problem. Therefore, the rate has to be reduced before using the method. For continuous systems, the data rate does not introduce any problems; however, a solution of the nonlinear Riccati equation must be supplied during the filtering. For complicated phase noise models it may not be easy to obtain this solution.

ACKNOWLEDGEMENTS

We thank our colleague Professor S. C. Dutta Roy for his critical reading of the manuscript. We also thank the Engineering Research Institute for support during this investigation.

REFERENCES

Clark, B.G., Kellermann, K. I., Bare, C.C., Cohen, M. H., and Jauncey, D.L.: 1968, "Astrophysical Journal," 153, pp. 705-714.

Fitzgerald, R. J.: 1966, "AIAA Guidance and Control Conference Proceedings," pp. 500-509.

Moran, J. M.: 1973, "Proceedings of IEEE," 61, pp. 1236-1242.

Mutel, R. L., Broderic, J. J., Carr, T. D., Lynch, M., Warnock, W. W., and Klemperer, D.: 1974, "Astrophysical Journal," 193, pp. 279-282.

Okatan, A.: 1977, Unpublished Ph.D. dissertation, Electrical Engineering Department, Iowa State University, Ames, Iowa.

Moran, J.M.: 1976, "Methods of Experimental Physics", 12, part C, pp. 174-197, M.L. Meeks, ed., Academic Press, New York.

Note.
The data of fig. 1 were collected on August 19, 1971, with a 26.3 MHz, 950 km baseline, interferometer operating between Boulder, Colorado and Ames, Iowa. The antennas at each site consisted of 160 pairs of crossed full-wave dipoles. The data were recorded using Mark I VLBI tape recording terminals on loan from the National Radio Astronomy Observatory.

DISCUSSION

Comment D.B. SHAFFER
1) As I understand it, the size you measured for 3C144 was too large because your visibility values were too low, and your technique removes ionosphere phase fluctuations which caused loss of coherence. Correct?
2) You can also go to shorter coherent integration intervals.
Reply A. OKATAN
1) Answer to the first question is yes. 2) If we want to keep uncertainties small, using shorter integration time is not useful, because the uncertainty on the estimate of the coherence function is proportional to the inverse square-root of the integration time.

Comment T.A. CLARK
As a collaborator in the acquisition of the 3C144 data presented, I would like to note that fringe amplitudes obtained by assuming partial coherence, e.g. $\sim$ 18 separate samples of 10 sec. each, gives visibility amplitudes quite similar to this approach and imply a 26 MHz apparent source size $\sim 1\overset{''}{.}5$.

Comment P.N. WILKINSON
The closure phase is still a good observable at 26 MHz!
Reply A. OKATAN
Yes. But, if you don't have three or more stations operating at the same time, the closure phase cannot be observed.

PART III

PROCESSING TECHNIQUES AND DISPLAY METHODS

DIGITAL PROCESSING METHODS FOR APERTURE SYNTHESIS OBSERVATIONS

(Invited paper)

B. G. Clark
National Radio Astronomy Observatory. Operated by Associated Universities, Inc. under contract with the National Science Foundation.

1. INTRODUCTION

The rapid growth, over the last two decades, of the amount of digital arithmetic capability available for a given cost is obvious to all, and is apparently not yet at an end. Therefore, any discussion of the best equipment and even of the best algorithms with which to attack a given problem will soon become obsolete as decreasing costs bring things previously considered inconceivable within economic feasibility. The general trend is that, as the cost of hardware decreases, the discussions of algorithms and procedures becomes simpler, as fewer approximations have to be made, and the mathematically simple correct formulae may be directly implemented.

This paper discusses two disjoint subjects. The first is simply a discussion, much limited by the state of the art of computer construction, of the least costly way of doing Fourier transforms for producing images from correlation data. The answer of course depends on the size and other particulars of the problem. The second is a very brief discussion of the relation of the map made by means of the Fourier transform to the sky brightness.

2. FAST FOURIER TRANSFORMS WITH GENERAL PURPOSE COMPUTERS

The discussion in this section is predicated on the use of the algorithm of Cooley and Tukey (1965). Occasionally, this is not the most appropriate algorithm; the discussion of this point is found in the following section.

The cost of doing a Fourier transform is comprised of two elements: arithmetic elements and memory. They are essentially non-interacting, and can be discussed separately.

In evaluating the cost of doing arithmetic operations, it is conventional to make a distinction between several classes of computers:

C. van Schooneveld (ed.), Image Formation from Coherence Functions in Astronomy, 113-120.

microprocessor systems, minicomputers, "main frames" and array processors. The distinction between the various classes is somewhat a matter of opinion; they are reviewed here. Perhaps the most useful distinction between "microprocessor systems" and "minicomputer systems" is that modern mini's have hardware floating point arithmetic units (at least optionally) and microprocessors do not. The most common (and most primitive) microprocessors, e.g., Intel 8080, Zilog Z80, National SC MP, do not even have hardware multiply.

The distinction between minicomputers and the class of device best given the computer jargon term "main frame" is even fuzzier. The traditional division is that a minicomputer is built around a 16, 18 or 24 bit word length, or is an upward compatible member of a family of 16 bit word length computers. Main frames were conceived and planned around a longer word. The distinction is not useful in comparing the arithmetic capabilities needed for the FFT; some main frames are many times slower than some minicomputers. The chief practical distinction between main frames and minicomputers is that the former are provided by the manufacturer with vastly superior software systems.

The term "array processor" has accumulated a meaning not adhering to it four or five years ago. The most successful devices now sold under this name are not, like the ones marketed half a decade ago, devices equipped to perform a few simple operations on linear arrays of numbers. They are instead general purpose minicomputers, with all accouterments not directed to fast floating point arithmetic stripped, to hold down costs, and the hardware optimized to do floating point arithmetic as fast as practicable. The manufacturers of these devices also hold down costs by providing, chiefly, a library of arithmetic routines, and only a minimum of resource management, device handler, and other expensive system software. The array processors do not currently support higher level languages (except, perhaps, as a device for calling low-level-coded subroutines), and the low-level languages require much care and sophistication to utilize the full capabilities of the device. This situation may be expected to change, and this will make a great improvement in their usability.

The times and system costs for doing Fourier transforms in systems of various classes is given in Table I. The reader is warned that this table is assembled from highly heterogenous sources, and reflects a variety of implementating, done by different people in different languages. The numbers are indicatory only, and cannot be taken seriously to a factor of 2 or so. All numbers quoted are Fortran programs, though by different authors with different details, except those for the 8080, the Cray I, and the array processors.

Memory costs are also rapidly decreasing. However, for large images it is still worthwhile to implement a hierarchy of memories. For making a large set of million point images one might have 16 K words of array processor memory (access time 300 ns, cost $20,000 per million bits), 1 million words of minicomputer storage (function:

TABLE I

TIMES FOR 1024 COMPLEX FFT

DEVICE	TYPICAL SYSTEM COST	TIME, SECONDS
Microprocessors		
Intel 8080	$ 4,000	27
Minicomputers		
Sperry-Univac 77-400 (no floating point hardware)	20,000	16
Modcomp II	50,000	1.6
DEC-PDP-11/70	100,000	0.8
Main Frame		
IBM 360/50	700,000	2.6
DEC KI/10	1,000,000	1.1
IBM 360/65	1,500,000	0.8
IBM 370/155	2,000,000	0.3
TI ASC/P2	4,000,000	0.004
Cray I	8,000,000	0.001
Array Processors		
FPS AP120B	150,000*	.005
CSPI MAP-300	150,000*	.006

*including host

transposing rows and columns, access time 1 μs, cost $12,000 per million bits), 200 million words of disk (working store, buffer to output; access time 25 ms, cost $3 per million bits) and an indefinite number of magnetic tapes (access 5 minutes, cost $0.25 per million bits).

3. FAST FOURIER TRANSFORMS WITH SPECIAL PURPOSE DIGITAL HARDWARE

Opening the door to consideration of special purpose devices introduces more options into questions of design than merely choice of the particular chip. For general purpose machines, for any really massive aperture synthesis, the only reasonable choice is to build the procedure around a Fourier transform using the Cooley-Tukey algorithm. Anything else is either grossly inefficient use of the equipment or is too difficult to organize. In special purpose equipment, efficiency is not particularly a consideration. It must be fast enough to do its hardest job, but it doesn't matter if for easier jobs it does them efficiently or not; being a special purpose device there is probably nothing useful it could be doing in the time saved by doing easy jobs efficiently.

For some special cases, the use of the Cooley-Tukey algorithm is not the most efficient solution either. Some corrections for instrumental problems can conveniently be done by modifying the nature of the transform relationship, which may involve a much slower algorithm than that of Cooley and Tukey, but which may still be feasible for a special device. Corrections which may be done this way include the non-coplanar baseline effect and the correction for instrumental bandwidth, both mentioned later. If one is interested in making, in real time, a map that is not too large with input data rates not too high, one can build a classical discrete Fourier transform device to do it. The expenditure of a few thousand dollars on arithmetic elements will result in a capability approaching 10^9 multiplies per second. If one is constructing an output image of, say, 10^6 points, one could support an input data rate of the order 10^3 correlation points per second. The possibilities have been explored extensively by Frater (1978). The limitation of this method for real-time use is that one is often interested in collecting data for several images at the same time. The memory for several images of this size is costly if implemented as semiconductor store, and inconvenient if implemented as a rotating magnetic memory. The approach is less attractive for non-real-time use because, once the data have been introduced into a general purpose computer system for management and recording, the question of efficiency again raises its head, and one has hopes of saving on the cost of the device by using the most efficient algorithms possible, and using the time saved on easy problems to catch up on work fallen behind during hard problems.

In cases where the machine can be built to do one job only and need not, for instance, do transforms of different lengths, Winograd's algorithm (Winograd, 1978) may be superior to that of Cooley and Tukey. It involves many fewer multiplications and a similar or smaller number of additions. It does seem to require about one additional bit of precision (Patterson and McClellan, 1978), and the structure of the algorithm does not lend itself to the beautiful symmetry of the Cooley-Tukey algorithm; therefore the wiring to accomplish the transform in hardware will be a much larger chore to design. This last consideration is the primary reason why the algorithm is not rapidly displacing the Cooley-Tukey algorithm in general purpose computers - the overhead required may eat up most or all of the time saved by the fewer multiplies, and the result is an opaque program that offers only small speed increases over the Cooley-Tukey algorithm.

Even in special purpose hardware implementations, the Cooley-Tukey algorithm is not unattractive. Using only a few chips of LSI, with parallel multipliers, it is possible to build a hardware "butterfly" (a butterfly on a single chip is being designed by TRW, Inc.) which should result in a transform speed of about .1 μs times $\log_2 N$ if they are provided in a parallelism of N and of 1 μs if in a parallelism of $N\log_2 N$, for a transform of length N. With these extremely large data rates all the difficult problems fall in storing, managing, and assimilations of the data. Arithmetic operations have become the easier part of the job.

Parallel multipliers grow in complexity as the square of the datum length, so the Winograd algorithm increases in attractiveness for higher precision operations.

The usual rule of thumb is that one loses half a bit of precision, in the Cooley-Tukey algorithm, per butterfly stage. In floating point one should carry mantissa lengths equal to the desired output accuracy plus $\frac{1}{2}\log_2 N$, where N is the image size. For fixed point operations, the convention is to represent the input data to its inherent accuracy, and then add a bit on the left every butterfly stage and drop a bit on the right alternate butterfly stages. In most, but not quite all, images, this ends up with many empty bits on the left, but simplifies the management greatly.

4. SOME THEORETICAL PROBLEMS IN FOURIER TRANSFORM ANALYSIS

Although the most exciting areas for theoretical research lie with the non-FFT algorithms of image formation, which are the topic of most of the discussion at this symposium, there are questions in the simple case of Fourier transforming which are not without interest.

Perhaps the simplest of these is the correction for the non-coplanar baseline effect. For earth rotation synthesis with non-east-west baselines the loci of the antennas as the earth rotates do not describe a plane in space, but occupy a solid. The relation between the sky brightness and the measured correlation functions involves all three direction cosines of the brightness element, and therefore the image is not recoverable by a two-dimensional transform. For an interferometer element located at (u,v,w) relative to a reference element (u and v conventionally lie in a plane perpendicular to the "phase tracking center"), and a brightness element at (s_x, s_y, s_z) the phase is $us_x + vs_y + (s_z-1)\,w$. The -1 in the last term arises because the phase tracking center $\{s = (0, 0, 1)\}$ has been removed from the calculation. Since s is a unit vector, $s_z - 1 \approx -\frac{1}{2}(s_x^2 + s_y^2)$. It is easy to verify that this is an important effect for many practical cases. Its relative importance is measured (roughly) by the product of the size of the synthesized image in radians and the size of the synthesized image in beamwidths. If this product approaches one the effect must be corrected.

There are three practicable cures for this problem. One may deal only with small areas at a time, so that the eventual synthesized image is generated from a mosaic of images with differing phase tracking centers. One may make maps at frequent intervals, short enough that the rotation of the earth does not destroy the two-dimensional nature of the array. Or one may make a three-dimensional transform of the correlation data, and sample it along the spherical surface

$$s_x^2 + s_y^2 + s_z^2 = 1$$

which automatically aligns the phases. This last solution appears to be the most attractive computationally. It does raise the interesting question of how many samples are needed in the third dimension. The answer is roughly the number of samples needed in the other dimensions times the image size in radians. However, the effects of sampling are well understood only for a uniform random distribution of data within the sampling intervals, but in this case there are usually sufficiently few points called for that the distribution of points within the cells is far from uniform, and will vary greatly from place to place. A good calculation of this effect for practical cases apparently has not been done.

A second interesting problem is the breakdown of the quasi-monochromatic assumption. The usual effect of different frequencies is simply a scale change of the image with frequency. The data at various frequencies within the band of reception are, however, added together before the image is produced, and the net effect is a radial smearing that becomes more serious at the edges of the image. This smearing effect may be reduced to a simple convolution by merely reinterpolating the image onto polar coordinates logrithmic in the radius. The full power of conventional convolution theory is then available for correcting for the effect.

Perhaps the most interesting problem in conventional image formation is the question of weighting to compensate for incompleteness. To see how the problem arises, suppose we have a randomly located collection of samples of the correlation function plane. If we take a very coarse grid, and average the points within each cell (or convolve in some other fashion and resample at the cell center), we will have a nearly complete sampling of the plane, and the synthesized beam (after a suitable taper is applied) is well behaved, has low side lobes, and causes a minimum amount of trouble. If we try to do the same thing on a very fine grid, however, we shall find that we have only one sample within each grid cell, and our synthesized beam will have side lobes at the mercy of accidental or systematic fluctuations in point density about the plane. In some sense this is fundamental; a side lobe of the synthesized beam, taken in this manner, reflects a real difficulty in distinguishing between radiation coming from two locations separated by the separation of beam and side lobe. On the other hand, the difficulty must be to some extent apparent, since these side lobes due to uneven sampling vanish when one interpolates to a coarse grid. The difference lies in the assumption, implicit in resampling on a coarse grid, that the radiation comes from a restricted area of the sky. For many purposes, one desires to have a large image produced under the assumption that the radiation comes from one or a few relatively isolated regions. This is achieved, in a very rough approximation, by making the fine grid map with the weighting of each sample set to that appropriate for the coarse grid map. The implication is that, for each emitting region, a tight cluster of points in the correlation function plane carry much the same information, and therefore are not weighted as heavily as isolated points. A more elegant approach would be to convolve the unit weights of the points with a suitable smooth function, and to weight the individual points with the reciprocal of the convolved weighting function at their location. This has the effect of suppressing the side lobes of the synthesized beam within an area whose size is the reciprocal of that of the convolving function.

A yet more elegant procedure is to choose weights which cause the synthesized beam to approach most closely a desired beam (say a Gaussian) within a specified area, or approach most closely with errors weighted by a grading function of distance from synthesized beam center. One can't carry this approach too far; for instance it is easy to show that if one asks to have the least squares fit of the synthesized beam to a Gaussian over the whole image, the answer is to simply Gaussian weight the sample points.

5. CONCLUSION

The computation of a Fourier inversion of the correlation data is nearly always the first step in any formal image formation procedure, and is part of an iterative step in many of them. For this reason, and because the transform map is an easily produced case of something that is at least consistent with the data, the Fourier transform technique will never be without interest. The facilities for producing transforms have dramatically dropped in cost, so that the limitations of the devices are more strongly met in providing or organizing storage for the output than in arithmetic elements. With the possibility of building, at attainable costs and complexities, devices which can produce a Fourier transformed output approaching 10^9 numbers per second, we may say that the arithmetic of the Fourier transform is not a conceptual difficulty for image formation.

REFERENCES

Cooley, J.W. and Tukey, J.W.: 1965, "Math. of Computation" 19, 297.

Frater, R.: 1978, "Astron. and Astrophys.", in press.

Patterson, R.W. and McClellan, J.H.: 1978, "IEEE Trans. on Acoustics, Speech and Signal Processing", in press.

Winograd, S.: 1978, "Math. of Computation" 32, 175.

DISCUSSION

Comment A.T. MOFFET

The problem of bandwidth smearing may be a red herring. In cases where $\Delta\nu/\nu$ is appreciable it is usually necessary to split the data into a number of narrow-band channels. Otherwise the "delay beamwidth" would be smaller than the primary beamwidth; in many cases this multi-channel processing is also necessary in order to selectively reject interference. Westerbork and the VLA are both implementing this scheme, I believe.

Reply B.G. CLARK

Even if you split the band into pieces, this correction can be important within each piece.

Comment R.H. HARTEN

The bandwidth problem is noticeable even in the present Westerbork data. Comparisons of source fluxes at several frequencies require accurate fluxes and source sizes. The large number of sources per field and the large number of fields makes detailed corrections prohibitive. Thus some care must be taken to restrict the bandwidth or the field of view, to minimize the source smearing.

Comment P. DEWDNEY

In relation to the previous remarks I can cite an extreme example of a large field aperture synthesis telescope. I have a 22 MHz instrument which covers a field 40 degrees wide with 15 arc minute beam in which these radial bandwidth distortions are apparent if not corrected for. However, if fan beam analysis is used, it is possible to produce a circular beam on all positions of the map.

Comment T.W. COLE

The suggested procedure of converting three dimensional data back into two by Fourier transforming the sky brightness, sampled along the sky sphere, back to the U,V plane and resampling it at the original U,V points, to obtain "corrected" data, needs classification. One is likely to run into trouble if one tries to use these data for other algorithms; for clean, for instance, since the true beam shape is still varying with position in the field.

Reply B.G. CLARK

Yes, the effects occur, and the data differ from that that would have been obtained had the geometry been such as to allow the direct sampling of the points in question, but it is not clear that these effects will be a practical limit to a reconstruction process.

DIGITAL HARDWARE PROCESSING OF ROTATIONAL SYNTHESIS DATA

R.H. Frater
School of Electrical Engineering,
Sydney University, Australia.

ABSTRACT

Digital techniques for the direct transformation of rotational synthesis data are discussed with particular emphasis on current cost and technological limits. A prototype system designed to test various approaches will be described and its application to "clean" and maximum entropy reconstructions considered.

C. van Schooneveld (ed.), Image Formation from Coherence Functions in Astronomy, 121.

ANALOGUE PROCESSING METHODS FOR SYNTHESIS OBSERVATIONS

(Invited paper)

T.W. Cole
Division of Radiophysics, CSIRO, Sydney, Australia

1. INTRODUCTION

In aperture synthesis the formation of an image involves the two steps of spatial correlation across an aperture and transformation to the image. This is closely related to conventional imaging with a lens (Cole, 1977a), which Abbé interpreted as two successive transformations at the surfaces of the lens. With the simple lens the image is the light intensity in the output plane (Figure 1(a)). In aperture synthesis (Figure 1(b)) the image is the transform of the correlation but no detection takes place. The image corresponds to the 'light' amplitude rather than intensity.

The two classes of instrument, direct imaging and aperture synthesis, exist in radio astronomy but they are rarely as simple as depicted in Figures 1(a) and (b). Dilute aperture imaging arrays such as crosses or rings require processing in order to simulate an ideal aperture. One form of processing modifies the amplitude and phase distribution across the aperture by a filter. The corrected image is then the difference between images obtained when the aperture has been modified by each of two filter functions.

Similarly, in aperture synthesis the spatial correlation samples rarely have the distribution one would have from an ideal aperture. Correction in this case however is often achieved by a single weighting filter across the (spatial frequency) aperture as in Figure 1(d).

The similarities in Figure 1 form the basis of most of the analogue processing techniques to be discussed. The differences are however important, as will be seen.

2. AN APERTURE SYNTHESIS SYSTEM

A typical aperture synthesis system (Brouw, 1975; Fomalont and Wright, 1973; Pooley, 1976) and the stages involved in the processing

C. van Schooneveld (ed.), Image Formation from Coherence Functions in Astronomy, 123-141.

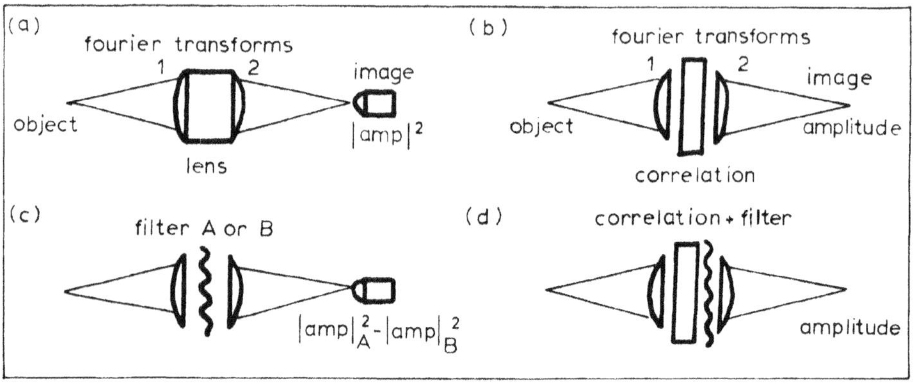

Figure 1. Conventional imaging with a lens (a) is diagrammatically compared with spatial coherence imaging (b). Processing for dilute apertures involves modifying the aperture with weighting filters and taking an image difference (c). Processing in aperture synthesis (d) involves a simple weighting filter.

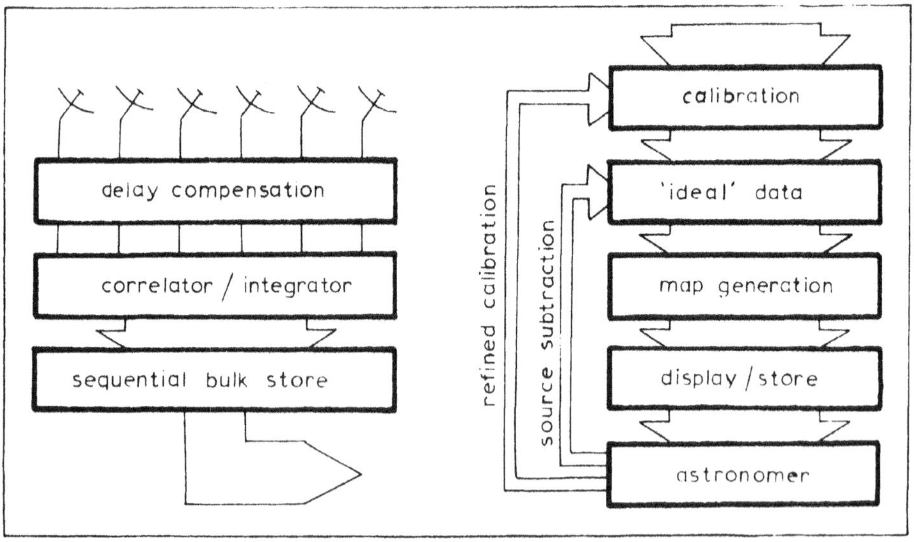

Figure 2. The basic steps in collection (left) and reduction (right) of aperture synthesis data involves several feedback loops.

of the data are illustrated in Figure 2. The antenna structures, and in many cases the feed lines, the delay compensation, and the correlators, are analogue devices, but from there on the data are almost invariably digitized. In an extreme case digitization could occur at the intermediate frequency right at the antenna elements of the array, and this will probably be the future trend. Only for observing bandwidths of hundreds of megahertz or more will analogue systems be used.

The sequentially stored data are processed in several stages. Calibration observations determine the fixed deviations in amplitude, phase and position of the array. The measurements are corrected to those which would have come from an array more nearly 'ideal'.

A map is obtained by transformation and is both stored and displayed. A number of options are then open. Calibration errors and deviations from ideal behaviour during the measurements can often be discerned in the point source responses in the map. Corrections to calibration can then be re-applied before making a more ideal map.

Sidelobes and grating responses occur in the response from even an ideal array. Strong sources in the map can therefore confuse weak ones with their (accurately predictable) sidelobes and grating responses. The most accurate way to remove the effects of strong sources is to subtract from the spatial frequency data. A new map would then have a lower peak value and show the weak structure more unambiguously.

In practice these two processing loops of calibration correction and source subtraction could be repeated many times before all the information in the data is clearly available in maps. Spectral line observations differ only in increasing the complexity of reduction by a factor which is somewhat larger than the number of frequency channels.

An important consideration emerges from this description. So long as one can repeat the map-making process the accuracy of the generated map need not be high. With sidelobes and grating responses usually 5% or more of the peak response one requires an accurate source subtraction step in order to see detail much weaker than 5%. But as one subtracts sources the dynamic range of the map decreases. It would be uncommon to require a dynamic range in the map-making stage of better than, say, 100:1, even though the original data might have a much higher accuracy.

Two approaches can be compared. Current practice, and as far as I understand it the projected routine practice for the Very Large Array (VLA), is to produce an accurate map of a field and the 'ideal' beam. Processing such as 'cleaning' (Hogböm, 1974) is then done in the map plane. But if one has multiple access to the (u,v) plane data, subtractions are more accurate, and it is then quite sufficient to generate maps of only 1% or 2% accuracy.

I want to concentrate on alternative means of generating maps from the spatial frequency plane data with an accuracy of 1%. If it is possible to do this quickly and economically, then truly interactive data analysis is possible. The quest for high dynamic range will force one to this interactive approach. It is not possible in all cases to produce a map which is strictly the convolution of the true distribution of the sky and a single, ideal beam. The variation of the beam shape across the map will form a fundamental limit to processing when only the map is available.

3. HYBRID RATHER THAN ANALOGUE

To generate the map from spatial frequency data I will consider special-purpose hardware rather than the various software approaches.

Software processing is very flexible. Programs can be quickly modified, and at the expense of speed and at a price the result can be obtained with as many significant figures as one wishes. However, the digital computer load of these methods is becoming too great.

In a special-purpose device, flexibility is sacrificed. There is a trade-off of flexibility (and perhaps accuracy) for economy and/or speed. Processing steps which are rigidly defined, and which are repeated many timés, are the ones to implement in hardware. From Figure 2 the data reduction after the calibration correction contains just such steps.

The input and output to the special-purpose device is therefore seen to be digital. It is almost always in the form of a serial access to magnetic tape, disc or memory.

Special-purpose digital hardware can be devised and several approaches are discussed by other speakers at this meeting. Instead, we shall consider analogue components.

The antenna elements, like any analogue component, have fixed errors (of amplitude, phase and position) which can be calibrated and corrected for in the digitally recorded data. They also have variable errors which limit the accuracy of the final map.

Optical systems, ultrasonic, electron beam or co-axial systems are all analogue, with basically identical accuracy limitations and characteristics. A special-purpose analogue system is an assemblage of these analogue devices which performs the required mathematical processing. Analogue processing would convert the digitized spatial frequency data to light, sound, electric charge and so on, and generate the desired image as a distribution of light, sound or electric charge. The problem is to identify the mathematical analogues, study the generation and detection of the analogue, and compare accuracy, speed and cost.

The analogue device will have fixed errors which can be compensated for digitally. The input and output will usually be digital. Therefore the special-purpose device may be considered as a hybrid - that is, as an analogue peripheral to a digital computer.

4. THE CLASS OF PROBLEM

One of the difficulties of hybrid processing is in choosing the optimum implementation for the particular aperture synthesis system. I identify three broad classes of aperture synthesis system, whose common basis is a Fourier transformation and, equally important, a coordinate transformation from the essentially polar coordinates of the measurements to the rectangular coordinates of the map. The differences between the systems will imply that the optimum processing is different for each application. The conclusion will be reached that the most desirable hybrid systems involve optical processing.

4.1 The VLA-type system

The most general Earth rotation aperture synthesis system is the Very Large Array (Heeschen, 1975).

With north-south components to the baselines, the spatial coherence function is sampled in a three-dimensional (u,v,w) space. This and the varying height along the arms of the array produce a complex grating structure which varies markedly with source position. This will make it difficult to detect calibration and other errors by eye, will restrict the use of map plane subtraction, and will probably limit the dynamic range. However, the instrument has the distinct astronomical advantage of almost equal resolution over the major part of the sky.

The mathematical description of the array requires a three-dimensional Fourier transformation of the data. An alternative analysis approach would be to project the data on a number of two-dimensional planes appropriate to small sub-sections of the field of view and to perform two-dimensional Fourier inversions for these sub-sections.

In either case the software approach would need to weight the measurements for a reasonable beam, convolve the data to a grid, and use the fast FFT algorithm. However, the optimum weighting varies across the field because of the differing projections.

The general geometry of the VLA places enormous loads on the data processing if conventional, linear methods are to be used and high accuracy is demanded.

4.2 The east-west array

In contrast, Earth rotation synthesis with east-west baselines samples a two-dimensional spatial frequency plane along concentric ellipses. The data are therefore recorded in polar coordinates as sequential sets of samples along a radial line. A two-dimensional transformation produces a map in which the beam varies by the simple scaling sec δ with declination (δ). The usual software approach is to apply a weighting, convolve to a grid, and use the FFT. The symmetries of the spatial frequency sampling make it easy to distinguish by eye features due to real sources from elliptical or radial artefacts due to calibration or other errors. The result is a system in which high dynamic range is possible but which cannot effectively map low declination regions.

4.3 The Culgoora array

The middle ground between the two previous extremes is epitomized by the Culgoora array (Labrum et al., 1975) with 96 elements around a 3-km-diameter circle. The elements are in a plane although the plane is tilted with respect to the equatorial plane. In Earth rotational synthesis this makes the beam and the grating responses position-dependent. The circular symmetry is a powerful factor in simplifying the description of the spatial frequency sampling. The 96 elements are sufficient to form an image at each instant, and indeed in the solar application up to about 20 images per second might be produced. Rotational synthesis could therefore be processed as the sum in the map plane of sub-images produced at different times of the synthesis. Geometrical symmetries will be seen to make this approach attractive. However, the complicated grating responses due to the north-south components and the limited antenna tracking range will make high dynamic range difficult, while the relatively wide bandwidth will restrict the field of view and thereby simplify processing.

5. SOME APPROACHES OTHER THAN OPTICAL

I will concentrate on just three general approaches in this section. All of them could perform the processing required in aperture synthesis. But their disadvantages illustrate the major factors involved in a desirable hybrid system.

5.1 Ultrasonic imaging

A piezoelectric transducer readily converts a radio signal to an ultrasonic wave in the radio frequency region. An array of such transducers can generate waves which on propagation to the acoustic far field yield a Fourier transform image. The ultrasonic imager has been described by McLean and Wild (1961), analysed by Oliver and Billingham (1973), and used by Briggs and Holmes (1973). In the spatial frequency application, pairs of transducers generate spatial frequencies in the output plane and arrays of transducers in that plane measure the acoustic image amplitude. The problem is that detector arrays are rather limited in both number of elements and to the extent that each element must be connected to its own amplifier chain. There are developments in the medical field, but for parallel processing, ultrasonic imaging is still badly limited in its scope.

5.2 Radio frequency imaging

There are a number of forms of the radio equivalent of a lens. One is a free-space propagation system similar to the ultrasonic system (Oliver and Billingham, 1973) but this suffers, as do all the other radio frequency systems that follow, in the complexity of the detector system at each output point.

Radio frequency lenses can compress the size of such an imager but the most highly developed lenses are only one-dimensional and are limited in the number of input and output ports (Rotman and Turner, 1963).

Co-axial cable systems, such as Butler matrices (Shelton and Kelleher, 1961) and branching networks, perform the required transformation between aperture and image plane. The 96 × 48 cable system at Culgoora (Fourikis, 1967) probably represents the practical limit of a cable system.

5.3 Electron beam processors

A somewhat more attractive processor uses stored charge as the output parameter and electron beams to both form and read the charge distribution. In the charge storage tube electrons can be deposited at any of up to about 1000 × 1000 locations and the charge distribution can be progressively integrated. The device is readily interfaced to a computer and is basically a coordinate converter in which polar data can be recorded and integrated before being read-out in rectangular coordinates (Silver and Luedicke, 1971).

There is no Fourier transform performed in the device so that integration must be of data already processed in the computer. For individual spatial frequency samples, the processor would generate the corresponding fringe pattern. Strip scans could be produced by the digital one-dimensional Fourier transform of radial lines of spatial frequency samples and processing via strip scans simplifies the source subtraction step. The charge storage tube would integrate these with the correct orientations and scale. In the more general array geometries instantaneous maps could be integrated. In all cases, the final, integrated distribution would be read back to the computer.

The major restriction of the device is its inability to write negative as well as positive values. Two solutions are possible. Either the values are floated around an offset bias charge or two sequential integrations are performed, one for positive and one for negative values. In either case a subtraction is needed for the final map, but the latter, the sequential approach, is clearly far superior. The final subtraction removes effects of fixed background variations due to leakage current and a maximum dynamic range is achieved. One notes however the restriction that the electron beam writes point by point. The painting of the whole screen could take several milliseconds if one assumes 2 μs per line and 1000 lines to the fringe or strip scan.

Typically one could then expect to complete a picture within 1 s each for positive and negative or 2 s for the complete 512 × 512 or more picture. At 2 μs per line, the required data rate of 500 kHz would probably be limited by digital disc transfer rates.

Linearity and dynamic range need still to be determined but experiments are taking place in the medical computerized tomography (CT) imaging field as well as in the radio field at Dwingeloo. The electron beam approach is, to my mind, the best of the non-optical approaches, even though it requires data to be transformed to strip scans or sub-images.

6. INCOHERENT OPTICAL PROCESSING

Optical processing of information has been discussed, analysed, and in some cases applied for a number of years (Tippett et al., 1965; Preston, 1972; Stroke, 1972). Its appeal lies in the availability of a wide range of optical components, light sources, light modulators and light detectors. A more subtle appeal is the possible speed of processing. An optical image can contain millions of picture points to be transformed by lenses in parallel and in nanoseconds.

For aperture synthesis processing it is not difficult to envisage an incoherent optical implementation of the charge storage tube processor described above. In this case a two-dimensional, optical, integrating detector (discussed below) accumulates sub-images generated by scanning a light source modulated by the digital data.

Other incoherent approaches could be derived (Cole, 1976), but all would suffer from two basic drawbacks. Offset light levels are needed, and to avoid diffraction effects, picture elements need to be much larger than the wavelength of light. This limits the processing capacity while the almost inevitable inclusion of mechanical movement in the processor can be a limit to speed. Their redeeming characteristics are probably cost, combined with a simplicity of design when compared with the use of coherent optics.

7. THE COHERENT OPTICAL PROCESSOR

The principles of coherent optical processing are now well established (e.g. Goodman, 1968) and some radio astronomy applications have been discussed (McLean and Wild, 1961; McLean et al. 1967; Cole, 1975). The processing relies on the Fourier transform properties of a lens. Near the optical axis where $\sin\theta \approx \theta$, the light amplitude distribution across an input aperture one focal length from a convex lens is Fourier-related to the light amplitude distribution across the output plane, one focal length on the other side of the lens.

Variations of the basic approach include cascaded transforms, where the input can be recovered after modification of its Fourier transform in the intermediate plane. Cylindrical lenses permit parallel transformation of a number of one-dimensional light distributions. Lasers and lenses are readily available. Lens accuracies can be one-twentieth of a wavelength or better r.m.s. and processing accuracies of better than 1% are common (Preston, 1972).

The implementation of a coherent optical processor for aperture synthesis data is now clear. From the computer-stored data, one needs to modulate the laser light distribution across the processor input plane. A simple lens provides a Fourier transform of this distribution. But from Figure 1(b) we realize that the light amplitude in the output plane needs to be measured. Detectors measure intensities, so that the amplitude needs to be determined by a further two-stage process somewhat akin to holography. As outlined by Bulabois (1976), a uniform reference light beam (R) firstly in-phase and then out-of-phase with the signal light amplitude distribution (S) in the output plane would give intensity distributions $|S+R|^2$ and $|S-R|^2$. On expansion of these terms and formation of the difference between them, we have 4RS, proportional to the desired amplitude S.

The practicality of the processor depends critically on the light modulators and detectors. They can be assessed in relation to several broad classes of configuration. The input and output digital stores are serial devices whereas the optical processor is parallel. A two-dimensional light modulator presents a two-dimensional output to be read either line by line with a linear detector array or in parallel by a two-dimensional detector array. The use of a two-dimensional integrating detector array will be seen to permit the sequential use of a one-dimensional modulator. In most cases however one is limited by the speed of serial access to digital stores and the optimum configuration depends critically on the detector and modulator.

8. LIGHT DETECTOR ARRAYS

Light detector array technology is advancing rapidly and a number of suitable components exist. The most studied one-dimensional array is the sensitive, accurate and reliable solid-state, self-scanned photodiode array (Fry, 1975; Vogt et al., 1978). With up to 1800 or so photodiodes delineated on a silicon substrate, incident light is integrated as stored charges to be read out on a single video output line by shift registers and control circuitry integrated on the silicon chip. The response is linear, and low-noise processing of the output produces a dynamic range between light saturation and r.m.s. readout noise of typically one or two thousand to one. To achieve this performance one removes fixed effects of leakage currents and fixed patterns by image subtraction.

Two classes of device exist. On one the photodiodes are sampled sequentially by switching each to the output video line. In the other all diodes are sampled at the same instant and read-out along an analogue charge coupled device (CCD) shift register. This second approach is preferred, since lower output video line capacitance and on-chip pre-amplifiers should give lower readout noise while the simultaneous sampling more closely matches the processing problem of comparing consecutive images. The so-called CCPD combines photodiodes (P) and CCD technology (Reticon, 1977).

In two dimensions CCD arrays already exist, and again those arrays which simultaneously sample all photodiodes are preferred (Barbe and Campana, 1977). Commerically available devices include the 120 × 150 point CCD camera from GEC currently being used at CSIRO. Laboratory devices exist with 10 or more times as many picture points.

Amongst the more traditional image detectors the silicon vidicon stands out as offering a very large number of output points and (with care) a high dynamic range (McCord et al., 1975). However, all the vidicon-type devices suffer in readout speed. The electron beam readout is sequential and must be complete before the next image can be displayed for integration.

The literature contains reviews and discussion of performance of most of these devices. There is however one point which needs to be stressed.

All of the detectors use stored charge as the parameter representing integrated light energy. Each storage area has a maximum charge typically of the order of 10^6 electrons. One has to consider the effects of shot noise, which for 10^6 electrons has an r.m.s. value of 10^3 electrons or one part in one thousand for the saturation charge. The other source of noise is readout noise, which is independent of the charge amount being read-out and which can be made much less than 10^{-3} of saturation charge. If the image being read-out is of low contrast (and it will usually be in optical processing), then shot noise will become the dominant source of processor noise. If a dynamic range of more than 30 or 33 dB is needed, then care is required in choice of detector.

For high-contrast images, dynamic range is determined by readout noise and could approach 40 dB. One notes however that this is up to 10 dB worse than might be achievable in a radio-frequency square-law detector!

Nevertheless, there is available a range of suitable detectors matched to the optical processing problem, especially where processing involves forming the difference between two images.

9. LIGHT MODULATORS

For speed the light needs to be modulated across an input plane. Therefore only those modulators which are multi-channel will be considered. Table I summarizes their properties.

Parameter \ Device type	Elastomeric		Pockels effect	
	Oil film	Thermoplastic	Optically written	Electron-beam written
Aperture size (mm^2)	25 x 25	50 x 50	25 x 25	50 x 50
Limiting resolution (line pairs/mm)	50	50	80	30
Space-bandwidth product	$< 10^6$	10^6	10^5	$10^5 - 10^6$
Cycle time	30 ms	>0.5 s	10 ms	30 ms
Storage time	∿100 ms	Months	Minutes to hours	Minutes to hours
Erasure time	∿100 ms	∿1 s	∿1 ms	<1 ms
Lifetime	∿3000 h	∿10^5 cycles	Long	Long

Table I. General characteristics of some current spatial coherent light modulators.

The most traditional modulator is the silver-halide photographic transparency. Varying density distributions modulate the transmitted light amplitude. In order to cope with negative values the modulation function needs to be floated about a bias (grey) level. In order to modulate light phase more complex arrangements are needed. The most usual is to employ a spatial frequency carrier so that each element of the input distribution is in the form of a diffraction grating. The phase of the light diffracted from this grating element depends on the relative position of the element centre and the grating maxima.

Photographic modulators of a million or more picture elements are possible but in practice they have great drawbacks. Accurate exposure of the modulating pattern is difficult, especially in the aperture synthesis case, where a large peak occurs at low values of spatial frequency. Surface irregularities on the film require compensation. Chemical development and the non-reusable nature are further drawbacks. The main positive factor is the 'low technology' content.

As an alternative I wish to summarize developments in three broad classes of modulator; elastomeric, Pockels effect, and electro-optical.

The elastomeric modulator (Thomson, 1977; Casasent, 1977)

relies on the deformation of the surface of a layer through which, or reflected from which, the light passing is to be modulated. The surfaces include thermoplastics or oil films and the required pattern is formed by writing an electrical charge distribution with an electron beam. The general properties of several elastomeric devices are listed in the table and a number of disadvantages emerge. In general they are expensive and of limited life. The decay time is either too slow or too fast. The optical quality is not always high, and like photography, these devices require a bias level and spatial carrier for complex light modulation.

A more attractive class of modulator relies on the Pockels effect whereby the linear polarization angle of the light is locally rotated by a charge distribution across an electro-optical substrate (Marie et al., 1974). Crossed polarizers convert the rotation to an amplitude modulation.

With 1000 × 1000 points or more, up to one hour storage of pattern, less than 1 ms to erase, long life, and high optical quality, the devices are very attractive. They do require a bias level and a spatial frequency carrier but the required modulating pattern is written either by a scanned electron beam or by a modulating blue light. The commercial development of these devices is sure to continue.

There is also an electro-optical modulator specifically derived in response to the aperture synthesis processing problem. It is an advance on acousto-optical light modulators for antenna array processors (Lambert et al., 1965), which are difficult to arrange in large numbers. The new device is a variation of the total internal reflection electro-optical modulator (Scybor-Rylski, 1977, 1974). An interdigitated metal finger pattern is deposited on a suitable substrate surface from which light is internally reflected (Figure 3(a)). For electro-optical and piezo-electric materials such as lithium niobate a voltage across the fingers can produce a combined surface deformation and electro-optical effect which phase-modulates the reflected light distribution. Some light is diffracted from the main beam by the applied voltage (several tens of volts) and both positive and negative amplitude modulation is possible.

General phase modulation requires a more complex electrode pattern (Cole, 1978a,b). As shown in Figure 3(b), the pattern contains two electrodes whose modulation is 90° apart in phase. These COSINE and SINE electrodes permit fully complex light modulation at megahertz rates. The deposition of a number of such patterns permits the optical simulation of an antenna array (Cole, 1978a), as shown in Figure 3(c).

With standard microcircuitry techniques up to a hundred or so patterns could be accurately deposited on a polished substrate. Their use in the aperture synthesis problem is discussed below.

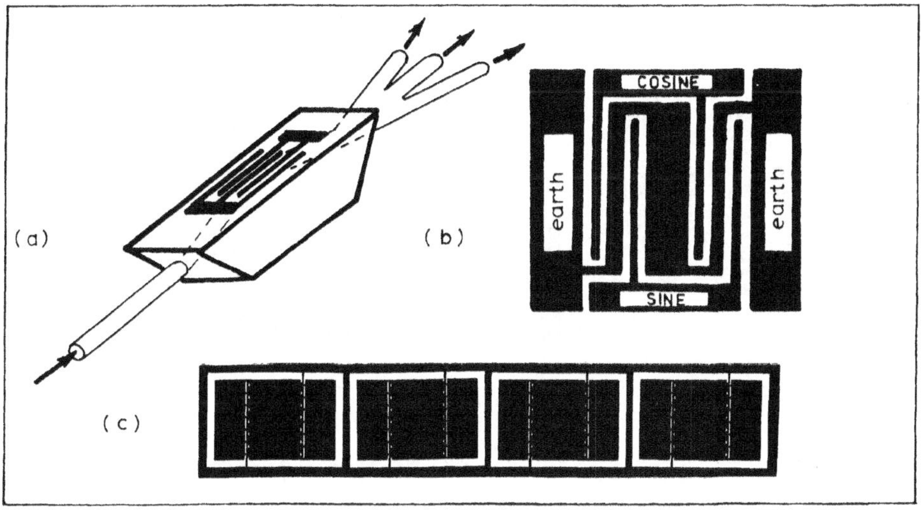

Figure 3. The total internal reflection electro-optical light modulator (a) uses interdigital metal finger patterns on lithium niobate. The light is reflected internally and the modulation diffracts light from the main beam. To modulate phase as well as amplitude a more complex pattern (b) with in-phase and quadrature elements as well as earth common is needed. A number of such modulators (c) forms an array processor.

10. SOME SYSTEM DESIGNS

Having defined the problem and assessed the components available for its solution we can consider those solutions. I concentrate on three systems. The first, the computer-generated hologram, is the most general, and in many ways the least attractive. The other two exploit the symmetries of particular antenna geometries.

10.1 The computer-generated hologram

The optical system can be required to fulfil only the Fourier transform function. A photographic transparency of the spatial frequency plane samples, suitably modulated on a spatial carrier and suitably embedded in a refractive index matching oil, forms the input, and photodiode arrays read the output distribution (Figure 4). The use of a switched reference light beam extracts the light amplitude (Somers and Burns, 1976; Bulabois, 1976; ERIM, 1977; Cole, 1977b, 1978c).

This is essentially the proposed optical processor for the VLA. I do not intend to go into great detail, but several comments are appropriate. The undesirable photographic input could be changed to

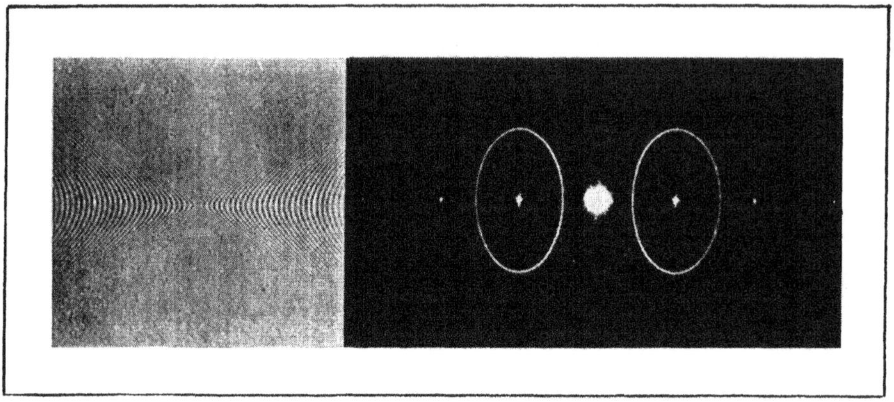

Figure 4. A computer-generated hologram of the Westerbork antenna beam in (a) can be optically transformed as in (b) to two images displaced about the undiffracted light beam. Several defects are discernible in the reconstruction (see Cole, 1978c).

any of the two-dimensional continuous modulators discussed in Table I. The approach is general in that any geometry can be processed as transformations of planes of spatial frequency samples. However, a not insignificant load is presented to the host computer because of the VLA geometry. The data would probably be placed on a grid (although not strictly necessary), the optimum taper would need to be applied, and most importantly, the data would need to have been projected to a plane. The north-south geometry strictly requires a number of planes for subsections of the field of view. The processor is not called upon to transform the entire field of view. The high parallel processing speed of the optics is not fully utilizable in the VLA optical processor.

However, it is clear that the system would work and produce outputs with 1% accuracy or better. It is not clear though that it could form part of a fully interactive data reduction system. Its generality is probably a disadvantage when compared with systems exploiting symmetries in the spatial frequency sampling. At best, the holographic approach could process east-west array data by recording along concentric ellipses for which the weighting factor is constant. This would greatly ease the load on the host computer and avoid aliasing due to gridding.

10.2 Strip scan processing

Earth rotation aperture synthesis with an east-west array measures sequential, radial lines of spatial frequency samples. The Culgoora array spatial frequency sampling can also be described in polar coordinates (see below). In these situations the transformation is describable as the sum of the one-dimensional transforms (or strip scans) of the properly weighted, radial lines of samples. A two-dimensional integrating array is required as the output plane.

An optical strip scan processor would use a two-dimensional detector and an effectively rotating, one-dimensional input system of light modulators.

Two-dimensional detectors exist. The one-dimensional modulator array is easily within current technology. In all antenna arrays there is only a finite number of spacings. For example, at Westerbork the minimum spacing increment of 18 m implies 167 or so possible spacings over 3 km. The rigid geometry of Culgoora also strictly defines the possible spatial frequency sampling. Therefore, it is feasible to contemplate individually addressed modulators aligned with a spacing pattern which closely matches the array whose data is to be processed.

For example, I would contemplate a one-dimensional line of total internal reflection modulators as in Figure 5. Each is addressable via

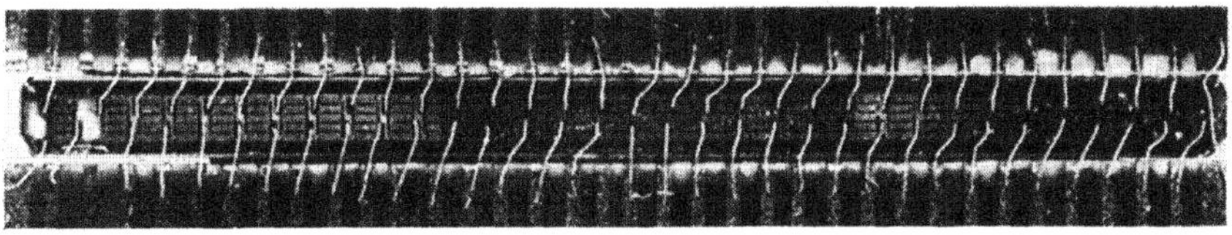

Figure 5. A complex internal reflection modular array with the geometry of one central arm of the Fleurs array is shown in the centre. The finger widths are 10 μm and the 33-element array measures 25 × 1 mm.

a multiplexer and charge-hold circuit from the computer. At each hour angle the data are loaded and the two-dimensional strip scan is integrated on the output plane. The particular geometry of Culgoora requires one to use two types of radial sampling patterns (Figure 6(a) and (b)). In all cases one must determine light amplitude by a process of image differencing between outputs with two states of a reference beam. It is easy to insert a reference modulator at the spatial frequency origin. (One also notes that with careful control of the reference phase, it is possible to transform just the positive half of the spatial frequency plane. This eases modulator design and speeds processing.) The requirement to subtract in a two-pass imaging system not only obtains light amplitude but also cancels fixed irregularities in the detector.

The processing speed would now be limited by the computer. The total internal reflection modulator responds at megahertz rates. At, say, 20 μs per strip scan, picture processing rates of many tens of pictures per second appear not unreasonable. The use of the fast, re-usable, solid-state modulators completely alters the processing

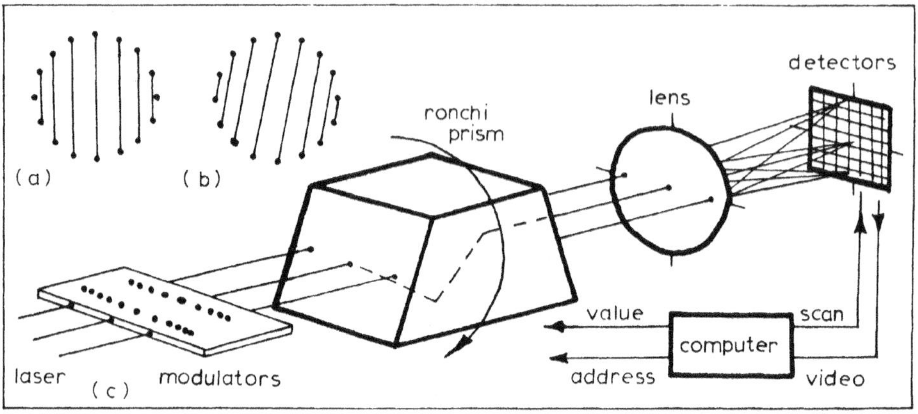

Figure 6. For the Culgoora array, those spatial frequency samples with the same orientations are shown in (a); (b) for a 16-element array. Two sets (odd and even) emerge with an almost twofold redundancy. In (c) is sketched one form of coherent optical processor for Culgoora (or east-west) spatial frequency data. The ronchi prism rotates the transformed image of the modulators on the two-dimensional detector array.

concept. To rotate the strip scans on to the detector a simple Dove prism system suffices (Figure 6(c)). With such a prism (or mirror equivalent) system, the image rotates twice as fast as the prism (or mirrors) so that mechanical constraints are not high.

10.3 Integration of instantaneous images

Strip scan processing for east-west arrays is a special form of a more general approach. Aperture synthesis is merely the summation of images obtained at a number of sequential times. The only requirement is that the spatial frequency sensitivity of the sub-images match the requirements of the complete, synthesized, spatial frequency sampling. A general case is Culgoora, capable of imaging a region of sky at any instant. Indeed, in its solar application, one is interested in producing up to 20 images per second. Earth-rotation aperture synthesis is therefore the addition of a large number of images which have been appropriately rotated and scaled to match the array rotation and foreshortening due to Earth rotation.

Similar arguments apply to the VLA, except that the VLA is not exactly in a plane. Simple rotation and foreshortening of a rigid planar array therefore do not exactly match the VLA behaviour. Nevertheless a device for generation of instantaneous images from the current samples of the spatial frequency plane could form the basis of a synthesis processor.

Without going into details, we can say that a two-dimensional distribution of light modulators in the same pattern as the array can be used to generate fringes appropriate to each measured spatial frequency (Cole, 1976). Acousto-optical and total internal reflection modulators could be used. However, it is difficult to see any advantage when this use is restricted to processing spatial frequency plane measurements from east-west arrays and from Culgoora. For these, strip scan processing appears simpler.

11. CONCLUSIONS

Undoubtedly, hybrid processing could undertake much of the repetitive digital processing stages in Earth rotation aperture synthesis. It has a distinct role in interactive reduction systems where the 20 dB or so limit to dynamic range is not itself a limitation. Hybrid processing is somewhat more attractive in those situations where it can perform the major part of the reduction. That is, those geometries for which weighting, coordinate conversion and transformation can be performed in the analogue components are more relevant than those for which only the Fourier transformation is performed.

As a result, the choice of processing depends critically on the antenna geometry. I would, in fact, prefer to see the reverse occur: where the processing would influence the initial antenna design. If this had occurred I am sure that several present arrays would have been constructed slightly differently.

The basic components of hybrid systems have all been studied, often in great detail. What then is the inhibiting factor to their use? The main problems appear to me to be human. So often it appears as an 'easy option' to just buy more digital computing power and take a software solution. Hybrid systems break interdisciplinary ground and do have accuracy limitations. Few of the traditional radio astronomy centres have the people prepared to face up to the challenge and possible risk. But if careful consideration is given to the computations to be done and careful selection made of the special purpose implementation, there is a significant role to be played by faster and cheaper hybrid processing in Earth rotation aperture synthesis.

ACKNOWLEDGEMENT

My attendance at the meeting was due to the generous support provided by the Netherlands Foundation for Radio Astronomy and the Max-Planck-Institut für Radioastronomie, Bonn.

REFERENCES

Barbe, D.F., and Campana, S.B.: 1977, *Advances in Image Pickup and Display* (Academic Press, New York) Vol. 3, p. 171.
Briggs, B.H., and Holmes, N.: 1973, *Nature, Phys. Sci.* 243, p. 111.
Brouw, W.N.: 1975, *Methods in Computational Physics* (Academic Press, New York) Vol. 14, p. 131.
Bulabois, J.M.: 1976, *National Radio Astronomy Observatory VLA Computer Memorandum* No. 128 (Charlottesville, Va.).
Casasent, D.: 1977, *Proc. IEEE* 65, p. 143.
Cole, T.W.: 1975, *Opt. Acta* 22, p. 83.
Cole, T.W.: 1976, *Netherlands Foundation for Radio Astronomy Technical Report* ITR 145 (Dwingeloo, Netherlands).
Cole, T.W.: 1977a, *Progress in Optics* (North-Holland, Amsterdam) Vol. 15, Chap. 4.
Cole, T.W.: 1977b, *Proc. Astron. Soc. Aust.* 3, p. 105.
Cole, T.W.: 1978a, *Appl. Opt.* October issue (in press).
Cole, T.W.: 1978b, submitted for publication.
Cole, T.W.: 1978c, *J. Opt.* (in press).
ERIM: 1977, Study for an Optical Processor for VLA Radiotelescope Data, ERIM, Michigan Report 123400-8-F.
Fomalont, E.B., and Wright, M.C.H.: 1973, in *Galactic and Extra-Galactic Radio Astronomy* (Springer-Verlag).
Fourikis, N.: 1967, *Proc. IREE (Aust.)* 9, p. 315.
Fry, P.W.: 1975, *J. Phys. E: Sci. Instrum.* 8, p. 337.
Goodman, J.W.: 1968, *Introduction to Fourier Optics* (McGraw-Hill, San Francisco).
Heeschen, D.S.: 1975, *Sky Telesc.* 49, p. 344 (June).
Hogböm, J.A.: 1974, *Astron. Astrophys. Suppl.* 15, p. 417.
Labrum, N.R., McLean, D.J., and Wild, J.P.: 1975, in *Methods in Computational Physics* (Academic Press, New York) Vol. 14, p. 1.
Lambert, L.B., Arm, M., and Aimette, A.: 1965, *Optical and Electro-Optical Information Processing* (MIT Press, Cambridge, Mass.) Chap. 38, p. 715.
McCord, T.B., Bosel, J.P., and Frankston, M.J.: 1975, *Image Processing Techniques in Astronomy* (D. Reidel, Dordrecht) p. 91.
McLean, D.J., and Wild, J.P.: 1961, *Aust. J. Phys.* 14, p. 489.
McLean, D.J., Lambert, L.B., Arm, M., and Stark, H.: 1967, *Proc. IREE (Aust.)* 9, p. 375.
Marie, G., Donjon, J., and Hazan, J.-P.: 1974, *Advances in Image Pickup and Display* (Academic Press, New York) Vol. 1, p. 225.
Oliver, B.M., and Billingham, J. (eds.): 1973, *Project Cyclops*, NASA publication CR114445.
Pooley, G.: 1976, *Methods of Experimental Physics* (Academic Press, New York) Vol. 12, p. 158.
Preston, K.: 1972, *Coherent Optical Computers* (McGraw-Hill, New York).
Reticon Corporation, California: 1977, Data sheets.
Rotman, W., and Turner, R.F.: 1963, *IEEE Trans.* AP-11, p. 623.
Scybor-Rylski, M.T.V.: 1973, *Electron. Lett.* 9, p. 309.
Scybor-Rylski, M.T.V.: 1974, *Electron. Lett.* 10, p. 4.
Shelton, J.P., and Kelleher, K.S.: 1961, *IEEE Trans.* AP-9, p. 154.

Silver, R.S., and Luedicke, E.: 1971, *IEEE Trans.* ED-18, p. 229.
Somers, L.E., and Burns, W.R.: 1976, *National Radio Astronomy Observatory VLA Computer Memorandum* No. 129 (Charlottesville, Va.).
Stroke, G.W.: 1972, *IEEE Spectrum* p. 24 (December).
Tippett, J.T., Berkowitz, D.A., Clapp, L.C., Koester, C.J., Vanderburgh, A. (eds.): 1965, *Optical and Electro-optical Information Processing*, (MIT Press, Cambridge, Mass.).
Thomson, B.J.: 1977, *Proc. IEEE* 65, p. 62.
Vogt, S.S., Tull, R.G., and Kelton, P.: 1978, *Appl. Opt.* 17, p. 574.

DISCUSSION

Comment J.P. HAMAKER
I want to put on record that, following Trevor Cole's original idea, we are constructing a storage tube-based Fourier transform machine of the type he just described, for getting a quick-and-dirty look at Westerbork maps. Results on its performance are not yet available.

Comment R.H. HARTEN
Isn't the whole idea of an interactive system dependent on the response time of the user (astronomer)? This is much slower than any disk head movement or I/O access time.
Also, the implemention of the 'digital' corrections to the data is the biggest problem: every data point must be handled, and at its observed coordinates, not at the gridded ones. This does not lend itself too well to special purpose devices.
Reply T.W. COLE
I consider one to ten seconds per cycle as matching the user response time and this probably matches the imaging systems under discussion. Digital correction is not as big a problem as you intimate when the antenna geometry is symmetric. Then it exactly matches special purpose devices.

EXPLORING METHODS OF EFFECTIVE DATA DISPLAY IN AN INTERACTIVE ASTRONOMICAL DATA-PROCESSING ENVIRONMENT

(Invited paper)

R.J. Allen
Kapteyn Astronomical Institute
University of Groningen

1. INTRODUCTION

Many of the image restoration algorithms discussed during these past days work best on fields which are virtually empty except for a few discrete sources. But the sky also has faint regions of extended emission; these regions invariably turn out to be of great astrophysical interest and must be represented as accurately as possible in the maps. The methods chosen to present the data in pictorial form can have an important effect on the speed and efficacity with which the astronomer can extract the useful information. I would be afraid of boring all of you with this discussion of display methods if I did not know that, besides being well versed in subjects of probability, statistics, and applied mathematics (and in some cases philosophics and polemics), you are also astronomers with a strong motivation to sift out of your data the useful information on the physics of cosmic radio sources, and to do it in the most efficient way possible.

I should say at the start that I shall not discuss either the structure or the implementation of interactive computer image-processing systems, nor shall I describe the various enhancement and processing algorithms applied to images. My concern here is restricted simply to the problems of displaying the images. Although I am convinced that careful attention should be paid to displaying the data at every step of the reduction, most of my own experience has been with maps of the radio sky at the latest stages of the data reduction process.

2. DISPLAYING SINGLE-IMAGE DATA SETS

When the area to be displayed contains only a modest number of telescope beams (10^1-10^3), the traditional representations as contour and ruled surface maps are generally adequate (e.g. Fig. 1). Indeed, the addition of shades of grey or colour to the contours in this case is usually viewed as unnecessary embellishment, of use possibly only for public relations purposes. Contour diagrams have the great advantage of providing a

C. van Schooneveld (ed.), Image Formation from Coherence Functions in Astronomy, 143-155.

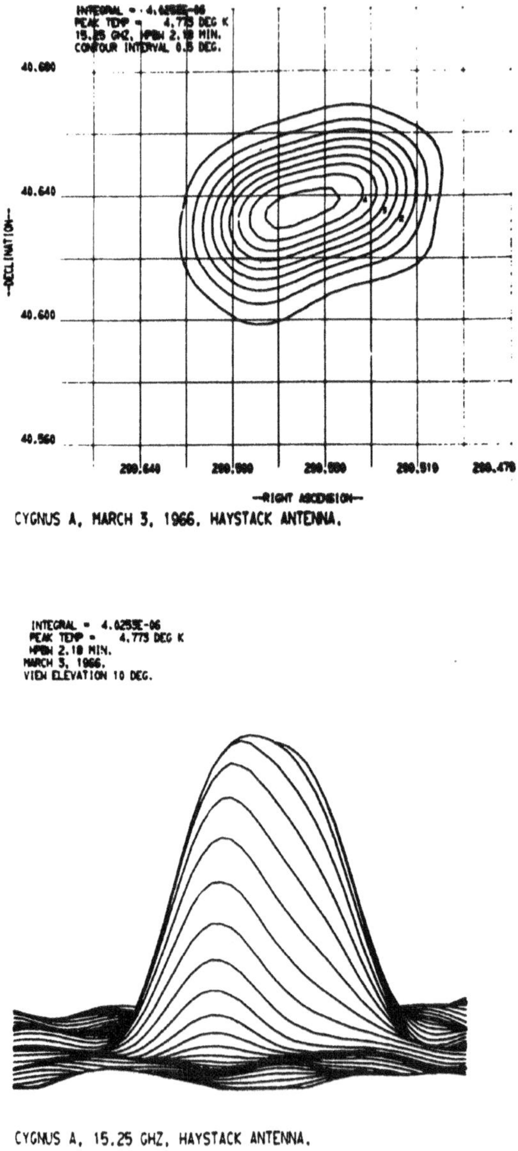

Figure 1. Contour (upper) and ruled-surface (lower) representations of a radio image of Cygnus A. This is a familiar way of displaying fields which contain only a modest number of beam areas.

method for quantitative display, in particular for negative intensities if some means of flagging negative contours is used.

Difficulties with contour and ruled-surface displays begin to appear when the area to be represented covers very many telescope beams (10^3 or more). In this case, grey shading and the use of colour can indeed be useful and may even be essential.

a) Contour Display

Although retaining the important advantage of being reasonably quantitative, contour plots of high-resolution images give a poor visual impression; gradients are emphasized rather than intensities (e.g. Fig. 2), and if the S/N ratio is furthermore not very high the particular choice of contours and contour interval has a great effect on the appearance of the picture (e.g. Fig. 3). Reducing the number and choosing the level placement of contours in order to improve the visual impression inevitably results in a loss of information which can severely limit the utility of the display.

b) Ruled Surface Display

This type of display provides a reasonable visual impression for high-contrast features (e.g. Fig. 4) such as discrete sources and bright ($S/N > 10$) extended regions. It is essential that the "hidden lines" be deleted from the display; otherwise the impression of viewing a solid surface disappears and the picture dissolves into a jumble of wiggly lines. An important shortcoming is that weaker features of objects can literally be hidden behind stronger ones.

If the S/N ratio per beam area is only of order unity, extended sources remain invisible to the eye on ruled surface plots. Grey-scale displays are better for detecting such features.

c) Grey Display

The right panel of Figure 5 is a high-contrast grey display of the same image from which the ruled surface display of Fig. 4 was made. The existence of faint extended emission from the disk of the galaxy (left panel of Fig. 5) is unequivocal even though the S/N per beam area is only about 0.5. This type of display presents a more familiar kind of picture to the viewer, and the highly-developed pattern-recognition capacity of the eye and the brain is more readily utilized.

Grey displays are, however, generally not very quantitative; they also require arbitrary zero level shifts in order to represent negative intensities. The available dynamic range of the final display is an important limitation; representing an intensity range more than about 30 on photographic printing paper is hardly feasible. Transparent film is better ($\sim$300) in this respect. For a given level of background illumination the range of intensity discernable to the average human eye is of the order of 1000.

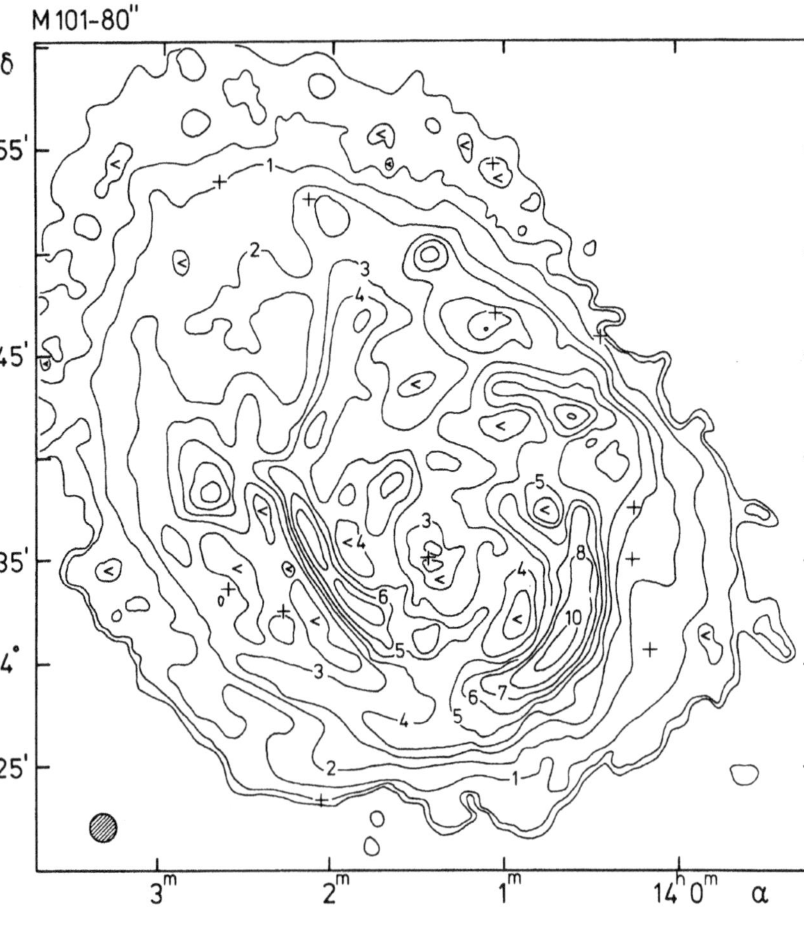

Figure 2. Contour display of the total HI in M101. A selection of contour levels to display has been made; for example there are no negative contours, the zero contour is omitted, and the contour intervals are not equal. Although the visual presentation is thereby improved somewhat, the overall impression is still confusing, and gradients are emphasized rather than intensities. The region of interest contains about 500 beam areas.

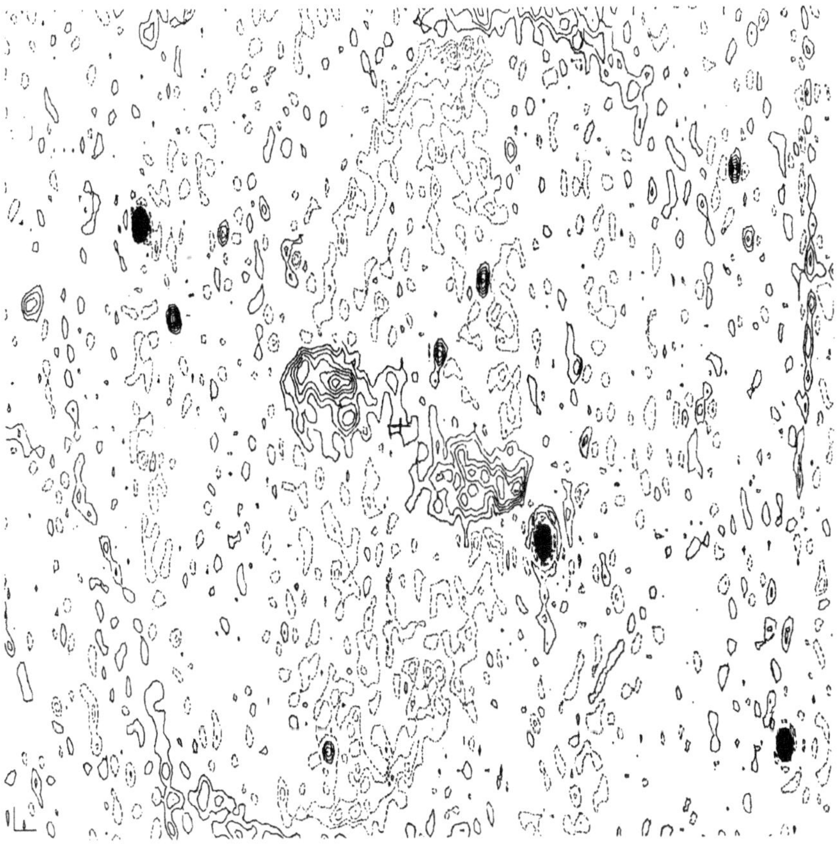

Figure 3. Contour display of a region containing about 5000 beam areas. The general first impression is one of confusion, even though the zero contour has been omitted to improve the presentation. Negative contours are dashed.

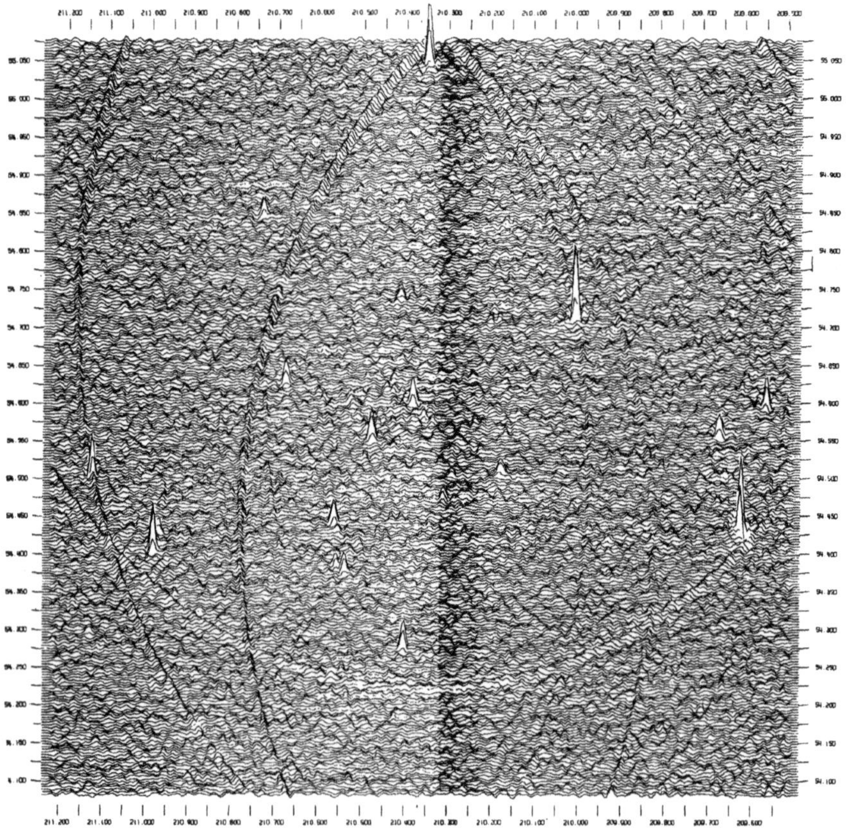

Figure 4. Ruled surface display of a region containing about 11,000 beam areas. High-contrast features stand out here, such as the discrete sources and their associated grating rings. Faint extended regions of low surface brightness remain invisible (cf. Fig. 5).

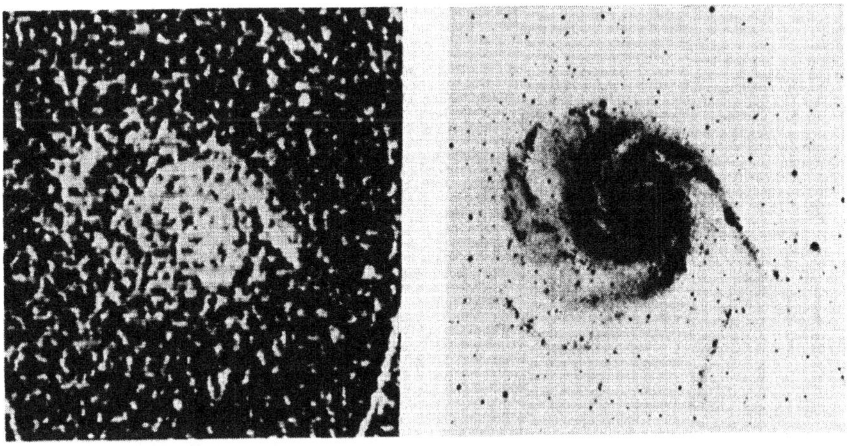

Figure 5. The right panel is a high-contrast representation of the central regions from Fig. 4; all positive areas are simply set white. The faint (S/N ≈ 0.5 or less) extended continuum emission from the disk of the galaxy is clearly recognized here, although it remains invisible in Fig. 4 even if one knowns exactly where to look. The optical image of the galaxy is shown in the left panel.

d) Pseudo-Colour Display (colour contouring)

For this type of display each data value is used to determine the colour (lightness L, hue H, saturation S) of the image at that point. Usually a small range of data values are all assigned the same colour; the effect is the same as when one colours the areas between two contours on a contour plot drawing. Besides the pattern-recognition capabilities of the eye we can here make use of the capacity to detect small changes in hue.

Pseudo-colour displays also offer a method for obtaining higher dynamic range in the final representation than is possible with grey displays. Successive contours can be assigned colours which cycle through the spectrum, effectively providing a high-contrast display at every contour level. Furthermore, negative intensities can also be easily distinguished by making a judicious choice of intensity level colour coding around the zero level. Most important is that colour printing methods can preserve the integrity of a pseudo-colour image in a final publication.

One difficulty is that the visual impression created by a pseudo-colour image may sometimes be too strong, such as when large areas of the picture are filled with bright saturated colours or when the colours for successive contours are chosen at random. Even more serious is that the visual impression may also be incorrect if unimportant features are inadvertently over-emphasized. It would seem that there is room here for some controlled and systematic experimentation in order to arrive at an effective presentation.

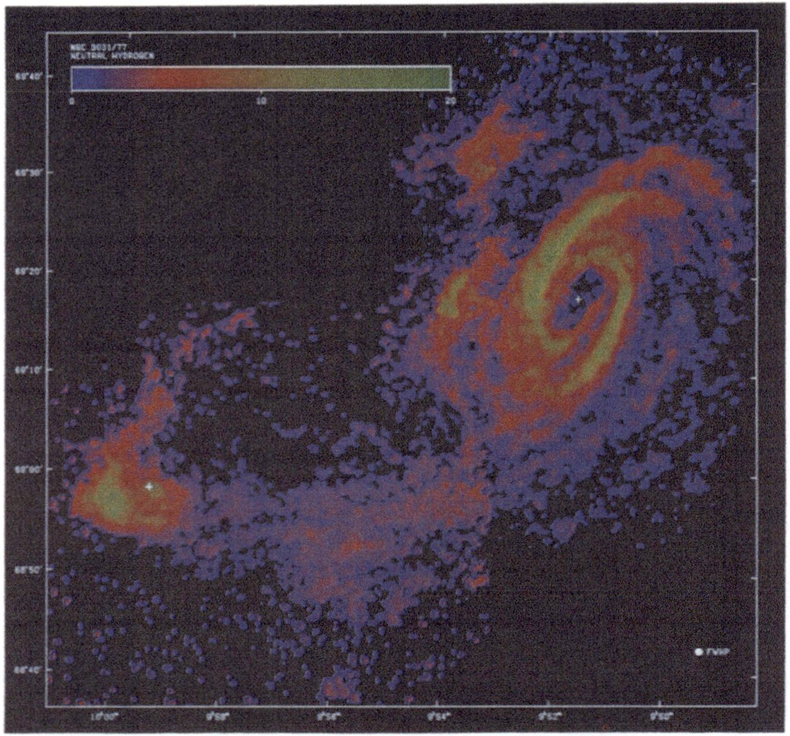

Figure 6. A pseudo-colour (colour-contour) display of the HI distribution in the region of M81 and NGC 3077. Care has been taken to represent contours covering larger areas of the display in more subdued tones.

Figure 10. False colour image of the HI distribution and kinematics of M101. There are more than 100,000 independent data values represented in this picture.

Figure 6 shows an example of colour contouring in which a useful choice of colours has been made.

3. DISPLAYING MULTIPLE-IMAGE DATA SETS

Two methods have been explored for representing a third independent variable such as polarization or frequency in an image data set; these are "false" colour and cinematography. Before describing an example of a false colour display it may be useful to examine Table 1, in which the different kinds of colour displays are described in the terms commonly used in the electronic-image-display industry.

a) False-Colour Display

If the data set contains only 3 images, a one-for-one assignment of L,H, S may be tried; two examples are for radio polarization (e.g. Intensity = L, Position angle = H, Percent polarization = S) and for the first three moments of spectral line profile images (e.g. Profile integral = L, Mean frequency = H, Profile width = S). As with the pseudo-colour display, some systematic experimentation is needed here.

If the data set contains $N > 3$ images, some $N \times 3$ matrix transformation can be tried. Figure 7 shows a grey display of two images from a set of 12 single-channel HI maps of a nearby galaxy. A possible matrix transformation on this set is shown diagrammatically in Figure 8 superposed over the total integrated HI profile of the galaxy. The three R,G, B output images are shown in Figure 9, and the final colour composite in Figure 10. The features best illustrated by this type of display are the overall rotation of the galaxy (i.e. the gradient of colour from blue on the lower right to red on the upper left), and the large-scale deviations from circular motion. Here too it is clear that further experimentation is necessary; one could, for example, try to enhance the smaller-scale velocity anomalies by first subtracting off a model of the circular rotation.

b) Cinematographic Display

This type of display seems especially useful for detecting correlations among the different images of the data set. It makes use of the highly-developed capacity of the eye and the brain for detecting motion. An experiment has been made by time-sequencing a set of spectrometer maps along the frequency axis and recording the sequence on video tape. A number of new phenomena could be seen in the data (the video tape was demonstrated at the meeting).

Another possibility would be to time-sequence the remaining spatial dimension in a data set organized as position-velocity images. There is here also much room for further experimentation. The cinematographic display probably is especially useful in the initial phases of reconnaissance of the data set.

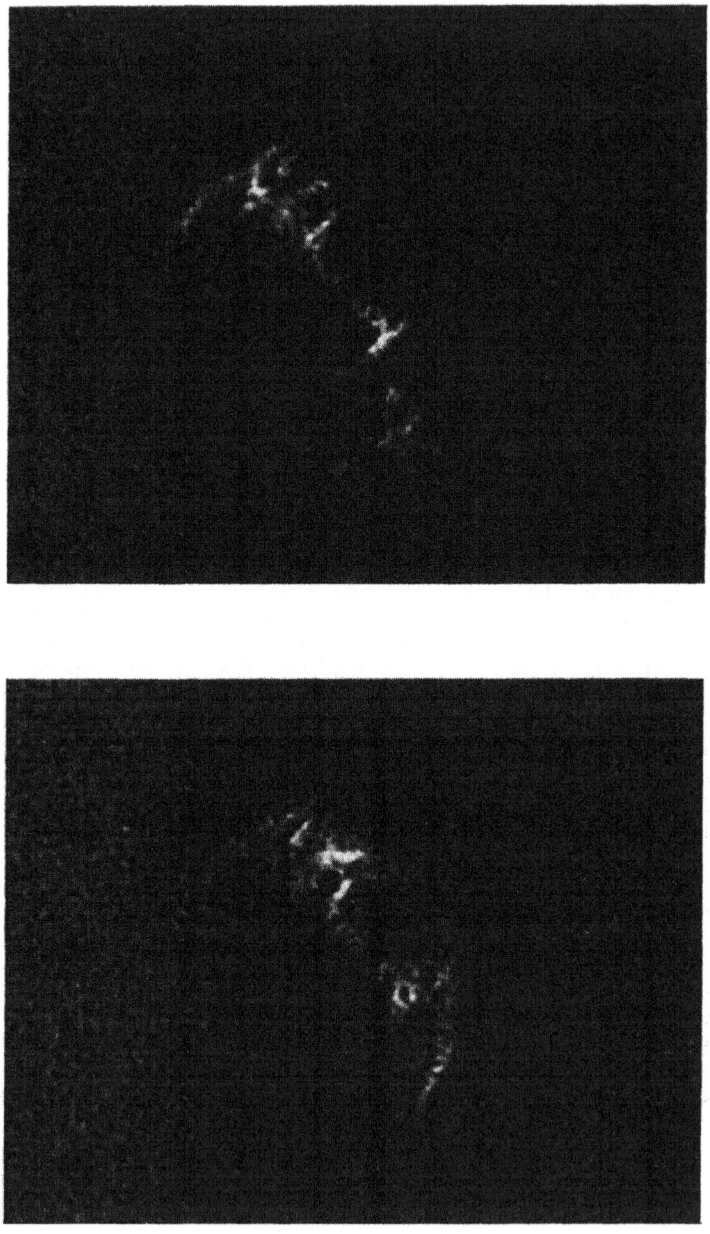

Figure 7. Two neighbouring channels (at velocities of 220 and 240 km s^{-1}) from a spectrometer synthesis survey of a large nearby galaxy are shown here as a grey display. The set of 12 such images was used as input to the "3-filter" matrix transformation shown in Fig. 8.

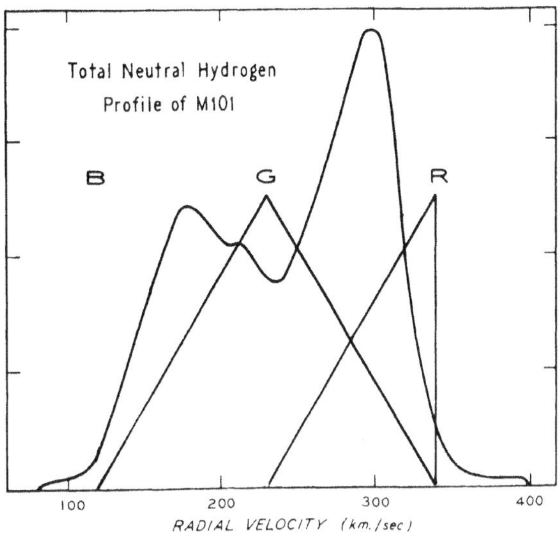

Figure 8.

Graphical representation of the R,G,B filters used to transform the 12-map set of images (cf. Fig. 7) to the 3-map set of Fig. 9.

---

1. Pseudo-Colour (colour contouring)

Use the numerical value of a pixel at a given point of a single image to compute an assignment of displayed R,G,B intensities at that point. The computation may pass through an intermediate step of choosing the lightness, hue, and saturation (L,H,S) which are then transformed to R,G,B intensities and additively displayed.

2. "True" Colour

Photograph a given real scene three times onto black-and-white film through R, G, and B filters successively. These three colour separation images are then projected by the same R,G,B filters onto a single screen to produce a "true" colour display.

3. "False" Colour

Subject a set of N $\geqslant$ 3 two-dimensional images to some computational procedure which produces 3 secondary R,G,B images; these 3 intensities are then additively displayed. As with the pseudo-colour display an intermediate step through L,H,S may be made.

TABLE 1. Operational definitions of the different types of colour displays as understood by the image display industry.

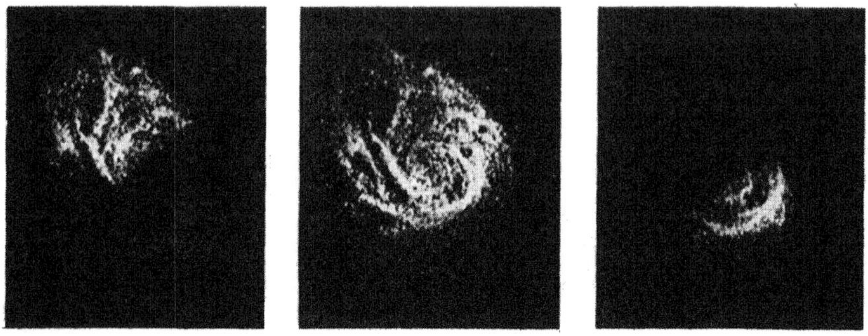

Figure 9. The R,G,B colour separation images produced by a 12×3 matrix transformation in the spectral dimension on the data set of Fig. 7.

4. POSSIBLE DIRECTIONS FOR FUTURE DEVELOPMENT

The need for experimentation in methods of effectively displaying astronomical data has been repeatedly mentioned in the previous discussion. In order to carry out such experiments it is necessary to have the hardware at hand. It would seem therefore profitable to set up a data analysis and display system which makes grey, colour, and cinematographic displays easily and conveniently available to the astronomer at the earliest stages of his image-processing effort; these facilities should be provided as an integral part of an interactive computer data reduction system. A means would then be available for:

- examining all the data taken;
- early detection of residual instrumental problems;
- rapid access to the astrophysically interesting parts of the data set (often only a very small fraction of the total);
- recognition of unexpected results.

A second area still to be explored is to try to bring the model-building phase of the analysis as much as possible into the same interactive picture-processing system. This makes available all the display facilities in order to compare models and observations in a similar way, and should provide the possibility for a rapid parameterization of the astrophysically interesting results in the data in terms of the models.

ACKNOWLEDGEMENTS

I am grateful for the collaboration of R.D. Ekers on several of the experiments described in this talk. J.M. van der Hulst and A.H. Rots kindly provided me with the pseudo-colour picture shown in Figure 5. R. Sancisi and U.J. Schwarz gave helpful comments on an earlier version of the manuscript.

DISCUSSION

Comment R.H.T. BATES

Have you attempted to use histrogram equalisation, which is now standard practice in remote sensing? Histogram equalisation significantly increases the visually apparent details in an image.

Reply R.J. ALLEN

No, I have not tried to do this on radio astronomical images, although I have observed the technique being applied to satellite photographs.

Comment D.C. BACKER

I have worked with false color displays of pulsar polarization observations and I have encountered a problem with color blindness: two of the pulsar experts involved were color blind. I would like to test the statement that a large fraction of the males in the audience is color blind. Could we have a show of hands of those who are partially color blind?

Reply by AUDIENCE

Less than a few percent of the 100 attendants indicated color blindness.

Comment B.F. BURKE

Color displays of data have absorbed large amounts of money and talent, with almost no results so far. Grey-scale representations seem to be adequate. I suggest that there is a psychological basis for the lack of usefulness of color displays. Perhaps black and white are good enough for the "gestalt", while color involves the emotions.

Reply R.J. ALLEN

You may very well be right. However, I think that far too little systematic experimentation with the color display of astronomical data has been done, perhaps because the presently available techniques are rather clumsy and time-consuming. I would prefer to regard your comment as a hypothesis to be tested.

Comment R. GORDON

As an application of appropriate interactive or automatic image processing, it should be possible to guide telescopes according to the maps gathered so far. This could substantially reduce the observing time.

Reply D.B. SHAFFER

There is an on-line example of imaging at the NRAO single dishes. Line and continuum observers can get a quick look at their spectra and maps, and decide on their next move. This ability has proved to be efficient in terms of telescope time.

Reply R.J. ALLEN

Yes. I'm sorry to have let my personal bias come through; I was not thinking primarily of single-dish spectral line data for this presentation, although it is clear that more space should have been devoted to it.

PROCESSING AND DISPLAY OF SPECTRAL LINE ABSORPTION MAPS

Melvyn Wright
Radio Astronomy Laboratory
U.C. Berkeley, California, U.S.A.

ABSTRACT

We describe the processing and display of aperture synthesis maps of the H I absorption against the radio source Cas A. Then data were obtained with the NRAO interferometer and constitute 384 narrow band maps of Cas A combined to produce a picture of the optical depth and mean velocity of the H I absorption due to the Perseus arm.

Both digital and analogue data techniques have been used to Fourier transform, combine and display the data and the merits and usefulness of each are described.

C. van Schooneveld (ed.), Image Formation from Coherence Functions in Astronomy, 157.

A PROPOSED CORRELATOR BACK-END FOR THE CULGOORA RADIOHELIOGRAPH

D.J. McLean
Division of Radiophysics, CSIRO, Sydney, Australia
(Lecture presented by N.R. Labrum.)

1. INTRODUCTION

The original design of the Culgoora radioheliograph (Wild, 1967; Sheridan et al., 1973) incorporated a number of features, such as high time resolution and limited declination coverage, which were particularly suited to its primary function of observing the Sun. Although some of these features are disadvantageous for cosmic work, the instrument has nevertheless proved useful for a variety of non-solar studies as well.

At the present time design work has begun on a new correlator back-end for the Culgoora array which, it is hoped, will enhance the power of the instrument in both its solar and its cosmic roles. The array consists of 96 aerials equally spaced around a circle 3 km in diameter. The new plan is to correlate the signals coming from each pair of aerials and then to process all the measured correlations so as to synthesize an image of an extended field; for solar work, this synthesis will have to be done in real time.

For 96 aerials there are 4560 different pairs of aerials, although almost half the aerial spacings occur twice; if we do not attempt to record redundant spacings we will need a minimum of 2304 correlators. To limit the cost and complexity of such a large project we have chosen to build these as one-bit (or possibly 1½-bit) digital correlators and to do much of the processing digitally (delays, correlators, image synthesis).

2. ADVANTAGES

The advantages which we expect from this undertaking differ for the two cases of solar and cosmic observations.

For solar observations one important advantage is improved time resolution; in its present form the radioheliograph takes one second

C. van Schooneveld (ed.), Image Formation from Coherence Functions in Astronomy, 159–164.

to form a pair of images in two states of circular polarization at one frequency. Originally, the heliograph operated at only one frequency, 80 MHz; now however the available time is shared between three frequencies (soon four) and the time resolution at any one frequency (3 s) is poor compared with the fastest time scales for bursts (<1 s). With the new back-end the image rate will be increased to 8 or 10 per second - sufficient for one image per second in each state of polarization at each of the four frequencies. Although an even higher image rate would be desirable in some cases, we have accepted this limitation both because of the embarrassingly high data rate and because we need about 0.1 s integration to ensure a good dynamic range.

A second major advantage for solar observations is the possibility of using the considerable redundancy of the array, particularly at close spacings, to determine a much more accurate phase calibration. (The principles of this technique are discussed briefly later.) An accurate phase calibration is important for solar observations because it permits a synthesized beam with lower sidelobes, providing a larger dynamic range for observations of bursts and a more accurate image of the quiet Sun. The large dynamic range is required at the time of major solar events, when interesting weak activity (e.g. moving type IV bursts) may be confused by the sidelobes of much stronger sources. In the case of the quiet Sun, the need for a low sidelobe level is evident from the following example: if the r.m.s. sidelobe level were 1% of the main beam then with about 100 beam areas on the Sun, the r.m.s. uncertainty of the measurements of the quiet Sun would be about $\sqrt{100} \times 1 = 10\%$ of the mean brightness!

A third advantage for solar observations is that we will also obtain a modest gain in sensitivity compared with the present scanning system. Although sensitivity has not been a major problem in studies of the quiet Sun (because integration times can be long) or for bursts (because of their high intensity) the extra sensitivity will become necessary in studies of weak features that will be visible with reduced sidelobe levels. A search for wave phenomena on the quiet Sun is an example of an application where increased sensitivity is paramount.

For cosmic observations of known sources we expect a modest gain in sensitivity of about six times, made up of a factor of two, because the source will be observed all the time rather than 25% of the time as with the existing system, and a factor of about three, because it will no longer be necessary to use J^2-synthesis for beam forming.

For cosmic observations of weak sources the new back-end will open up the possibility of earth-rotation synthesis. The primary beam of the aerials of the array is large - about 34° diameter to the 10 dB level at 80 MHz, half that at 160 MHz and half again at 327 MHz. The spacing of the aerials, 100 m, ensures that the grating lobes fall outside a 2° field suitable for solar observations. In a short observation of weak cosmic sources the grating lobes associated with other sources will limit our ability to discern the source. Earth-rotation synthesis can reduce this confusion limit: Swenson and Mathur (1967) have shown that, with the limited hour-angle coverage of the existing aerials, ±2.4 h, the grating-lobes will be significantly reduced, although not eliminated (better results would be obtained with ±4 h coverage). For the study of weaker sources then it will be valuable to synthesize the full field of the primary aerial beam so that residual grating-lobes due to strong nearby sources can be recognized and possibly cleaned. Further, with the possibility of synthesizing such a large field in five hours' observations it becomes feasible to survey all the sky within the declination range of the aerials (roughly $-30° < \delta < 30°$).

3. PHASE CALIBRATION METHOD

Figure 1 shows the UV plane coverage for a circular array similar to the radioheliograph, but for 48 rather than 96 aerials. For a strong source such as the Sun we can assume that all the flux recorded originates within a limited field, and consequently we need only sample the UV plane at a moderate density; the density of sampled points about half way from the centre to the edge of Figure 1 is sufficient for a 2° field. Both at the centre and the outer edge of the figure the UV plane is heavily oversampled and this oversampling can be used in much the same way that the redundancy of an equispaced linear array can be used to correct phase errors (Jennison, 1958; Ishiguro, 1971; McLean, 1973). In practice, in solar work only the correlations from the inner few circles of spacings will be strong enough to be useful. The simplest way to use these is to compare the individual correlations with the running mean of about 16 complex correlations taken around the inner ring. The errors so determined can be assigned successively to the individual aerial channels. Because of the way the errors grow during this process the major question is whether the phases of the longer base lines will be improved; the preliminary study suggests that they will. If so, then new observations incorporating the first set of corrections can be used to refine the corrections, and repeated applications of this process will then converge to an accurate calibration.

4. PROBLEMS

We are currently studying a number of problems which might limit the success of this project.

(a) The limited sky coverage of the aerials of the array has already been mentioned as a major limitation of synthesis observations. Modification of the aerial mounts may be possible at some future time.

(b) Coupling between the open wire transmission lines and possibly radiative coupling between aerials will introduce spurious correlations between pairs of signals. Although the level of coupling is low it will become serious when we seek to process individual correlation measurements to extract phase calibration data. We believe that we can eliminate the effects of coupling using synchronous switching techniques.

(c) A third problem is caused by the complexity of the system that we are proposing and the large number of correlators required. Furthermore, the large rate of flow of data means a heavy, on-line, computing load. With a minimum of 2304 correlators, each recording two numbers each 0.1 s, we must process some 40,000 numbers per second. For solar observations we need images in real time and so we must have sufficient computer power to cope with this rate of data and to reduce it to an amount that can be recorded and stored.

(d) Although preliminary evaluation of the phase calibration system described above suggests that the process converges, if slowly, a more thorough evaluation has yet to be made to confirm this.

5. CONCLUSION

The Culgoora radioheliograph is already a powerful instrument (it is, as yet, the only fast imaging radio telescope in the world), and with the new back-end described here its usefulness will be further extended. Operating in the metre and decimetre bands at 43, 80, 160 and 327 MHz, with resolution in two dimensions ranging from about 7'.5 arc at 43 MHz down to about 50" arc at 327 MHz, it will be capable of studies of solar bursts and the quiet Sun with high time resolution and large dynamic range. It will also be capable of extensive studies of galactic and extragalactic objects with a sensitivity at 80, 160 and 327 MHz of about 0.1 Jy.

REFERENCES

Ishiguro, M.: 1971, *Proc. Res. Inst. Atmos., Nagoya Univ.* 18, p. 73.
Jennison, R.C.: 1958, *Mon. Not. R. Astron. Soc.* 118, p. 276.
McLean, D.J.: 1973, *Proc. IEEE* 61, p. 1318.
Sheridan, K.V., Labrum, N.R. and Payten, W.J.: 1973, *Proc. IEEE* 61, p. 1312.
Swenson, G.W. Jr. and Mathur, N.C.: 1967, *Proc. Inst. Radio Electron. Eng. (Aust.)* 28, p. 370.
Wild, J.P. (Ed.): 1967, *Proc. Inst. Radio Electron. Eng. (Aust.)* 28, No. 9.

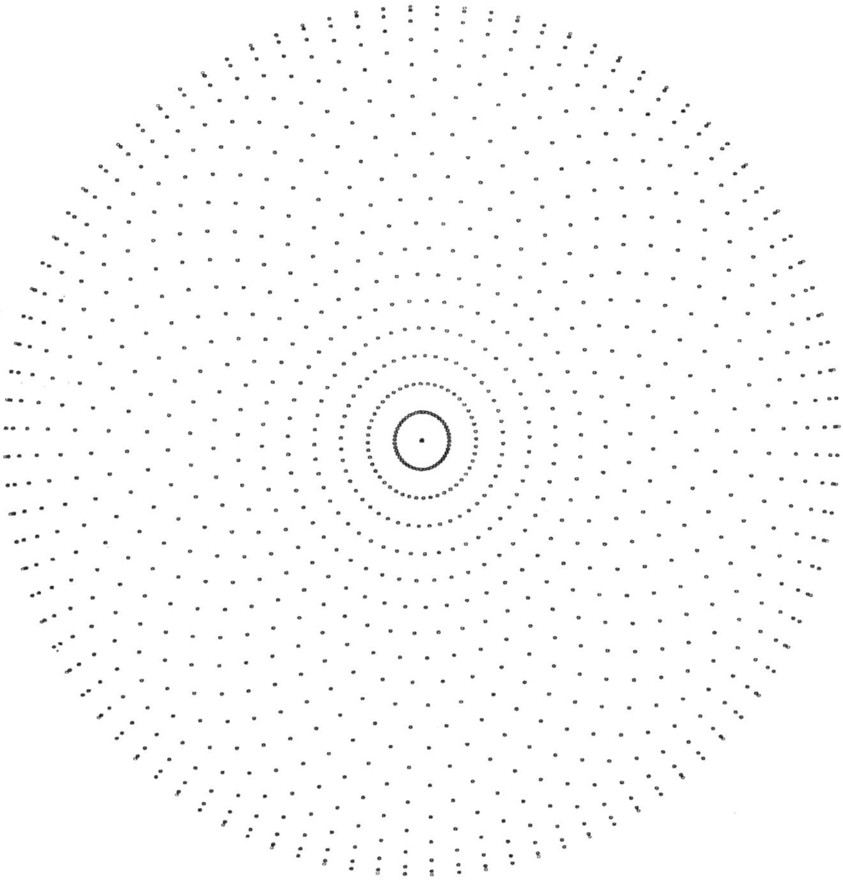

Figure 1. The UV plane for a circular array like the Culgoora radioheliograph. Each point represents the direction and spacing between a pair of aerials of the array. To simplify the diagram an array of 48 aerials has been represented, rather than the actual 96.

DISCUSSION

Comment W.M. GOSS
What are the representative sensitivities and resolutions for the correlation mode of the radioheliograph? Is the sensitivity limited by confusion?

Reply N.R. LABRUM
Beamwidth at zenith ranges from $7'.5$ at 43 MHz to $0'.9$ at 327 MHz. Sensitivity at 80, 160 and 327 MHz will be about 0.1 Jy. Rough calculations indicate that the confusion limit will be less than 0.1 Jy at all these three frequencies.

AN ACOUSTO-OPTICAL IMAGE PROCESSOR AND ITS APPLICATION TO A 160 MHz INTERFEROMETER

Keizo Kai and Takeo Kosugi
Tokyo Astronomical Observatory, University of Tokyo
Mitaka, Tokyo 181, Japan.
(Lecture presented by M. Morimoto.)

ABSTRACT

Radio emissions from the Sun are characterized by extremely rapid time variation. Although they are relatively strong, the signal to noise ratio becomes critical when we observe them with high resolution in space and time. To improve the S-N ratio, we have investigated the feasibility of an acousto-optical processing of interferometer images and applied it to the 160 MHz interferometer of the Nobeyama Solar Radio Observatory. We present here results of simulation experiments and preliminary solar observations with the new image processor.

I. INTRODUCTION

A 160 MHz interferometer of a compound type has been put in operation since 1970 at Nobeyama to investigate fine structure of radio emissions from the Sun (Takakura et al., 1967). The interferometer scans 64 min of arc centred at the Sun in the E-W direction at a rate of 10 Hz to follow rapid time variations. The scanning of the fan beam across the Sun is performed by a phase-controlled local oscillator system. The signal of a variable frequency (f_c) oscillator (VFO) is transmitted from one end of the array via each antenna to the other end of the array where the signal is mixed with the signal of an another oscillator of a fixed frequency (f_0). The converted signal of the frequency of (f_0-f_c) is transmitted back via each antenna. Under each antenna these two signals are mixed to produce the fixed frequency (f_0) signal which is fed to the mixer of the pre-amplifier as the local source. The phase of the local signal changes with the frequency of the VFO. The amount of the phase change is proportional to the distance of an antenna in question from the antenna at the east end.

The beam scanning system has a disadvantage as to S-N ratio. Signals which come from a region of the sky are missed unless the beam is directed there. Let θ be the angular extent (in one dimension) of the field of view of an interferometer and $\Delta\theta$ the width of the fan beam.

C. van Schooneveld (ed.), Image Formation from Coherence Functions in Astronomy, 165-172.

Then, the S-N ratio for the interferometer would be deteriorated by a factor of $(\theta/\Delta\theta)^{1/2}$ for such a beam scanning system as described above than for a system of multi-beams which wholly cover the field of view. The S-N ratio of the 160 MHz is not good enough for weak bursts to see fine structures in 0.1s resolution. To improve the S-N ratio we have planned to modify the image processor either to a multi-channel ($\sim$ 50 in the present case), a correlator mode or an acousto-optical processor. For some reasons we have chosen the last system for the image processor of the 160 MHz interferometer.

An acousto-optical processor for a radar array was first suggested by Lambert et al. (1965). The application of such a system to the Culgoora radioheliograph was proposed subsequently by McLean et al.(1967). However, eventually 48 multi-beams which scan the Sun in the E-W direction were constructed for the radioheliograph image processor. The principle of an acousto-optical processor is briefly given below (for detailed descriptions, see Lambert et al. (1965) and Cole (1965)). A RF signal is applied to a transducer attached on the top of an acousto-optical medium. The transducer converts the RF signal to the acoustic wave which travels through the medium and damps into an absorber at the bottom of the medium. The refractive index of the medium varies periodically as the acoustic wave travels. If a light beam is applied to the medium with a large angle to the direction of the wave propagation, part of the light beam is diffracted and focused on the image plane. The intensity of the diffracted light is proportional to the power of the applied RF signal, and the phase relative to that of the RF signal is maintained. Therefore, when coherent RF signals are applied to two separate acousto-optical modulators (Ao), the light diffracted through one of the AO_S can interfere with that diffracted through the other. If signals from two separate antennas which are directed to the sky are applied to the two separate AO_S, the image of the interference pattern is just the Fourier component of the sky which the two antenna interferometer would give. Any different signals which come from different directions produce independent diffracted images simultaneously. Consequently the S-N ratio is improved by a factor of $(\theta /\Delta\theta)^{1/2}$ than that of an interferometer of a conventional beam scanning type. In the present case, $\theta = 64'$ and $\Delta\theta = 2'$ so that $(\theta / \Delta\theta)^{1/2} = 5.7$

We have made fundamental experiments on an acousto-optical image processor since 1975 with a special application to an interferometer in mind. In this report we present results of the experiments and preliminary observations of solar bursts recorded with this new image processor.

2. ACOUSTO-OPTICAL IMAGE PROCESSOR FOR THE 160 MHZ INTERFEROMETER

The antenna arrangement of the 160 MHz interferometer is illustrated in Figure 1. Signals from 8 short-spacing antennas are multiplied by those from 3 long-spacing antennas to form a sharp fan-beam of approximately 2 min of arc in the E-W direction. Eleven AO_S, each of which

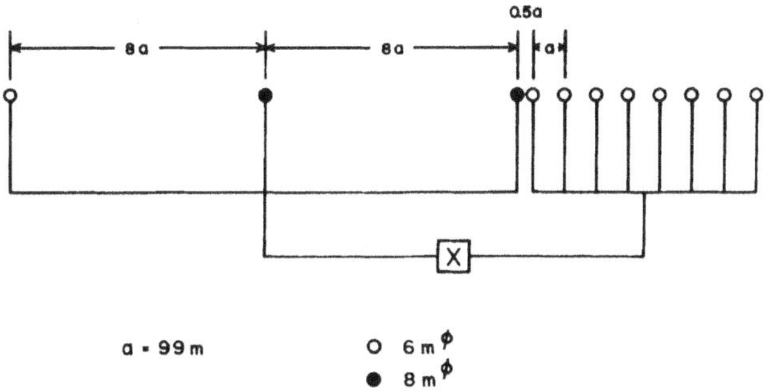

Figure 1. The antenna arrangement of the 160 MHz interferometer at Nobeyama.

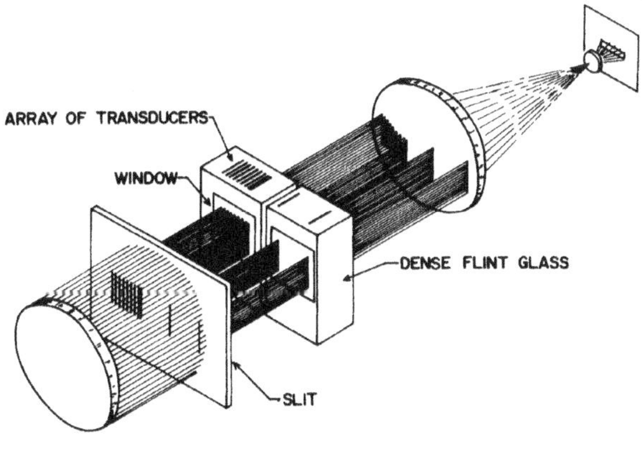

Figure 2. Acousto-optical light modulators and an array of slits placed in front of the modulators.

is to be connected with the corresponding antenna, are arranged to form an array analogous to the antenna array in a reduced scale. The fundamental spacing of the AO array is 4 mm. To keep the spacings very accurately an array of slits (width and length are 0.5 and 2.0 mm res-

Table 1. Specification of Acousto-optical Modulators

centre frequency ---------	37 MHz
band width ---------------	$\pm$ 5 MHz
transducers --------------	$LiNbO_3$ (2 x 14 mm^2)
medium -------------------	dense flint glass
sound velocity -----------	3720 m/s
deflection efficiency ----	30 % max

Figure 3. A photograph of the acousto-optical image processor to illustrate the arrangement of components. A He-Ne laser, a beam expander, a mirror (top : left to right), a mirror, a collimating lens, an array of slits, acousto-optical modulaters, a focusing lens, an expanding lens and a 256 photo-diode array (bottom : right to left).

pectively) is placed in front of the modulators (see Figure 2). Physical parameters of the AO are shown in Table 1. A 15 mW He-Ne laser is used as a light source. The beam is expanded by a combination of a microscope lens and a pin hole and then collimated by a 15 cm ϕ lens (f = 700 mm). The collimated light is diffracted by the AO_s and focused by a 15 cm ϕ lens (f = 70C mm). The first-order diffracted light only is expanded to form an image on a 256 photo-diode array (Figure 3). The optical components including the $AO's$ are set on an optical bench in a dark box. Two tire-tubes support the optical bench to isolate it from disturbances from floor. Tire-tubes are proved to be good absorbers of vibrations over a wide frequency range.

RF signals are amplified and converted to 15 MHz under individual antennas and transmitted to the observing room. IF signals are again

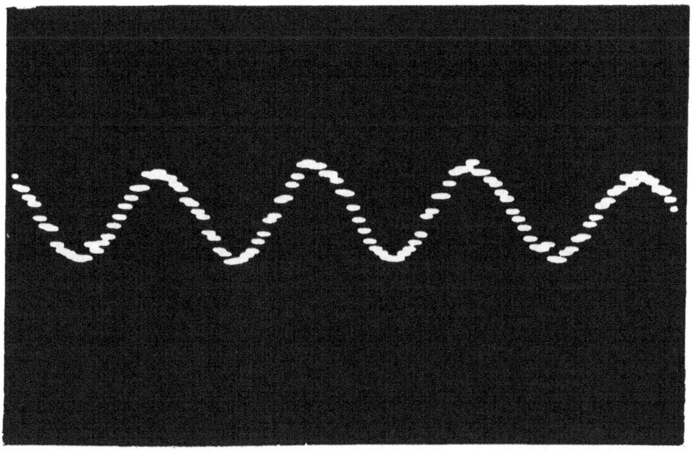

Figure 4 (a)

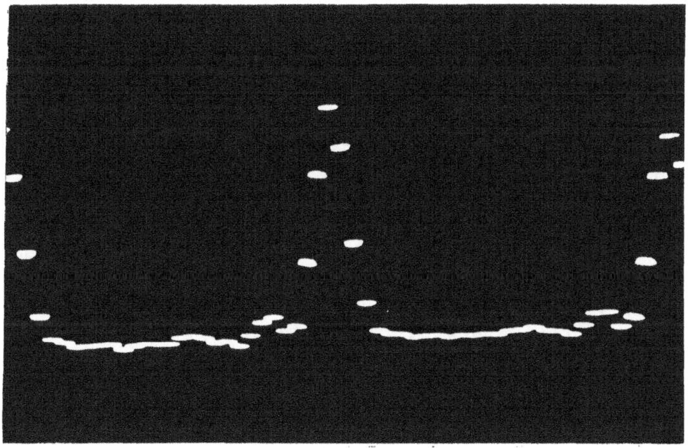

Figure 4 (b)

Figures 4 (a) and (b). Interference patterns due to adjacent two modulators (a) and eight modulators of short-spacing (b). Signals from SG are applied to modulators through power amplifiers. Different scales are used in (a) and (b).

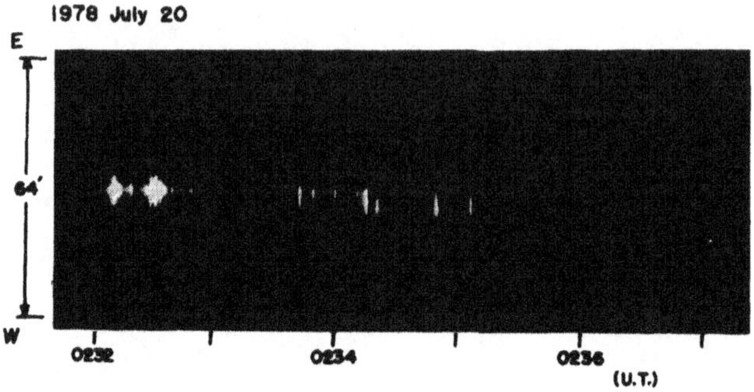

Figure 5. An example of the east-west distribution of solar bursts recorded with the acousto-optical image processor.

converted to 37.4 MHz and applied through power amplifiers to the transducers. The phase of signal from long-spacing antenna is changed by 0 and 180 degrees alternately in order to perform the multiplication of signals from long-spacing antennas with those from short-spacing antennas. In the first cycle when the two signals are added in phase, the charge accumulated in the photo-diode array is picked up sequentially and stored in an analogue memory. In the next cycle when the two signals are subtracted, both the signal from the photo-diode array and the signal stored in the memory are fed synchronously to a differential amplifier to get the signal resulted from the multiplication. A computer processing would be more accurate. However, since two small computers are almost fully occupied to control various kinds of equipments, to process images of a 17 GHz interferometer of a correlator-type in real time and so on, there is no extra ability left to make further processings. Only the output of the differential amplifier is stored in magnetic tapes.

3. RESULT

Simulated patterns are shown in Figures 4(a) and (b) for 2 and 8 element interferometers respectively. Signals are fed to two adjacent modulators (a) and to eight modulators of the fundamental spacing (b) directly through power amplifiers. Sensitivities of individual photo diodes have not been calibrated there. The accuracy of the sensitivity is approximately 5%.

Solar bursts were observed tentatively with the 8 element grating interferometer of short-spacing antennas. Part of the records is shown in Figure 5. The output of the photodiode array was displayed on CRT and photographed on a film which moves continuously. The result as shown above is only preliminary. Observations of the Sun with the compound system will work soon.

As far as the present system works well, it is not difficult to extend the single frequency observation (say, 160 MHz in the present case) to a simultaneous multi-frequency observation. When two signals of different frequencies are applied to the modulators simultaneously, two diffracted images form on two separate horizontal lines. Therefore, with the use of an area array or two different arrays of photo diodes in parallel we would obtain brightness distributions of the Sun at two frequencies simultaneously.

The authors thank Dr. Fukumoto, the Matsushita Research Institute Inc. for generous cooperation, Prof. J. Tsujiuchi, the Tokyo Institute of Technology, for valuable suggestion, Mr. H. Nakajima for stimulated discussion, Mr. Y. Shiomo and Mr. S. Aiba for help in construction of optical and electronic parts. This work was supported by the Toray Science Foundation (75 - 1602).

REFERENCES

Cole, R. W., 1965, _Optics Technology_, 1, p. 31.
Lambert, L. B., Arm, M. and Aimette, A., 1965, in "Optical and Electro-optical Information Processing", ed. Tippet et al., MIT press., p. 715.
McLean, D. J., Lambert, L. B., Arm, M. and Stark, H., 1967, _Proc. I. R. E. E., Australia_, 28, p. 375.
Takakura, T., Tsuchiya, A., Morimoto, M. and Kai, K., 1967, _Proc. Astron._ Soc. Australia, 1, p. 56.

DISCUSSION

Comment M.S. EWING
If you were to use a cylindrical transform lens, you would have an output plane having a spectrum in one direction and fringe-rate spectrum in the other direction. This would concentrate the laser power into one or a few fringe rate cells (at least for a 2-element interferometer). Thus less laser power might be required.

Reply M. MORIMOTO
Because of the large distance (typically 5 mm or 10^4 λ) between the piezo-transducers fringe separation tends to get too small. Thus the problem is how to expand the beam in the fringe axis rather than concentrating it.

Comment T.W. COLE
Do you make your own modulators and is it a problem to obtain modulators made to your own specifications at a non-infinite price?
Reply M. MORIMOTO
As I mentioned the modulator was made under a generous cooperation of the Matsushita Research Insitute Inc. However, the accuracy of the modulator geometry is determined by the slits in front rather than by the setting accuracy of the transducers. Thus it does not absorb costs of an infinite amount.

PART IV

OPTICAL INTERFEROMETRIC METHODS

IMAGING WITH A COHERENCE INTERFEROMETER IN OPTICAL ASTRONOMY

C.RODDIER and F.RODDIER
Département d'Astrophysique de l'I.M.S.P.
Equipe de recherche associée au C.N.R.S.
Université de Nice
Parc Valrose
06034 N I C E CEDEX - F R A N C E

I. INTRODUCTION

In optical astronomy, images are highly degraded by atmospheric turbulence. The diameter of the seeing disk may vary from 0.2 arcseconds to 10 arcseconds or more according to weather conditions. A typical diameter is of the order of 2 arcseconds which is the angular resolution of a 6cm lens at the diffraction limit.

Large telescopes have been built for their light collecting capabilities, not for their resolving power which is, in general, of the order of 0.2 arcseconds, that is far from the diffraction limit.

Up to now, there have been several approaches to overcome this situation. Three kinds of approach may be distinguished :

a) Postdetection processing of the image : Usual deconvolution procedures have been unsuccessful for two reasons : the exponential decrease of the modulation transfer function for long exposures and the random properties of the short exposures. A considerable improvement has been brought by Labeyrie[1] who applied the speckle-interferometric technique to the statistical analysis of short exposure sequences. However such a procedure gives only the autocorrelation or power spectrum of the image. Obtaining an image is not straightforward in the general case.

b) Imaging through pupil masks : The double aperture of the Michelson stellar interferometer[2] was the first mask to be used. Russel and Goodman[3] have investigated the use of more elaborated non redundant masks. The main disadvantages of these methods are that the image Fourier components are measured sequentially and most of the light is lost.

c) Active optics : this is certainly the most promising technique at high light levels. However Goodman[4] has shown that at low light levels, such as those encountered in Astronomy, photon noise introduces errors which make passive methods superior to active ones.

C. van Schooneveld (ed.), Image Formation from Coherence Functions in Astronomy, 175-185.

2. IMAGING WITH A COHERENCE INTERFEROMETER

Our approach is of a fourth kind. It consists in postdetection processing of interferograms in the pupil space. It is somewhat similar to Michelson interferometry but allows us to observe all the Fourier components of the image at the same time. It is equivalent to a great number of Michelson interferometers working in parallel. This technique is also called incoherent holography[5,6]. It consists in observing the interference pattern produced by two superimposed images of the telescope pupil, one being rotated with respect to the other. In order to do this, we use a rotation shearing interferometer[7] also called coherence interferometer[8]. It is schematically represented on fig. 1.

The mechanism of this interferometer is best explained when prism A is rotated by an angle of ninety degrees which corresponds to a one hundred and eighty degrees rotation of the image. In this case, the interferometer gives two images of the object symetrically displayed around the axis. Each point source in the object gives rise to fringes of equal thickness with different period and orientation. Since the object is incoherent, all these fringe systems will add in intensity producing a Fourier transform of the intensity distribution in the object.

The maximum resolution corresponding to the pupil frequency cut off is obtained in this case. For a smaller rotation angle, a magnification of the hologram is observed and the cut off frequency is reduced[9]. By demagnifying the image a higher luminosity is obtained at the expense of the resolving power so that the best compromise between luminosity and resolution can always be achieved.

In order to get maximum contrast fringes for any rotation angle, the polarisation states of the two interfering beams must be matched. This has been successfully achieved by cementing appropriate phase plates on the entrance face of our roof prisms[10].

Under usual conditions, the bandwidth must not exceed 50 Å. We have been able to use a bandwidth 10 times larger by putting a chromatic optical system which gives a magnification proportional to the wavelength before the interferometer.

Incoherent holograms behave like usual holograms since a restitution of the image can be obtained by observing their Fraunhofer diffraction pattern. Fig. 2 shows an example of such a restitution. In fact, in this case, the squared modulus of the object intensity is restored.

One of the main problems encountered with incoherent holography is the decrease of the signal to noise ratio as the square root of the number of pixels in the image[11]. However, when dealing with highly degraded images we shall see that the signal to noise ratio can be higher in the Fourier plane than in the image plane.

3. IMAGING THROUGH FIXED ABERRATIONS

We shall first examine the problem of restoring images degraded by fixed aberrations due to bad optics. This is done by using a reference point source within the isoplanatic patch.

When done in the image plane, the procedure is known as deconvolution. The optical transfer function is obtained by computing the Fourier transform of the recorded point spread function. In the case of strong aberrations, this transfer function is highly attenuated everywhere except near the origin and goes down to zero at several frequencies where the information is definitely lost (Fig. 3). The average attenuation factor is known to be of the order of $\sqrt{N}$, where N is the number of speckles in the point spread function[12]. The signal to noise ratio is affected by the same factor. The net result is that deconvolution is difficult and the image can only be partially restored.

On the other hand, incoherent holography allows direct recording of the complex optical transfer function without any attenuation at any frequency (Fig. 3). Indeed the fringe amplitude is not affected by optical aberrations as we shall see in a simple example.

Fig. 4 shows a hologram of a simulated double star (two point sources). The two associated fringe systems give rise to a Moiré pattern. The beat frequency and direction give the space between the stars and their orientation.

Fig. 5 shows a hologram of the same source taken through strongly aberrated optics. The fringes are highly distorted. Under these conditions the double star was not resolved. However, since the two fringe patterns are distorted in the same way, the resulting Moiré pattern remains unaffected.

Such information is lost in the image space as shown in fig. 6 obtained by computing the 2-dimensional Fourier transform of the intensity recorded in the hologram presented in fig. 5. However, the amount of fringe distortion can be calibrated by recording a hologram of a single star (reference point source) under the same conditions. After subtraction of the phase errors a good image is obtained (Fig. 7). Evidently, such a perfect restoration could not have been done from photographs taken in the image plane.

We have seen that the signal to noise ratio in our incoherent holograms decreases as the square root of the number M of resolved pixels in the image. With more complicated objects the restoration would not have been as good. Comparison with the average signal to noise ratio in image plane recording shows that incoherent holography remains better than usual deconvolution as long as the number of resolved pixels is smaller than the number of speckles in the point spread function, that is

$$M < N$$

This is obviously the case for a strongly aberrated image of a double point source.

4. IMAGING THROUGH TURBULENCE

Let us now examine the case of random aberrations such as those produced by atmospheric turbulence.

By standard speckle interferometry techniques, one obtains only the amplitude of the Fourier transform of the object. However McGlamery[13] and more recently Knox and Thompson[14] in the United States, have shown that the phase can also be estimated by computing the average phase over all the Fourier transforms of each image. However, one must take care because the phase excursions due to turbulence increase with spatial frequency and may become much larger that 2π. Therefore the phase must be followed with continuity from zero spatial frequency.

The same procedure can be used with a sequence of turbulence distorted holograms. It consists in estimating the average fringe position in order to produce a restored hologram, the Fourier transform of which gives the restored image.

We have first taken holograms of a point source in our laboratory through artificial turbulence generated by a toaster. In order to freeze the fringe distorsion due to turbulence, the exposure time was of the order of a millisecond.

Fig. 8 shows the holographic image of a point source taken without turbulence. Fig. 9 shows a typical holographic image obtained through turbulence. From a sequence of nine distorted holograms, we have computed a restored hologram. Fig. 10 shows the corresponding restored image which is almost as good as the original.

We did the same thing with a slightly more complicated object : a double point source. Fig. 11 shows a typical holographic image from a sequence of 5 holograms. None of the images show a resolved structure. However, the double point source is clearly resolved on the restored image with a correct ratio of intensities between the two point sources (Fig. 12).

A stellar hologram will be presented at the meeting.

REFERENCES

[1]A.LABEYRIE (1970), Astron. Astrophys., 6, pp 85
[2]A.A. MICHELSON (1920) Astrophys. J., 51, pp 257
[3]F.D.RUSSEL and J.W.GOODMAN (1971), J. Opt. Soc. Am., 61, pp 182
[4]J.W.GOODMAN and J.F.BELSHER (1977) Joint Automatic Control Conference San Francisco (June 22-24)
[5]G.W.STROKE and R.C.RESTRICK (1965) Applied Physics Letters, 7, pp 9
[6]G.COCHRAN (1966), J. Opt. Soc. Am., 56, pp 1513
[7]J.D.ARMITAGE, Jr and A.LOHMAN (1965), Optica Acta, 12, pp 185
[8]J.B.BRECKINRIDGE (1972), Applied Optics, 11, pp 2996
[9]S.LOWENTHAL, J.SERRES et C.FROEHLY (1969), C.R. Acad. Sc. Paris, 268, pp B 841
[10]F.RODDIER, C.RODDIER and J.DEMARCQ (1978), J. of Optics (to appear in July)
[11]S.LOWENTHAL, J.SERRES and H.ARSENAULT (1970), Optics Commun., 1,pp 438
[12]D.KORFF, G.DRYDEN and M.G.MILLER (1972), Optics Commun., 5, pp 187
[13]McGLAMERY, B.L. (1971), N.A.S.A. Tech. Report SP-256, pp 167
[14]K.T.KNOX and B.J.THOMPSON (1974), Astrophys. J., 193, pp L45

DISCUSSION

Comment J.C. DAINTY
Will the technique work for long baseline interferometry with two separated telescope pupils?
Reply F. RODDIER
Yes, but the phase must be followed by continuity in time rather than in frequency space. It can also be followed by continuity in wavelength (color) as suggested by Laurent Koechlin (CERGA Observatory - France).

Comment J.P. HAMAKER
As you said, you must be careful to avoid 2 π ambiguities in locating your fringes. This then means that you have to track the movement of individual fringes over the sequence of short exposures. Do you do this "manually" or have you developed computer algorithms to take care of it?
Reply F. RODDIER
We track the distortion of our fringes from the center of the hologram (zero spacing) to the edge (maximum spacing). We have developed a computer algorithm to do it, based on interpolations.

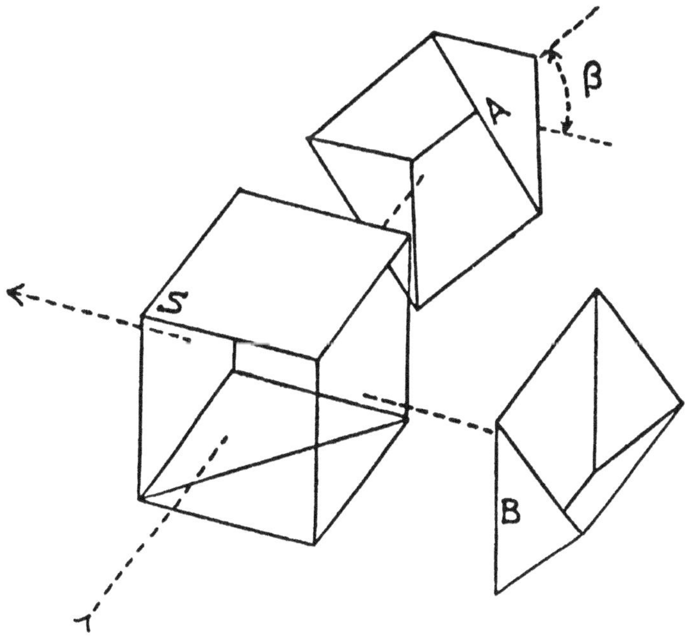

Figure 1. Perspective view of a rotation shearing interferometer - A: Rotatable roof-prism; B: Fixed roof-prism; S: Beam splitter.

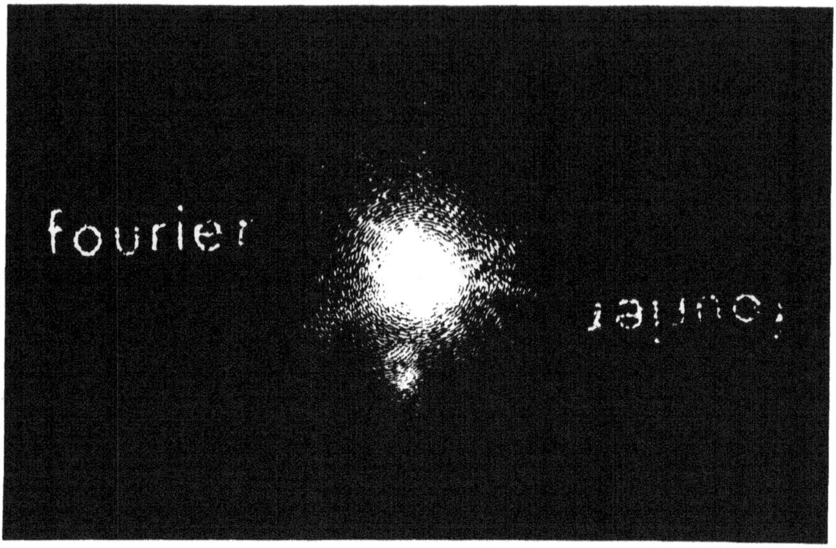

Figure 2. Diffraction pattern of an incoherent hologram of the word "Fourier".

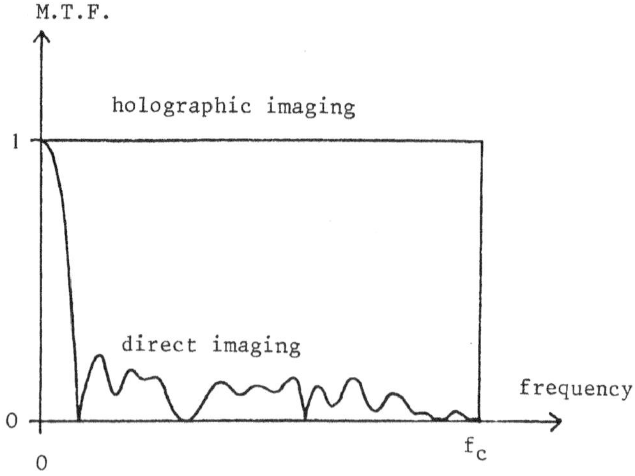

Figure 3. Modulation transfer function (M.T.F.) for direct and holographic imaging through aberrated optics.

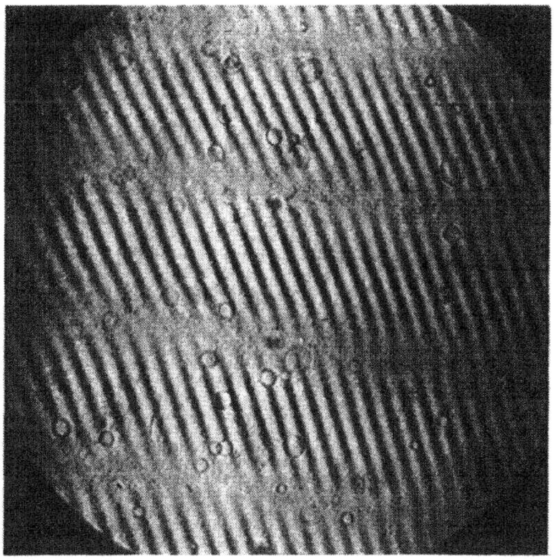

Figure 4. Incoherent hologram of a double point source through good optics.

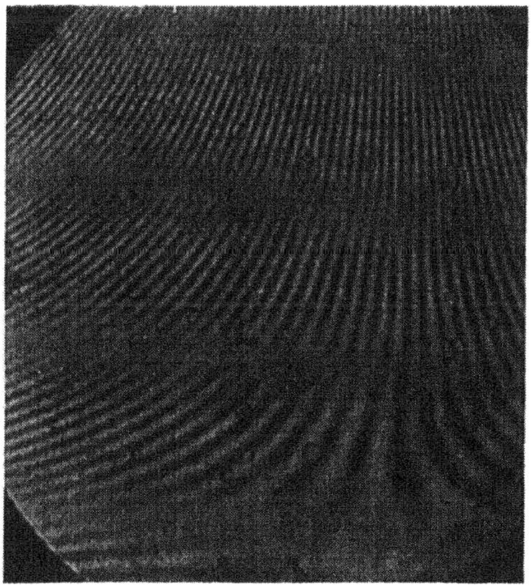

Figure 5. Incoherent hologram of a double point source through strongly aberrated optics.

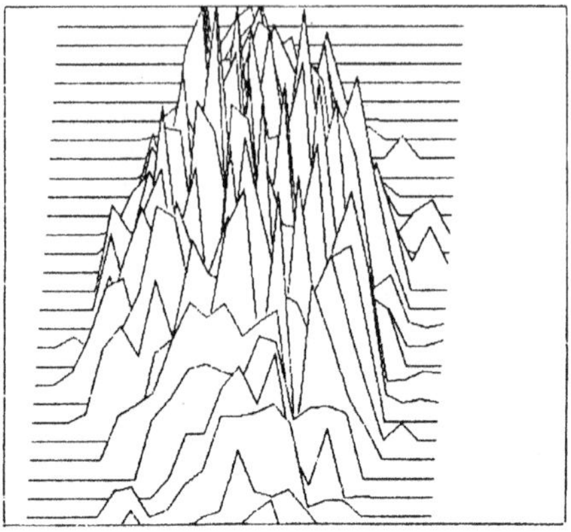

Figure 6. Image obtained by computing the 2-D Fourier transform of the hologram presented on Fig. 5.

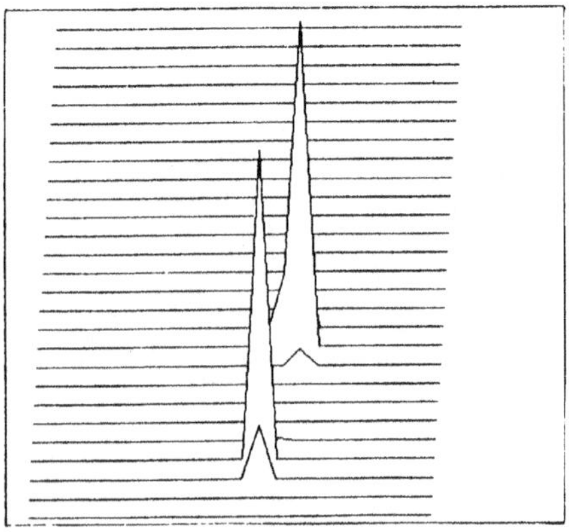

Figure 7. Restored image obtained from the hologram on Fig. 5 after subtraction of the phase errors given by a reference point source.

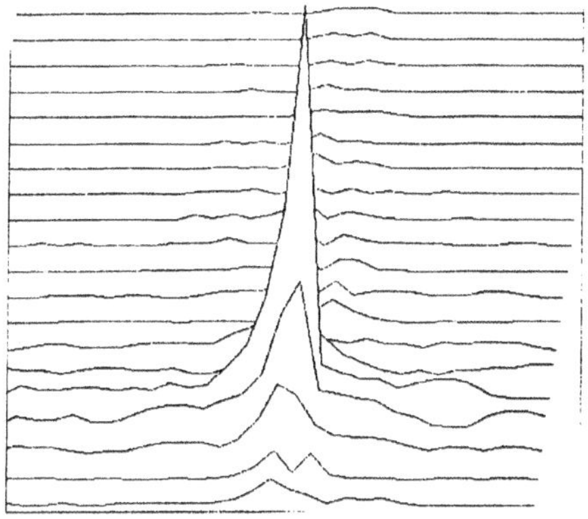

Figure 8. Holographic image of a point source taken without turbulence.

(See following pages for Figures 9-12.)

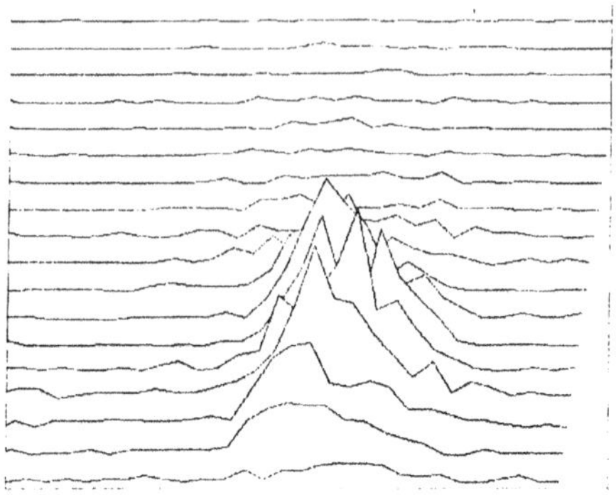

Figure 9. Typical holographic image of a point source taken through turbulence.

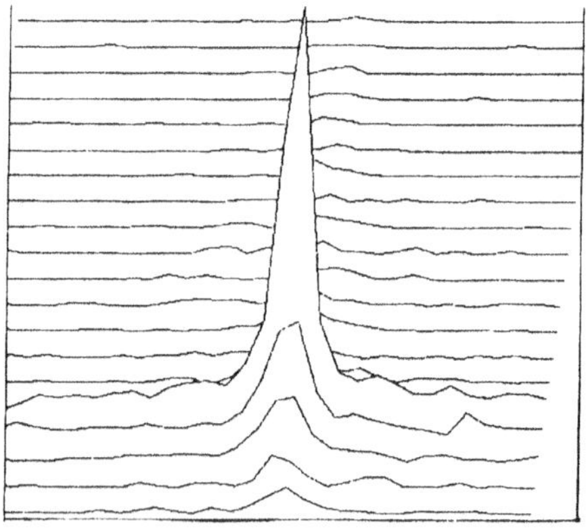

Figure 10. Restored image obtained by averaging fringe positions over 9 turbulence distorted holograms. Compare with the turbulence-free image on Fig. 8.

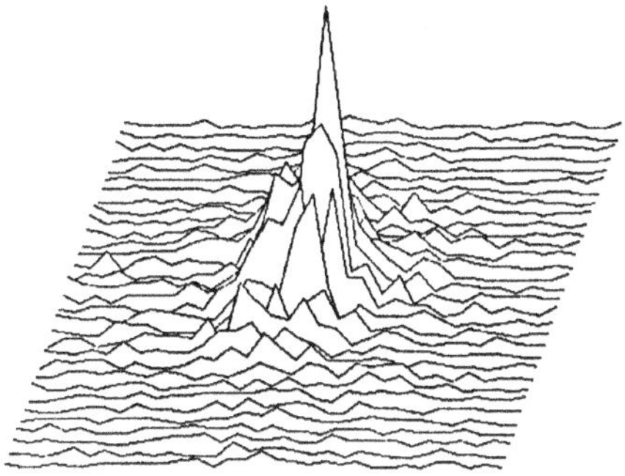

Figure 11. Typical holographic image of a double point source taken through turbulence.

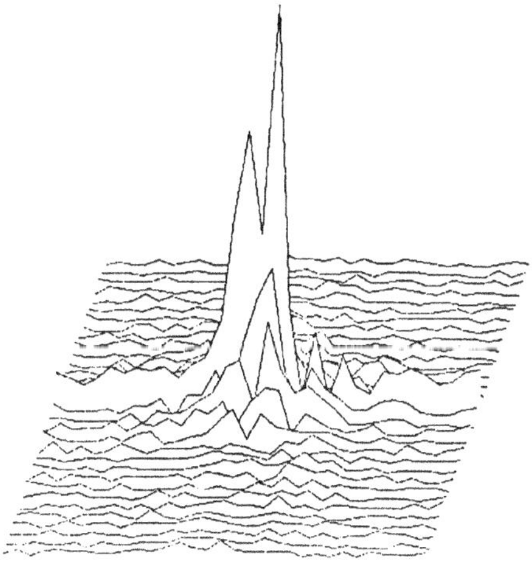

Figure 12. Restored image obtained by averaging fringe positions over 5 turbulence distorted holograms. The double point source is clearly resolved with a correct ratio of intensities.

TOWARDS IMAGING OF STAR CLUSTERS BY SPECKLE INTERFEROMETRY

R.H.T. Bates and M.O. Milner
Electrical Engineering Department, University of Canterbury,
Christchurch, New Zealand.

Labeyrie's (1970, 1976) speckle interferometry permits an astronomical object to be resolved to a limit approaching λ/D (where λ is the mid-band wavelength of the light accepted by a telescope having a pupil aperture of diameter D) in the presence of severe atmospheric seeing. As Dainty (1973) has pointed out, this resolution limit is virtually independent of the accuracy to which the telescope is figured - the seeing can actually improve a telescope's resolution! Even though a true image of the object cannot usually be constructed by Labeyrie's data reduction procedures, nevertheless the autocorrelation of (the distribution of brightness over) the object can always be reconstructed - provided the extent (width, apparent angular diameter) of the object does not exceed that of the isoplanatic patch (cf. Bates and Gough, 1975). The autocorrelation is useful because its extent is necessarily twice that of the object.

True images of certain objects can be formed straightforwardly by speckle interferometry - this is related to holography (cf. Bates and Gough, 1975). By superimposing the individual speckles apparent in a narrow-band, short exposure of an isolated object whose angular diameter is somewhat greater than λ/D, Lynds *et al.* (1976) have suggested that an actual image of the object is obtained. Unfortunately, when used unaided, such imaging techniques can only be applied to very limited classes of objects.

All but a very few individual stars are unresolvable in the largest telescopes, even when seeing is perfect, implying that diffraction-limited images of the great majority of star clusters are collections of Airy discs. The essential information in any such image can be represented by an array of two-dimensional delta functions, whose relative positions and intensities correspond to the relative positions and brightnesses of the Airy discs.

We have realised that by combining Labeyrie's original technique with the approach taken by Lynds *et al.*, and employing data reduction procedures developed by X-ray crystallographers (Baldwin and Warner,

C. van Schooneveld (ed.), Image Formation from Coherence Functions in Astronomy, 187–193.

1976, have already made interesting astronomical use of these), we should be able to reproduce accurate representations of those star clusters that could be imaged, if the seeing conditions could be perfect, with existing telescopes in the conventional way. We have carried out simulations in our optical laboratory, using a previously developed technique (Gough and Bates, 1974), and have already reported results of preliminary experiments in which only optical processing was used (Bates *et al.*, 1978).

To take full advantage of speckle interferometry we feel it is essential to use digital techniques. Accordingly, we have placed a CCD-camera (Cady *et al.*, 1978) at the focus of our simulated telescope, and have interfaced it to the EAI 590 Hybrid System in our Computer Laboratory (note that CCD stands for *Charge Coupled Device*). The theoretical foundations of our new method of imaging, the experimental apparatus and the computational procedures are described in detail elsewhere (Bates and Milner, 1978; Milner, 1978). Here, we explain the physical basis of our approach and delineate the essential steps of the data reduction procedures.

1. SPECKLE MASK PROCESSING

We call a narrow-band, short exposure of an astronomical object a *speckle image*. The short exposure of γ Orionis shown by Lynds *et al.* (1976) is typical of the many speckle images that have been recently reported in the literature. For our purposes, the essential characteristic of a speckle image is that many of its constituent speckles are distinct (i.e. they can be recognised as individual entities).

Suppose that the cluster shown in Figure 1 is being viewed (each circle in the figure represents an individual star, positioned at the centre of the circle and having a relative intensity denoted by the number immediately above the circle). Our CCD-camera has a dynamic range of 2^8, which is why the maximum intensity of each of our images is normalised to 255. We use the letter C to denote the pattern of delta functions corresponding to the circles shown in Figure 1.

The graphics facility in our computer laboratory incorporates a storage oscilloscope, so that the visual display has a low dynamic range. Consequently, when the display threshold is set to reveal most of a speckle image, neighbouring speckles are almost completely merged into each other. However, by applying a gradually decreasing threshold to the display, the brightest points in each of the brightest speckles are conveniently revealed.

When the cluster shown in Figure 1 is simulated optically and is viewed through simulated seeing, the brightest points in each of the brightest speckles in a typical speckle image appear as shown in Figure 2. The relative intensities of each of these brightest points (they are two-dimensional delta functions, in effect) are the same as the relative intensities of the corresponding speckles.

We call Figure 2 a *speckle mask* because, when we perform this processing purely optically (cf. Bates *et al.*, 1978), we place a transparent sheet over the speckle image, affix a black dot to the sheet over the centre of each of the brightest speckles (the diameter of the dot is proportional to the brightness of the corresponding speckle) and then photograph the sheet - the result is a mask suitable for performing an optical cross-correlation operation.

By cross-correlating the speckle image with the speckle mask, we obtain the *correlation image*, in which the brightest speckles are superimposed upon each other. The idea put forward by Lynds *et al.* (1976) is that, since the brightest speckles can be interpreted as distorted images of the object (when it is isolated and somewhat larger than λ/D), an improved image is formed by superimposing these speckles. Our extension of this idea is the suggestion that the brightest star in the cluster can generally be associated with the brightest point in each of the brightest speckles. We therefore expect the correlation image to be a distorted image of the cluster.

By gradually reducing the threshold applied to the display of the correlation image, we uncover individual stars in the cluster. The fainter stars are often hidden in the noisy background which permeates most correlation images. However, it is usually only necessary to recognise a few of the stars in the cluster, in order to be able to recover the remainder by the data reduction procedure described in the next section.

Figure 3 shows the recognisable stars abstracted from the correlation image which was formed by correlating the speckle mask shown in Figure 2 with its corresponding speckle image.

We give the name *basic star pattern* to the image consisting of delta functions having the same relative positions and intensities as the isolated stars that are recognisable in a correlation image. Figure 4 shows the basic star pattern derived from the correlation image shown in Figure 3.

2. CORRELATION PROCESSING

Figure 5 shows the autocorrelation of the cluster shown in Figure 1. The autocorrelation can be obtained in practice by processing a large number of speckle images according to Labeyrie's (1970) original prescription. Labeyrie's form of speckle interferometry has a sound and simple, yet widely applicable, theoretical basis. So one can be confident that the autocorrelation is as accurate as the observational conditions permit. However, the true image cannot be recovered unequivocally from the autocorrelation unless extra information is

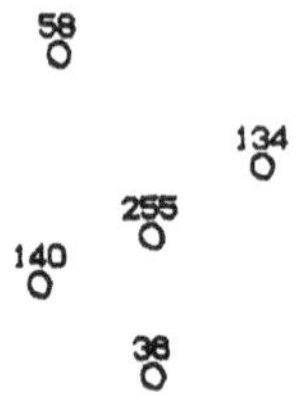

Figure 1. Star cluster C.

Figure 2. Speckle mask.

Figure 3. Recognisable stars abstracted from correlation image.

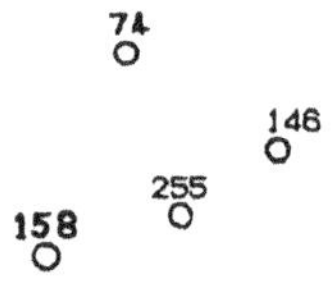

Figure 4. Basic star pattern $\bar{B}$.

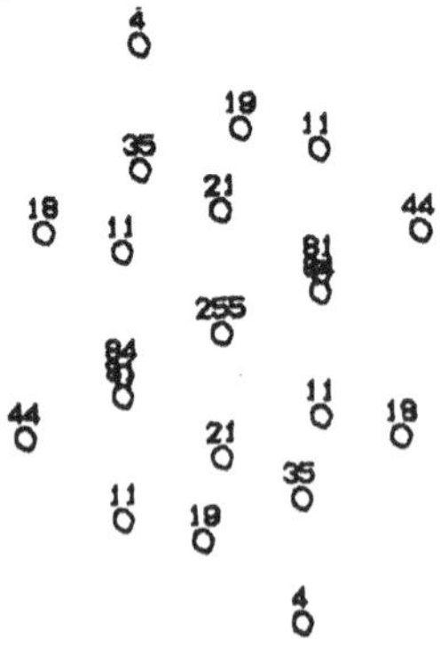

Figure 5. Autocorrelation A.

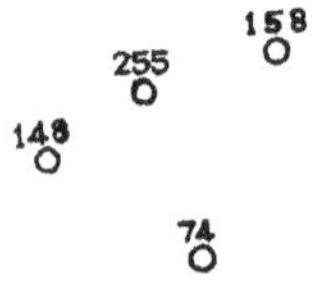

Figure 6. Inversion $\bar{B}_i$ of basic star pattern.

available (cf. Bates and Gough, 1975).

It is convenient to denote the autocorrelation and the basic star pattern by the symbols A and $\bar{B}$ respectively. It is also convenient to use a subscript i to denote the *inversion* of a pattern (e.g. for any pattern P, its inversion P_i is defined to be its reflection about its *origin* - its origin is always taken to be the position of the brightest star in P). Figure 4 shows $\bar{B}$. To make clear what we mean by inversion, $\bar{B}_i$ is shown in Figure 6.

The interpretation of the basic star pattern rests on much less firm theoretical foundations than does Labeyrie's original form of speckle interferometry. Consequently, $\bar{B}$ cannot be expected to be as accurate as A. In fact, the relative positions and intensities of the stars in $\bar{B}$ are likely to be significantly in error. However, as we argue below, $\bar{B}$ can often be expected to contain sufficient information to permit resolution of the ambiguities inherent in A.

The stars making up the pattern C of the cluster can be separated into two patterns, B and R say :

$$C = B + R$$

Pattern B contains those stars that have been recognised in $\bar{B}$ (but B is free of the errors present in $\bar{B}$). Pattern R contains the remainder of the stars in the cluster. Thus, A can be written as

$$A = B*B + R*R + B*R + R*B$$

where the asterisk denotes correlation. If there are n stars in the pattern R then R*B consists of n replicas of B_i. Each of these replicas is similar to $\bar{B}_i$. So, provided that $\bar{B}_i$ is not too distorted a version of B_i , each of these replicas (which we call *matching patterns*) can be identified by inspection of A.

Figure 7 shows A (same as Figure 5) with $\bar{B}_i$ (same as Figure 6) superimposed upon it (in the form of black dots) in such a way as best to show up a matching pattern in A. Careful scrutiny of Figure 7 confirms that there is no other matching pattern. So we deduce that n = 1. The single star in R is positioned (with respect to the origin of C) at the same point as the black dot marked 21 is positioned (in Figure 7) with respect to the origin of A.

Since the autocorrelation can be taken as accurate, the matching pattern identified in Figure 7 is an accurate version of B_i (see Figure 8). The basic star pattern has thus been corrected - it only had to be accurate enough to allow the matching pattern to be identified and to distinguish it from the inverted pattern - the latter consists of those circles marked by the horizontal arrows shown in Figure 7. The whole cluster C, as shown in Figure 1, is thus reconstructed.

Suppose that only the three brightest stars in the cluster had been identified in $\bar{B}$, which would then have been almost symmetrical about a line passing through the brightest star. It would have been impracticable to distinguish B_i from B. There would have been additional ambiguity if only two stars had been identified in the basic star pattern. However, it should now be clear that a basic star pattern need contain only two stars if the brighter stars in the cluster are arranged appreciably asymmetrically, as in general they are likely to be. It should also be clear that any possibility of ambiguity becomes apparent as the data reduction proceeds. If it is found that more stars are needed in the basic pattern, then many speckle images may be subjected to speckle mask processing so that many correlation images can be averaged (this improves the signal-to-noise ratio, thereby increasing the probability of identifying more stars in the basic pattern).

3. RADIO ASTRONOMICAL PROCESSING

At high radio astronomical frequencies, the visibility phase cannot be measured as accurately (using a synthesis telescope) as can the fringe visibility. By Fourier transforming the square of the latter, the autocorrelation of the observed radio sources is constructed. If these sources contain a cluster C of radio stars of small angular diameter, then the autocorrelation A of this cluster can be abstracted by inspection from the autocorrelation of the whole radio source field.

Provided the measured visibility phase is not too inaccurate, Fourier transforming the observed complex visibility provides an image of the cluster which may be sufficiently well defined that distorted versions of the more intense stars in the cluster can be recognised. This permits a basic star pattern $\bar{B}$ to be formed. An accurate version of C can then be constructed in exactly the manner described in the previous section.

4. CONCLUSIONS

The data reduction introduced here is a two-dimensional analogue of a procedure which we developed earlier (Bates and Napier, 1972). The present procedure represents an enormous computational simplification, although it is less general (cf. Bates, 1978). It has recently occurred to us that the principle of isomorphic replacement, as employed by X-ray crystallographers (cf. Lipson and Cochran, 1966), could be invoked to form a true image of a continuous distribution of brightness (e.g. a nebula) set as a background to a cluster containing at least one variable star (e.g. either a closely-orbiting pair that would be resolvable under perfect seeing conditions, or a spectroscopic binary).

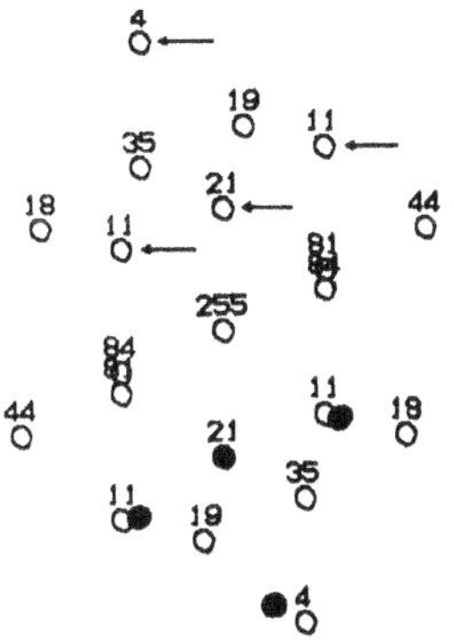

Figure 7. A with $\bar{B}_i$ superimposed; B identified by arrows.

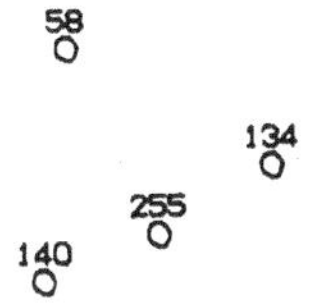

Figure 8. Pattern B.

REFERENCES

Baldwin, J.E., and Warner, P.J.: 1976, *Mon.Not.R.Astr.Soc.* 175, pp.345-353.

Bates, R.H.T.: 1978, *Optik* 51, in press.

Bates, R.H.T., and Gough, P.T.: 1975, *Trans.I.E.E.E.* C-24, pp.449-456.

Bates, R.H.T., and Milner, M.O.: 1978, *True imaging of star clusters by extensions of speckle interferometry*, Research Report (10 July 1978), Electrical Engineering Dept., University of Canterbury, Christchurch, New Zealand.

Bates, R.H.T., Milner, M.O., Lund, G.I., and Seagar, A.D.: 1978, *Opt. Commun.* 26, pp.22-26.

Bates, R.H.T., and Napier, P.J.: 1972, *Mon.Not.R.Astr.Soc.* 158, pp.405-424.

Cady, F.M., Hodgson, R.M., Ireland, A.J., and Duncan, G.D.: 1978, *N.Z. Institution of Engineers Proc.Tech.Groups*, in press.

Dainty, J.C.: 1973, *Opt. Commun.* 7, pp.129-134.

Gough, P.T., and Bates, R.H.T.: *Optica Acta* 21, pp.243-254.

Labeyrie, A.: 1970, *Astron. Astrophys.* 6, pp.85-87.

Labeyrie, A.: 1976, *Progress in Optics* XIV, North-Holland, pp.49-87.

Lipson, H. and Cochran, W.: 1966, *The Determination of Crystal Structures* (Sir L. Bragg ed.), G.Bell, Vol.III, chapt.8.

Lynds, C.R., Worden, S.P., and Harvey, J.W.: 1976, *Astrophys.J.* 207, pp.174-180.

Milner, M.O.: 1978, *Ph.D. Thesis*, University of Canterbury, Christchurch, New Zealand.

PART V

MAXIMUM ENTROPY IMAGE RECONSTRUCTION

RESOLUTION ENHANCEMENT; THE "MAXIMUM ENTROPY METHOD" AND THE "HIGH RESOLUTION METHOD"

(Invited paper)

C. van Schooneveld
Leiden Observatory,
Leiden, the Netherlands
and
Physics Laboratory NDRO,
The Hague, the Netherlands.

1. INTRODUCTION

1.1 Resolution enhancement and a-priori knowledge.

In image formation from coherence functions we start with a set of observed coherences (visibilities) $R'(u_k, v_k)$, measured at locations u_k, v_k in the u,v-plane:

$$R'(u_k, v_k) = R(u_k, v_k) + r(u_k, v_k) \; ; \quad k = -M, .., M. \tag{1}$$

$R(u,v)$ is the object's coherence function (visibility function, covariance function) and $r(u_k, v_k)$ are random measurement errors. In conventional image formation we apply a weighted fourier transformation to eq 1 and we get

$$B'(l,m) = \{G(l,m) \circledast B(l,m)\} + b(l,m), \tag{2}$$

where $B(l,m)$ is the brightness distribution of the object and $B'(l,m)$ is the image. $G(l,m)$ is a beampattern depending on the locations u_k, v_k and on the weights applied to them. The component $b(l,m)$ is the random error in the image, corresponding to $r(u_k, v_k)$.

The locations u_k, v_k are usually limited to some central area A of the u,v-plane. Consequently the high-order fourier components of $B(l,m)$, which correspond to coherences outside A, are missing from the image $B'(l,m)$. The result is a loss of resolution.

The effect of resolution loss can be controlled - to some extent - by using suitably tapered weights, but this method does not recover any of the object's missing fourier components. A better job can be done if it is possible to operate on $B'(l,m)$ (or on $R'(u_k, v_k)$) in such a way that part or all of these components are retrieved.

At first glance the problem of inverting eq. 2, i.e. finding B( )

C. van Schooneveld (ed.), Image Formation from Coherence Functions in Astronomy, 197-218.

from B'( ), looks like an ordinary deconvolution problem, but there is a difference. In a standard deconvolution problem the object's high frequency components are attenuated by a finite factor and are subsequently corrupted by the measurement errors. A solution is then sought within the class of linear operations on B'(l,m) (Wiener filters or modified inverse filters). In eq. 2, however, the high-order fourier components of the object were never observed. They have therefore effectively been attenuated down to zero. As a consequence, any inversion scheme that performs the required task must be non-linear or at least linear space-variant. No linear, space-invariant operation on B'(l,m) can create fourier components that are not already present!

The key to the inversion of eq. 2 is to use a-priori assumptions concerning the object B(l,m) or, better, a-priori knowledge. The assumptions must provide a deterministic coupling between the missing high-order fourier components and the available low-order ones. Eq. 2 can then be solved, for example, by fitting an a-priori constrained object model to the data B'(l,m). Equivalently, in the language of eq. 1, our assumptions must establish a deterministic coupling between the observed coherences inside the area A and the unobserved ones, outside A, thereby allowing an extrapolation of the coherence function. Two questions are important for the success of such an attempt at resolution improvement:

1. Are the assumptions valid for the object under consideration?
2. Is the coupling mechanism sufficiently powerful to withstand the random measurement errors?

Some frequently used assumptions concerning B(l,m) are:

1. Non-negativeness : $B(l,m) \geq 0$
2. Limited extent : $B(l,m) = 0$ outside some closed contour in the l,m-plane.
3. A variety of assumptions about the structure of the object, all of them characterized by a parametrization of B(l,m) (i.e. the continuous function B(l,m) is known if the values of N parameters have been found).

Based on such assumptions a number of resolution enhancement methods has been proposed (Frieden 1975). We shall discuss two of the better developed ones and we shall try to analyze their underlying principle. They are Burg's Maximum Entropy Method (MEM) and Capon's High Resolution Method (HRM), also erroneously known as Maximum Likelihood Method (MLM). In ch. 2 we repeat an argument from temporal spectrum analysis to show that the MEM, sometimes said to be "maximally non-committal" with regard to the observations, relies in fact on a structural assumption about the object, an assumption which may or may not be valid. The HRM, discussed in chapter 3, was originally developed for adaptive detection of point sources against an unknown and non-uniform radiation background, but Capon realized that the method yields improved resolution when used as an estimator of brightness distributions. We show that the HRM can be interpreted as a positive-constrained reconstruction method.

At this point we comment briefly on methods which rely on the assumption of limited object extent. Historically they were the first enhancement methods proposed and they are known as "superresolution" or "superdirectivity". (The temporal counterpart (section 1.3) is known as "superselectivity".) The idea is that an object of limited size produces an analytic coherence function. This implies that a set of coherence observations, very densely spaced, but extending over only a small area A in the u,v-plane, can be extrapolated to an area many times larger. In this way we can obtain beampatterns G(l,m) with a peakwidth inversely proportional to the number of u,v-samples, rather than inversely proportional to the dimensions of the area A. Physically this is achieved by making the wavefront from the desired steering direction (l_o, m_o) cancel to a high degree, while obtaining an even higher degree of cancellation for wavefronts from the other directions. However, this cancellation mechanism is also the source of the method's weakness: an extreme sensitivity to internal noise like thermal noise in the array sensors, quantization and round-off noise in digital processors, etc. By sensitivity we mean that a superresolving system, designed under the assumption of no internal noise, is completely swamped by such noise if it is indeed present. If, conversely, the internal noise is taken into account in the design, the system automatically reverts to a near-conventional processor and almost no superresolution occurs. In fact the assumption of finite object size is too weak to cope with the measurement errors, when unsupported by other assumptions. Superresolution can be ruled out as a practical tool for high-performance imaging systems. Conclusions of this sort were reached in the fields of E.M. antennas (Gilbert and Morgan 1955, Taylor 1955), accoustics (Cox 1973, Schultheiss 1977), radio astronomy (Cole 1973) and optics (Frieden 1975). We note in passing that other enhancement methods, for example the HRM, have the same flaw in their character - cancellation type processing - and special precautions must sometimes be taken to suppress the danger involved (Cox 1973, Owsley 1973, Griffiths and Hudson, 1977).

1.2 Simplification to 1 dimension.

We shall base our discussions on the 1-dim. equidistant case, i.e. line-shaped arrays with constant baseline increment Δ. The advantage is that we obtain simple closed-form solutions which reveal the mechanism of the method. Apart from this, 1-dim. spatial imaging is of practical importance, for instance in underwater acoustics where long horizontal arrays are used to map the radiation intensity as a function of azimuth. One-dim. strip scans in radio astronomy are another example. Standard radio astronomy images, however, are 2-dimensional. Extension of our discussion to 2-dim. arrays is possible, as well as to non-uniform sampling. Qualitatively the properties remain the same but a good deal of the simplicity of the equations is lost.

We use a fourier notation adapted to a (1-dim.) object B(l) of finite extent: $B(l) = 0;\ |l| \geq W$. (This is merely a matter of notation; we do not exploit the limited size of the object in the sense of superresolution.) B(l) can then be written as a fourier series on an interval of length $1/\Delta = 2W$:

$$B(l) = \Delta . \sum_{k=-\infty}^{\infty} R_k . \exp - 2\pi jkl\Delta \quad ; \quad -W < l < W \quad ; \quad \Delta = 1/2W, \qquad (3)$$

where the R_k are samples of the coherence function R(u) at equal spacings Δ:

$$R_k = R(k\Delta) = \int_{-W}^{W} B(l) . \exp 2\pi jlk\Delta . dl \quad ; \quad k = -\infty, .., \infty . \qquad (4)$$

With an array of finite length the observed coherences are limited to $k = -M, .., M$. In conventional image reconstruction eq. 3 is then replaced by a truncated (and weighted) fourier transform:

$$B'(l) = \Delta . \sum_{k=-M}^{M} (R_k + r_k) . g_k . \exp - 2\pi jkl\Delta , \qquad (5a)$$

whereas image enhancement methods try to get hold of the missing coherences, $|k| > M$, via an a-priori coupling mechanism. The r_k in eq. 5 are the measurement errors and the g_k are the weights. Eq. 5a can finally be written in the form of eq. 2:

$$B'(l) = \int_{-W}^{W} G(n-l) . B(n) . dn + b(l), \qquad (5b)$$

where G( ) is the fourier transform of the weights $\Delta . g_k$.

1.3 The space-time analogy.

Resolution enhancement applies equally well to temporal spectrum analysis of random signals as to (pseudo - monochromatic) spatial imaging with a line array or a plane array. Some spatial enhancement methods were first conceived with an application to temporal spectrum analysis in mind, and conversely. Both techniques are analogous ; we briefly recall the analogy below (table 1, fig. 1).

In temporal spectrum analysis we have a random signal s(t) which is sampled at intervals Δ seconds. We try to measure the power density spectrum B(f): the distribution of the average signal power over the temporal frequencies f. In spatial imaging we start with the random aperture illumination $\Delta(x,t)$, the (complex) amplitude of the E.M. field along the aperture axis x at one time instant t. We measure B(l), the distribution of the average power of illumination over the spatial frequencies l. Spatial frequency l corresponds to angular source direction ϕ via $l = \sin \phi / \lambda$. Actually, l is the projected wavenumber of a wavefront from direction ϕ.

In spectrum analysis we store a block of M + 1 samples $s_o, .., s_M$ in a shift register and we calculate the cross-products $s_i \; s_{i+k}$. Averaging over many datablocks produces covariance estimates $R'_{-M}, ..,$ R'_M which are processed to give a spectral estimate $\hat{B}(f)$. In the spatial

TIME	SPACE
. random signal: s(t)	random aperture illumination: s(x) ("frozen" at one time instant)
. time series after sampling at intervals of Δ sec.: .., s_k, ..	space series after sampling at distances of Δ meter: .., s_k, ..
. covariance function: $R_k=E\{s_i s_{i+k}\}$	coherence function: $R_k=E\{s_i^* s_{i+k}\}$
. power density spectrum: B(f)	brightness distribution: B(l)
f: temporal frequency (sec^{-1})	l: spatial frequency ($meter^{-1}$) (projected wavenumber)
. fourier-relation $B(f) \leftrightarrow R_k$: "Wiener-Khintchine"	fourier relation $B(l) \leftrightarrow R_k$: "v. Cittert-Zernike"

Table 1, space-time analogy.

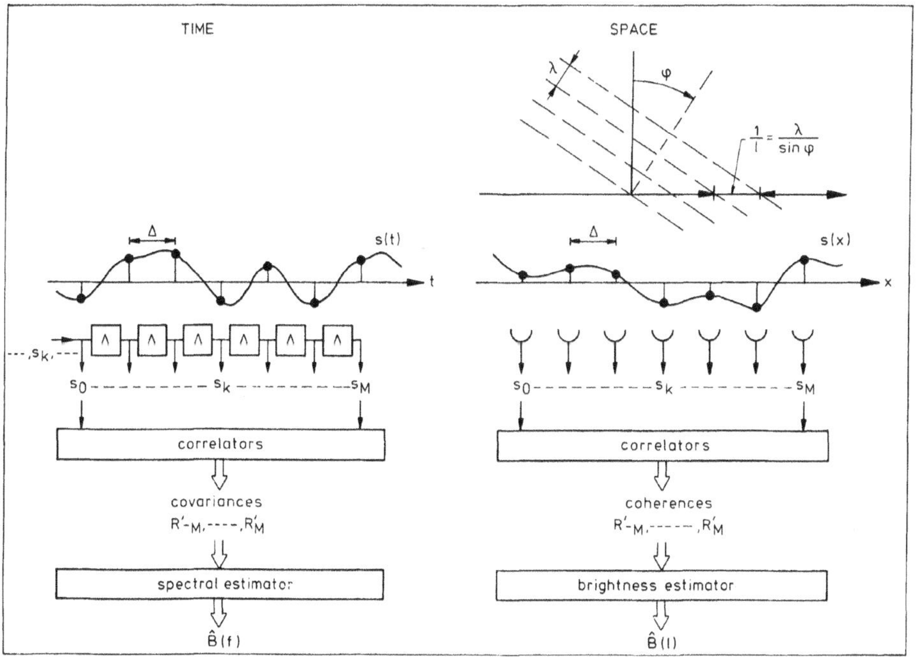

Fig. 1. Temporal spectrum analysis (left) and spatial imaging (right).

case the block of M+1 illumination samples, obtained at one time instant t_1, is multiplied to yield crossproducts $s_i^* s_{i+k}$. They are averaged over many successive times t_1, t_2, t_3, .. and the coherence estimates $\hat{R}_{-M}$, .., $\hat{R}_M$ are transformed to an estimated brightness distribution $\hat{B}(1)$.

There is one minor difference: power density spectra are symmetric functions of frequency; spatial objects are not. In order to distinguish positive and negative source directions in the spatial case, we must use a complex-valued illumination s(x). The result is a complex hermitic coherence function. In the temporal case the covariance function is real-valued and symmetric.

2. MAXIMUM ENTROPY METHOD (MEM)

The main purpose of this chapter is to repeat an argument from spectrum analysis (Burg 1967; van den Bos 1971) to show that MEM fits an a-priori constrained object model to the observed coherences.

2.1 MEM and Markov (or autoregressive) processes.

Following Burg (1967) we start in this section with an application of the MEM to spectrum analysis of a random time series .., s_k, .. . We have M+1 observed covariances R_k, k = 0, .., M (or k= -M, .., M since $R_k = R_{-k}$), spaced at Nyquist distances $\Delta = 1/2W$ (eq. 4). We neglect, until further notice, the measurement errors r_k. Burg selected as a spectral estimate the function $\hat{B}(f)$ which maximizes the entropy of the process .., s_k, .. under the condition of consistency with the observations, i.e.

$$\int_{-W}^{W} \hat{B}(f) \, . \, \exp 2\pi jfk\Delta \, . \, df = R_k \; ; \quad k = -M, .., M \tag{6}$$

The entropy of the process is a measure for the randomness, or the unpredictability, of .., s_k, .. . Maximum entropy means that a value s_k is maximally unpredictable from its entire past, s_{k-1}, s_{k-2}, .. , $s_{k-\infty}$, given the (M+1) observed covariances. This implies that the first M previous samples, s_{k-1}, .., s_{k-M}, are all that is needed to predict s_k and that the availability of earlier samples s_{k-M-1}, ..., $s_{k-\infty}$ does not improve predictability, despite the fact that they are correlated with s_k. Given the observed R_k's there is only one power density spectrum which belongs to such a "least predictable process"; this particular spectrum is taken as the MEM estimate $\hat{B}(f)$. Observe that the MEM criterion is a subjective one; it is equivalent to the a-priori-assumption of a "least predictable signal" .., s_k, .. . There is no evidence for this in the observed covariances themselves!

Solutions can be obtained in many different ways (Survey papers: Ulrych and Bishop 1975, Makhoul 1975). One approach is to express the entropy of the process in its spectrum (Shannon and Weaver 1949):

$$H \doteq \frac{1}{2W} \int_{-W}^{W} \ln B(f) \,.\, df, \tag{7}$$

and to maximize (7) subject to (6). A second approach (van den Bos 1971) is based on the theory of (wide sense) Markov processes. (Papoulis 1965). Our "least predictable signal", consistent with M+1 covariances, is a Markov process. Such processes can only be generated by an Mth order autoregressive filter (A.R. filter), with unit variance uncorrelated noise samples .., w_k, .. at the input (fig. 2). The output covariance function, although infinitely long, is determined by M+1 parameters, b_o, a_1, .., a_M. The relevant equations are:

$$\begin{pmatrix} R_o & \cdots & R_M \\ \vdots & & \vdots \\ R_M & \cdots & R_O \end{pmatrix} \begin{pmatrix} 1 \\ -a_1 \\ \vdots \\ -a_M \end{pmatrix} = \begin{pmatrix} b_o^2 \\ 1 \\ \vdots \\ 1 \end{pmatrix} \quad \text{and} \tag{8a}$$

$$R_k = \sum_{i=1}^{M} a_i R_{k-i} \; ; \quad k = M+1,\ M+2,\ \ldots,\ \infty \,. \tag{8b}$$

Given R_o, .., R_M we can solve eqs. 8a for b_o^2, a_1, .., a_M. This implies that we fit the Mth order Markov, or A.R., model to the data. We can then extrapolate the covariance function via eq. 8b and take the fourier transform to obtain the estimate $\hat{B}(f)$. Observe that eq 8b represents the coupling mechanism discussed in sec. 1.1. An equivalent method, less elaborate, is to solve for b_o^2, a_1, .., a_M and to take the AR filter's output spectrum as the estimate:

$$\hat{B}(f) = \frac{b_o^2}{\left|1 - \sum_{i=1}^{M} a_i \,.\, \exp - 2\pi j f i \Delta\right|^2} \,.\Delta \tag{9}$$

Eq. 9 has a (2M+1)-term fourier series in the denominator. Thus, $\hat{B}(f)$ contains fourier components up to infinity: many more than the conventional estimate of eq. 5.

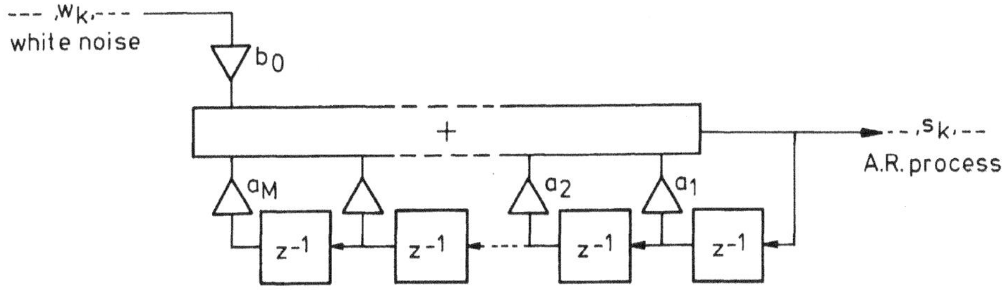

Fig. 2. Mth order AR filter. z^{-1} is the unit delay operator. This is the model, fitted to the data in a MEM spectrum analysis. The implicit assumption is that the spectrum consists only of resonances (poles).

The assumption of a "least-predictable" or A.R. or Markov process implies an a-priori assumption concerning the spectrum. Eq. 9 can be factored to give

$$\hat{B}(f) = \frac{b_o^{\,2}}{|(\exp 2\pi jf\Delta - p_1) \,..\, (\exp \; 2\pi jf\Delta - p_M)|^2} \cdot\Delta \quad , \qquad (10)$$

where $p_1, \ldots, p_M$ are the (real valued or conjugate paired) roots of $(z^M - a_1 z^{M-1} - \ldots - a_{M-1} z - a_M)$. The roots are the poles, or resonances, of $\hat{B}(f)$: a resonance occurs if a frequency f brings one of the terms $\exp 2\pi jf\Delta$ close to a pole p_i. Thus, the MEM estimate consists of resonances only ("all-pole-spectrum"). A general spectrum contains absorptions (zeroes) as well as resonances (poles). It would be obtained at the output of a filter with feed-forward branches in addition to feed-back branches (fig. 3).

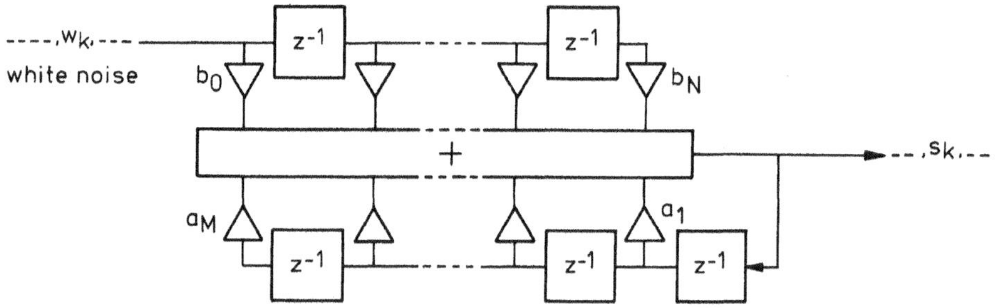

Fig. 3. General digital filter. This model should be used in the case of resonances (poles) and absorptions (zeroes).

For a mixed pole-zero spectrum the model of fig. 3 should be fitted to the data, rather than an A.R. model, and the solution would no longer correspond to an entropy maximization. Application, in this case, of the MEM leads at most to an approximation and it should be investigated whether this is a better one than the conventional fourier transform of eq. 5, or not. In the case of an all-zero spectrum the situation becomes surprisingly simple: no model fit is required since the covariance function R_k assumes a finite length. Under the proviso that all non-zero covariances have been observed there is no need for resolution improvement; the conventional fourier transform (eq. 5, unweighted) is optimum! Succesful applications of MEM to the detection and resolution of sharp spectral lines were reported in the past (references in Ulrych and Bishop 1975 and in Lacoss 1971). Spectral lines in temporal frequency analysis - and point sources in spatial imaging - are limit cases of resonances and are therefore marginally valid objects for the MEM.

2.2 Example

Fig. 4a shows an object spectrum B(f) of 20 poles (no zeroes). It was deliberately designed to have a narrow peak, some weak structure around the peak, and a dynamic range of > 100 dB. A piece of signal .., s_k, .. of 1024 points was used to determine the first 21 covariances, R_0, .., R_{20}. Fig. 4c shows the conventional estimate $\hat{B}(f)$ (eq. 5; g_k = gaussian taper): all spectral details are lost. The MEM estimate $\hat{B}(f)$, based on the same 21 covariances, is shown in fig. 4b. The deviations from the nominal spectrum B(f) are less than 2dB; they are due to the errors r_k caused by the finite (1024) signal duration (section 2.3).

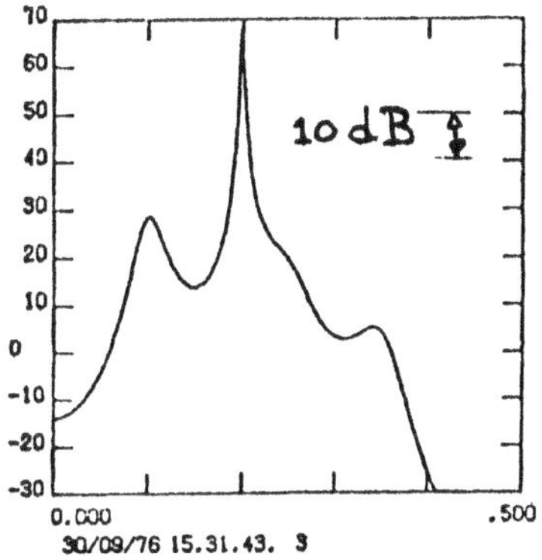

a) Spectrum B(f).

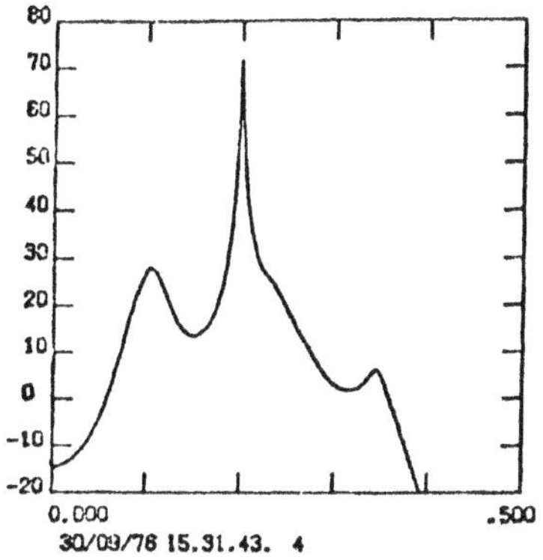

b) MEM estimate $\hat{B}(f)$(eqs 8,9).

Fig. 4. Maximum Entropy Method (MEM). The spectrum B(f) belongs to a 20th order Markov (or A.R.) process. The estimates $\hat{B}(f)$ are based on 21 observed covariances. The conventional estimate suffers an important resolution loss. The lost resolution is recovered by the MEM estimate.

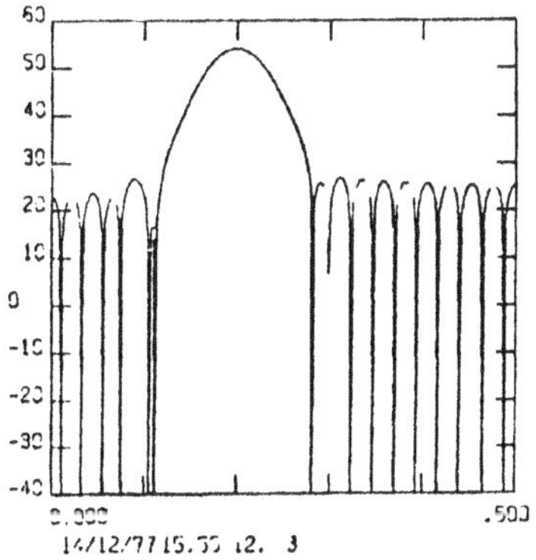

c) Conventional estimate $\hat{B}(f)$ (eq. 5).

Note that the conventional estimate (fig. 4c) could have been plotted on a wide f-grid without serious loss of detail. The MEM estimate, on the other hand, requires a dense plotting grid if the benefits of resolution enhancement are to be retained.

2.3 Measurement errors.

Actual observations of R_k are corrupted by errors r_k which propagate into the MEM estimate. The problem is particularly troublesome when B(f) has a large dynamic range. A general theory of error propagation is still failing but MEM does not seem to be unduly sensitive to random errors, provided that the problem of order estimation can be solved satisfactorily.

Up to now we have tacitly assumed that the order of the AR model is equal to the number of observed covariances, which implies that M+1 degrees of freedom are fitted to M+1 observations. Clearly we must not use too few degrees of freedom (resolution loss) but we must also avoid the danger of using too many (sensitivity to random errors) and we are therefore in need of algorithms for estimating the order of the model to be fitted. Various such methods are being developed, both for AR order estimation (Lacoss 1977, Jones 1976, Ulrych and Bishop 1975, Parzen 1974) and for model fitting in general.

Akaike's "Information Criterion" method (Akaike 1974) is an example from the 2nd class. An application is described by Ishiguro and Ishiguro (this volume). The method is a generalization of "naive" order estimation where one studies the behaviour of S, the sum of the squared residuals of a fitting procedure, as a function of N, the number of degrees of freedom. For small N the value S(N) drops rapidly as N increases. For large N the value of S decreases slowly and linearly until S = 0 when N becomes equal to the number of observations, M+1. The estimated order is the value of N where the linear behaviour sets in. The use of a value $N_{opt} < M+1$ leads, of course, to non-zero residuals. Consequently, the consistency conditions of eq. 6 must be relaxed.

Various authors have relaxed the consistency equations by maximizing eq. 7 under the condition

$$\frac{1}{M+1}\sum_{k=0}^{M}\left|R_k' - \int_{-W}^{W}\hat{B}(f)\,.\,\exp - 2\pi jfk\Delta\,.\,df\right|^2 = \sigma^2, \qquad (11)$$

where σ^2 is the variance of the r_k errors and R_k' are the observations (eq. 1). (Ables 1974, Wernecke and d'Addario 1977, Wernecke 1977, Gull and Daniell 1978). The advantage is that many constraints are replaced by a single one. In addition one obtains an "error-tolerant" estimate $\hat{B}(f)$ with an entropy value larger than obtained under the conditions of eqs. 6. Newman (1977) shows that the structure of the solution $\hat{B}(f)$ remains indentical to eq. 9, which suggests that the number of degrees of freedom is still equal to the number of observations. It would be interesting to compare the results of this approach with the results of order estimation methods.

2.4 One-dimensional spatial imaging.

Our discussion remains valid for 1-dim. spatial imaging with constant baseline increments Δ along a linear axis x, the aperture axis, provided that we use complex, hermitic coherences R_k. The underlying MEM assumption is, that the spatial series .., s_k, .. (E.M. field samples along the array axis) has a spatial AR- or Markov-structure, i.e. that M spatial samples are sufficient to predict the next spatial sample and that other samples, if available, would be useless for spatial prediction.

2.5 Two-dimensional spatial imaging.

The concept of maximizing eq. 7 under the constraints of eq. 6, or eq. 11, can be extended to 2-dimensional sets of coherence observations in the u,v-plane. Applications to radio astronomy were described by Ponsonby (1973), Wernecke and d'Addario (1977) and Gull and Daniell (1978). For simplicity we restrict ourselves to observations on a rectangular grid with spacings Δ in both dimensions:

$$R'_{ki} = R(k\Delta,\ i\Delta) + r_{ki}\ ;\quad k,i = -M,\ ..,\ M\ ;\quad R_{ki} = R^{*}_{-k,-i} \tag{12}$$

Newman (1977) showed that the estimated brightness assumes the form

$$\hat{B}(l,m) = \frac{b_o^{\ 2}}{\left|1 - \sum\limits_{\substack{g,h=0\\ g+h\neq 0}}^{M}\sum a_{gh} \cdot \exp - 2\pi j(gl + hm)\Delta\right|^2} \cdot \Delta^2\ , \tag{13}$$

i.e. the same structure as eq. 9. He derived eq. 13 in two ways: 1) maximization of the entropy $\int \ln \hat{B}(l,m)\, dl\, dm$ under consistency constraints, 2) via a minimum prediction error method. He concludes from the similarity of the results that the link between an AR model fit and the MEM extends to problems of 2 dimensions.

The implication is that the underlying a-priori assumption - a least predictable random process - remains valid in two dimensions. However, our process is a 2-dim. space series .., s_{gh}, .., the E.M. field amplitude at the aperture sample points and we must explain what is meant by a "least predictable process, subject to a set of observed coherences". This takes us into the area of 2-dim. Markov processes (Woods 1972, Jain 1977).

A process .., s_{gh}, .. is Markov if a sample value s_{g_o,h_o} can be predicted in terms of a limited number of samples in the

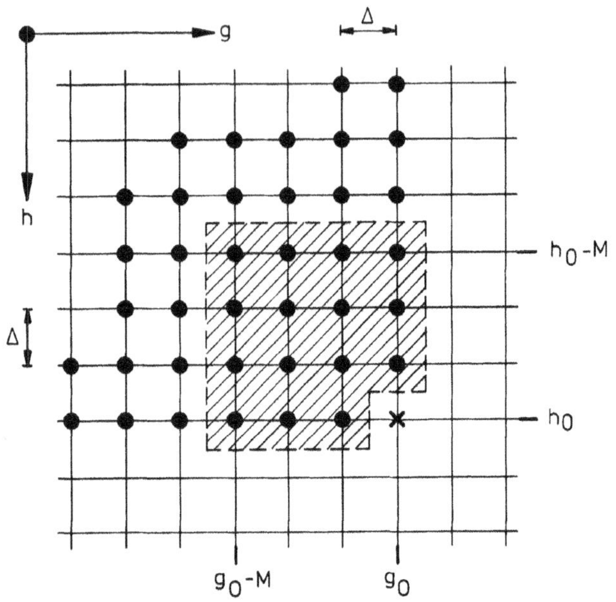

Fig. 5. Causal 2-dim. Markov process; order MxM. Prediction at g_o, h_o, based on all points $g \leq g_o$, $h \leq h_o$ $(g + h \neq 0)$ gives the same result as a prediction, based on the points $g_o - M \leq g \leq g_o$, $h_o - M \leq h \leq h_o$ $(g+h \neq 0)$. This is the implicit model used in a 2-dim. MEM.

vicinity of the point g_o, h_o. The other points are superfluous for prediction. Just as in the 1-dim. case, the MEM seems to be exclusively related to causal Markov processes. Fig. 5 illustrates a 2-dim. causal Markov process (Jain 1977). The vicinity, sufficient for prediction of s_{g_o,h_o} is "to the left and above" of the point g_o, h_o (hatched area); all other points in the upper left quadrant $g \leq g_o$, $h \leq h_o$ are superfluous. Causal Markov processes obey a difference equation analogous to the diagram of fig. 2, with coefficients a_{gh} (g, h = 0, .., M; g+h $\neq$ 0) and b_o. The coefficients can be solved from the observed coherences of eq. 12 via relations analogous to eqs. 8. As in the 1-dim. case, the brightness distribution is finally obtained as the power transfer function of the difference equation, which leads to eq. 13.

Thus, the conclusion seems justified that a 2-dim. MEM relies on fitting a 2-dim. Markov process to the observations.

2.6 Another entropy expression.

The preceding sections were based on the entropy expression of eq. 7: $H_1 \doteq \iint \ln B(l,m).dl\ dm$. We note that H_1 is a property of the random process .., s_k, .., related to its (un)predictability. It is not a measure for the likelihood of the estimate $\hat{B}(l,m)$, unless of course we have the a-priori certainty that .., s_k, .. is an all-pole-process.

More recently a second entropy expression was introduced: $H_2 = \iint \hat{B}(l,m).\ln \hat{B}(l,m)\ .\ dl\ dm$. Maximization of H_2 under consistency conditions is said to give a measure for the likelihood of $\hat{B}(l,m)$ (Frieden 1975, Gull and Daniell 1978, Kikuchi and Soffer 1977, various papers in this volume).

When using H_2 instead of H_1 the connection with Markov processes is broken and our preceding discussion does not apply anymore. Still, H_2 gives improved resolution, implying that the coherence function has been extrapolated, in one way or another. It would be interesting to find out if H_2 can also be interpreted as a model fit. In other words: what are the conditions on the object B(l,m) which make a H_2-coherence extrapolation a legitimate one?

3. HIGH RESOLUTION METHOD (HRM)

We analyze the mechanism of Capon's High Resolution Method (HRM) - also known, erroneously, as Maximum Likelihood Method (MLM) - and we indicate two areas which require further investigation.

3.1 Imaging with a beamformer-wattmeter combination.

We start with an array of M+1 sensors and we assume that all outputs s_o, .., s_M (aperture illumination) are simultaneously available for processing. The sensors may be distributed over a linear or a plane aperture, but for simplicity we shall discuss a 1-dim. line-shaped array with equal sensorspacings Δ (fig. 6). Note that the condition of simultaneous accessibility of all sensor outputs is not fulfilled in 2-dim. earth rotation aperture synthesis; we return to this point in sec. 3.6.

Fig. 6 shows the traditional array processor, consisting of a beamformer and a wattmeter. The beamformer adds the instantaneous sensor outputs s_i via weights $a_i^*(l_o)$ so as to steer an antenna beam in the desired direction l_o:

$$x = \sum_{i=0}^{M} a_i^*(l_o)\ .\ s_i\ . \tag{14}$$

The wattmeter estimates the power, incident upon the array from direction l_o, by squaring x(t) and averaging over a finite time T. Note that beams can be steered simultaneously in many different directions l_o by using different sets of weights for each beam. Neglecting the measurement error caused by a finite integration time T, the average output power is given below, where E denotes expectation:

$$\hat{P}(l_o) = E\ \{|\sum_{i=0}^{M} a_i^*(l_o)\ .\ s_i\ |^2\}\ . \tag{15}$$

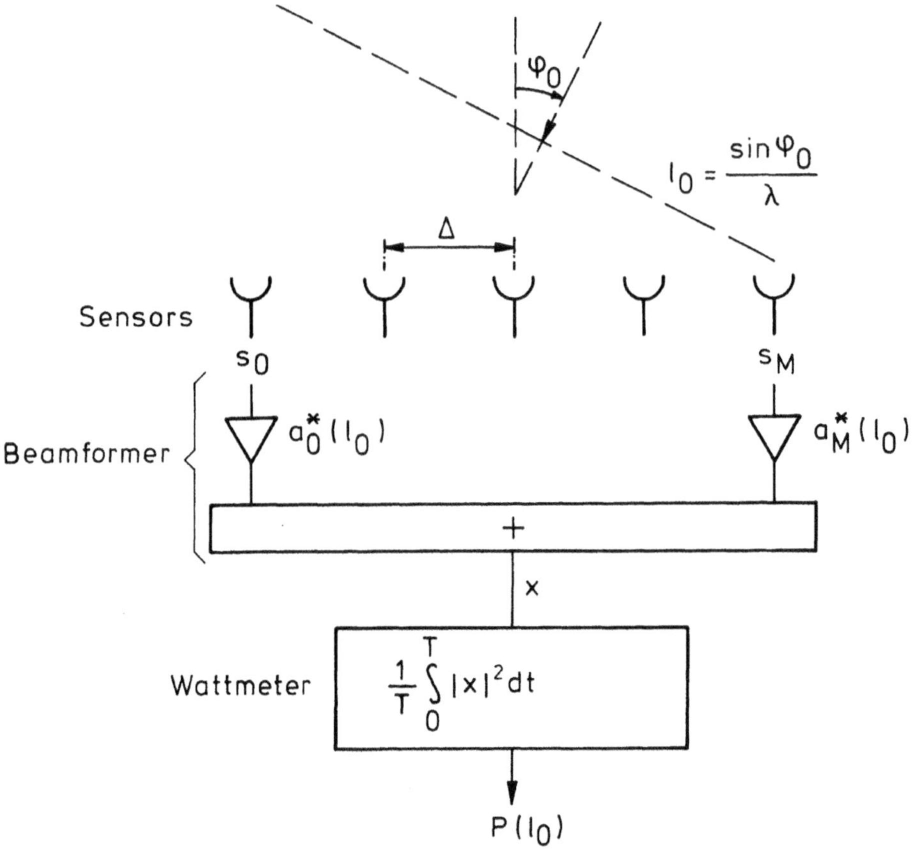

Fig. 6. Beam former + wattmeter; a traditional imaging system when all sensor outputs are simultaneously available for processing. l_o is the beamsteering direction. Using a parallel processor, beams can be steered in many different directions.

Invoking the definition of the coherence function, $E\{s_i^* \, s_{i+k}\} = R_k$, and the van Cittert-Zernike theorem, $R_k = \int B(l) \, . \exp 2\pi jkl\Delta \, . \, dl$, this can be written in various other forms:

$$\hat{P}(l_o) = \sum_{i,m=0}^{M}\sum a_i^* (l_o) \, . \, a_m (l_o) \, . \, R_{i-m} \tag{16}$$

$$= \sum_{k=-M}^{M} w_k^* (l_o) \, . \, R_k \tag{17}$$

$$= \int W(l,l_o) \, . \, B(l) \, . \, dl, \tag{18}$$

where

$$w_k^*(l_o) = \sum_{i=0}^{M} a_i^*(l_o) \cdot a_{i-k}(l_o) \tag{19}$$

$$W(l, l_o) = \sum_{k=-M}^{M} w_k^*(l_o) \cdot \exp 2\pi j l k\Delta = \left| \sum_{i=0}^{M} a_i^*(l_o) \cdot \exp 2\pi j l i\Delta \right|^2 . \tag{20}$$

All integrals above and below are over the source interval $[-W = -1/2\Delta$ $W = 1/2\Delta]$ (eqs. 3,4). Eq. 18 expresses the output power in the familiar form of the brightness distribution $B(l)$, convolved with $W(l, l_o)$, the antenna beampattern (BP).

Observe that eq. 18 allows the use of differently shaped BP's for different steering directions, by selecting specific sets of weights $a_i^*(l_o)$ (or $w_k^*(l_o)$) for each l_o. This opens the possibility to adapt the imaging system to the brightness distribution. Although $W(l, l_o)$ is limited to 2M+1 fourier components (eq. 20), we can realize quite arbitrary, and even asymmetrical, BP's. We shall take advantage of this feature by minimizing the power, picked up by our antenna beam, from directions other than l_o. Or, in the jargon of spectrum analysis, we shall minimize the "leakage" from object coordinates l to the steering direction l_o.

Eq. 18 holds also the potential of resolution improvement. Suppose that we divide eq. 18 by the effective width of the beampattern, say L, and that we regard $\hat{P}(l_o)/L$ as a brightness estimate $\hat{B}(l_o)$. In that case eq. 18 relates the estimate to the object via a shift-variant operation, contrary to a conventional image reconstruction which produces a shift-invariant operation, similar to eq. 5b:

$$\hat{B}(l_o) = \int G(l-l_o) \cdot B(l) \cdot dl. \tag{21}$$

(The conventional result can be obtained by putting $w_k^*(l_o) = g_k \cdot \exp - 2\pi j k l_o \Delta$ in eq. 17; compare with eq. 5a; the pattern G ( ) in eq. 21 is the fourier transform of g_k, $k = -M, .., M$.) In eq. 21 the number of fourier components in $\hat{B}(l)$ is limited by the number in $G(l)$, i.e. $k = -M, .., M$. In eq. 18, however, the range of fourier components in $\hat{B}(l)$ may exceed this number, i.e. it may exceed the limit set by the length of the antenna array, and we may hope for a resolution improvement.

3.2 Capon's High Resolution Method (HRM).

Capon (1969; Capon and Goodman 1970) imposed the following conditions on the beampattern:

1. $W(l, l_o) \geq 0$; all l, any l_o (22)

2. $W(1_o, 1_o) = 1$ (23)

3. $\hat{P}(1_o)$ = mimimum, (24)

and he proposed to regard the minimized output power $\hat{P}(1_o)$ as an estimate of the power radiated by the object B(1) in a narrow region around 1_o. What is the significance of these conditions?

1. By requiring a non-negative beampattern we have the certainty of a positive estimate $\hat{P}(1_o)$. Hence the method relies on the a-priori assumption of a positive brightness distribution (sec. 1.1). (For a processor of the type of fig. 6 the positivity assumption is already incorporated because of the presence of the wattmeter.)

2. Eq. 23 normalizes the beampattern's power sensitivity in the steering direction to unity. Radiated power from directions close to 1_o is measured correctly.

3. By minimizing the average output power $\hat{P}(1_o)$ under the constraints 22 and 23 we hope to minimize the power picked up via sidelobes of the BP (minimization of "leakage"; sec. 3.1). We shall see that this leads to a BP where the sidelobes have arranged themselves "inversely" with respect to the brightness distribution: low sidelobes where B(1) is large, and conversely.

Eqs. 22, 23, 24 are solved very easily for a beamformer system. No special precautions are required to satisfy eq. 22; the BP is always positive because of eq. 20. Further, eqs. 20 and 16 are used to express $W(1_o, 1_o)$ and $\hat{P}(1_o)$ in terms of the weights $a_i^*(1_o)$. The minimization is then performed via a Lagrange multiplier and we obtain an optimum set of weights $a_i^*(1_o)$ which can be resubstituted in eq. 16. We shall skip the calculations (Lacoss 1971, Cox 1973 (2), Capon 1969). The result is:

$$\hat{P}(1_o) = \frac{1}{\sum\sum_{i,k=0}^{M} Q_{ik} \cdot \exp - 2\pi j(i-k)\, 1_o\, \Delta}, \tag{25}$$

where Q_{ik} are the elements of the inverse, Q, of the coherence matrix R, which consists of the elements $R_{ik} = E\{s_i^* s_{i+k}\}$:

$$R = \begin{pmatrix} R_{oo} & R_{o1} & \cdots & R_{oM} \\ R_{10} & R_{11} & \ddots & \vdots \\ \vdots & & \ddots & \vdots \\ R_{Mo} & \cdots & \cdots & R_{MM} \end{pmatrix} ; \qquad Q = R^{-1} = \begin{pmatrix} Q_{oo} & \cdots & Q_{oM} \\ \vdots & \ddots & \vdots \\ Q_{Mo} & \cdots & Q_{MM} \end{pmatrix} \tag{26}$$

(In the present case of an equidistant linear array (fig. 6) we have $R_{ik} = R_{k-i}$). We make the following observations.

1. Although we started with a beamformer, requiring the simultaneous availability of all sensor outputs, we end up with an expression (25) that enables us to use observed coherences R_{ik}. In the equidistant case we only need the values R_{-M}, .., R_M since $R_{ik} = R_{k-i}$. Thus, we can start by measuring coherences in the usual way and then substitute them in eqs. 25, 26 to obtain $\hat{P}(1_o)$. Note that this implies the use of slightly incorrect values since the finite correlation time creates an error r_k (eq. 1).

2. Eq. 25 can be written as a fourier series of infinite length (finite length fourier series in the denominator). Hence $\hat{P}(1_o)$ contains fourier components missing from a conventional image (eq. 5a). Equivalently, the coherence function corresponding to eq. 25 has an infinite length and contains many of the unobserved coherences.

3. Unlike the MEM, the HRM solution (25) is not consistent with the original observations: the first M+1 coherences, calculated from eq. 25, differ from the observed R_o, .., R_M. This is simply checked by doing the calculations for a short series.

3.3 Example.

Fig. 7 shows an example of the HRM in temporal spectrum analysis, applied to the same 21 covariances as used in the MEM example of fig. 4. The nominal spectrum B(f) and the HRM estimate $\hat{P}(f_o)$ are depicted in figs. 7a and 7b, respectively. The resemblance is reasonable. Comparison with the conventional estimate of fig. 4c shows a significant enhancement of resolution. Note the discrepancy between the vertical scales of fig. 7a and 7b, caused by the fact that eq. 25 represents power rather than power density.

The plots of fig. 7c present beampatterns $W(f, f_o)$ for 4 different tuning frequencies ("steering directions") f_o, indicated by vertical lines. Observe that $W(f_o, f_o) = 1$, in accordance with eq. 23. Comparison of the BP's with the object spectrum B(f) illustrates the "inverse" behaviour of the sidelobes, in response to the required leakage minimization (eq. 24). The BP's have adapted themselves to the data. The effect is so strong that the main beam of the antenna pattern is no longer clearly recognizable.

3.4 Discussion.

In sec. 3.2 we derived the HRM estimate $\hat{P}(1_o)$ as the minimum power that can be measured with an all-positive beampattern, normalized to unit sensitivity in the steering direction. Thus, we interprete HRM as a positive constrained reconstruction method which minimizes leakage.

Yet it is amazing that a result of the quality of fig. 7b can be obtained solely on the assumption of a positive object function and one may wonder whether there is perhaps a hidden structural assumption behind the HRM, similar to the MEM assumption. This suspicion is fed by

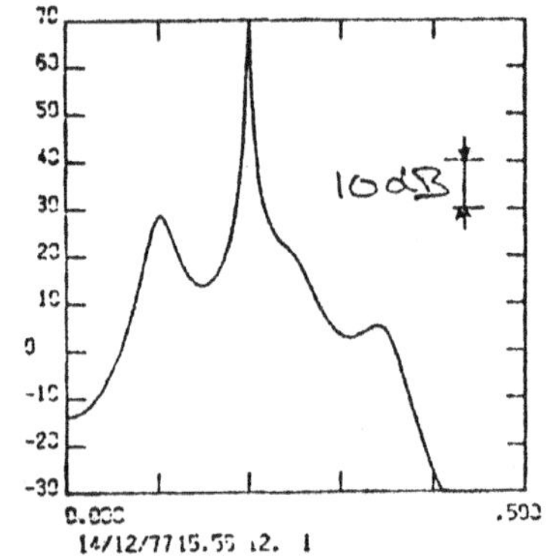

a) Spectrum B(f)

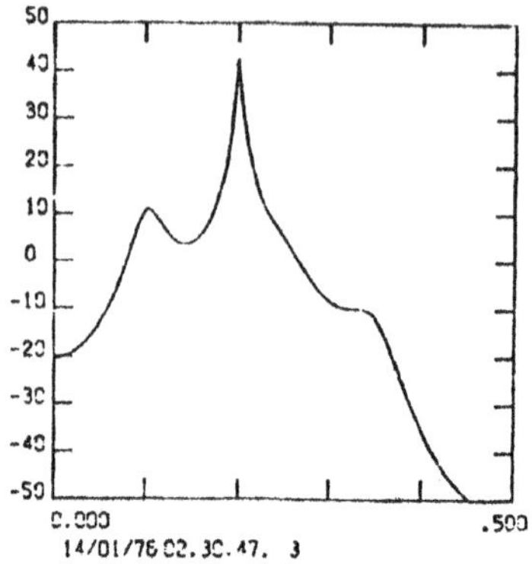

b) HRM estimate $\hat{P}(f)$

c) Beampatterns $W(f, f_o)$ for 4 different steering directions f_o.

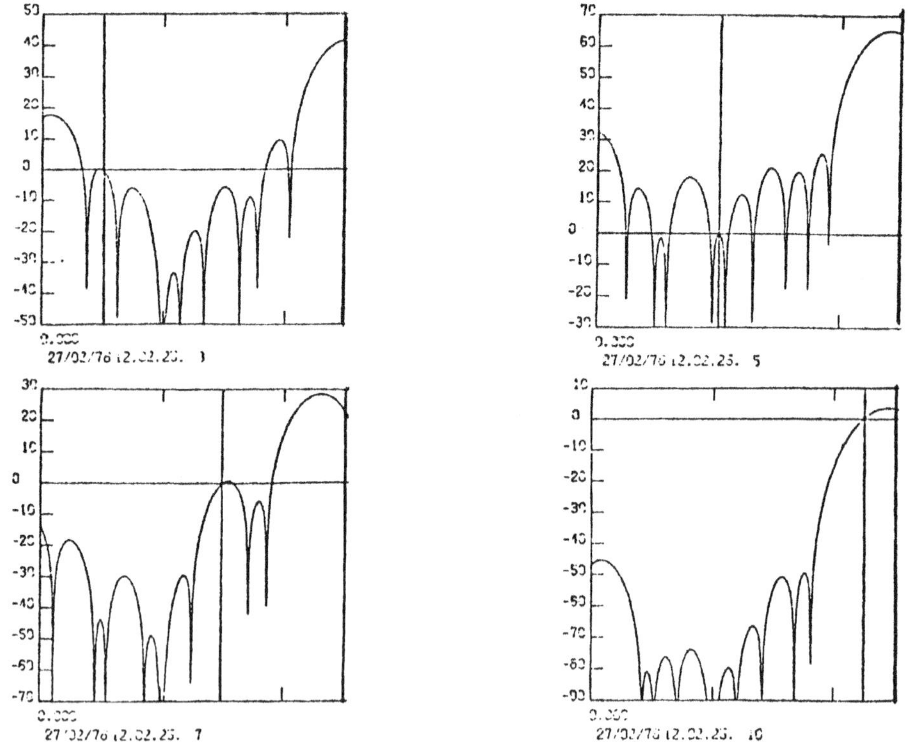

Fig. 7. High Resolution Method (HRM). Same spectrum B(f) as in fig. 4a. Comparison with the conventional estimate (fig. 4c) illustrates the resolution enhancement. Figs. 7c illustrate how the beampatterns adapt in an inverse sense to the object B(f), in order to minimize leakage.

the existence of an algebraic relation between MEM and HRM estimates, as demonstrated by Burg (1972).

3.5 Areas for further study.

There are at least two problems requiring further investigation: 1) intensity calibration and 2) incomplete coherence matrices. The 2nd point is important for application to 2-dim. earth rotation aperture synthesis.

Intensity calibration. We recall that eq. 25 is an estimate of the power radiated by the object within a narrow region around the steering direction; it is not an estimate of the brightness (i.e. density of power) B(1). The HRM has a long history in the area of underwater acoustics where the problem is to detect the presence or absence of a point source $P_s.\delta(1-1_o)$ against an unknown, continuously distributed background radiation $\bar{B}(1)$. The HRM, in the form of a power estimator, is perfectly suited to this purpose since the constraints 23 and 24 guarantee a maximum signal-to-noise ratio if a point source is indeed present in direction 1_o. (These applications go by the name of MLM: Maximum Likelihood Method, a misnomer since the method does not provide an ML estimate of power. Cox 1969, Cox 1973 (1), Cox 1973 (2), Lewis and Schultheiss 1970, Schultheiss 1977.)

Our present purpose is slightly different: we are interested in a density estimate. Hence eq. 25 must be divided by the effective width L of the BP's main beam. Since the BP's may assume strange shapes (fig.7c), and since the shapes differ from one steering direction to the next, the problem deserves some more attention. Various approaches can be taken.

1. An ad-hoc approach is to divide eq. 25 by the effective width L of a conventional unweighted beampattern G(1), to be obtained via eqs. 5.

2. Another option is to use the actual beampatterns $W(1, 1_o)$, to be derived from eq. 20, but this poses the problem of defining an effective width for BP's shaped like those of fig. 7c.

3. A 3d solution might be to replace constraint 23 by a modified one:

$$\int W(1, 1_o) \, . \, C(1-1_o) \, . \, d1 = 1, \tag{27}$$

where $C(1-1_o)$ is a positive pulse-type function (e.g. a gaussian), centered at 1_o, with an area 1 and an effective width L which could, for instance, be taken equal to a desired resolution width. Observe that eq. 27 reduces to the original constraint 23 if $C(1-1_o)$ is a delta function.

Incomplete coherence matrix. The preceding discussions extend immediately to non-equidistant arrays, distributed over a 2-dim. plane aperture. One can use a beamformer + wattmeter if all sensor outputs are simultaneously accessible or one can start from a previously measured coherence matrix R (eqs. 25, 26).

In the last case, however, the HRM requires that all cross coherences R_{ik} have been measured! This condition is not met in radio astronomy earth rotation aperture synthesis, where the usual observation scheme limits the coherences to those involving the sensor at the origin of the rotating array. This leads at best to the elements in the first row, the first column and the main diagonal of the coherence matrix R. Papadopoulos (1975) applied the HRM to radio astronomical data and he solved the dilemma by setting all unobserved matrix elements equal to zero.

A better solution for the case of an incomplete matrix is to break the link with simultaneously accessible arrays and to rephrase the problem. Suppose that a set of coherences R_{-M}, .., R_M has been observed, distributed arbitrarily in the u,v-plane. The HRM conditions could then read:

$$1.\ \text{Minimize}\ \hat{P}(l_o) = \sum_{-M}^{M} w_k^*(l_o).R_k \quad \text{under the constraints}$$

$$2.\ W(l, l_o) \geq 0 \qquad (28)$$

$$3.\ W(l_o, l_o) = 1\ ,$$

where $W(l, l_o)$ is the fourier transform of the weights $w_k^*(l_o)$. Unfortunately the solution of this problem is less straightforward than the one of eqs. 22, 23, 24 since a convenient relation of the type 20, automatically satisfying the positivity constraint, is no longer available.

ACKNOWLEDGEMENT

The author is grateful to P.J. Prinsen of the Physics Laboratory NDRO for many dicussions and for the preparation of figs. 4 and 7.

LITERATURE

Ables, J.G.: 1974,"Astron. Astrophys. Suppl." 15, pp. 383.
Akaike, H.: December 1974, IEEE Trans. AC-19, 6, pp. 716.
Burg, J.P.: 1967, Proc. 37th Meeting Soc. Explor. Geophys..
Burg, J.P.: April 1972, "Geophysics", 37, 2, pp. 375.
Bos, A. van den: July 1971, IEEE Trans IT-17, pp. 493.
Capon, J.: August 1969, Proc. IEEE, 57, 8, pp. 1408.
Capon, J., Goodman, N.R.: October 1970, Proc. IEEE, 58, 10, pp. 1785.
Cole, T.W.: 1973, "Astron. and Astrophys." 24, pp. 41.
Cox, H.: 1973, 1, in "Signal Processing", eds. Griffith, Stocklin, van Schooneveld, Academic Press.
Cox, H.: September 1973, 2, "Jour. Acoust. Soc. Am." 54, 3, pp. 771.
Frieden, B.R.: 1975, in "Picture Processing and Digital Filtering", ed. T.S. Huang, Springer.

Gilbert, E.N., Morgan, S.P.: 1955, "Bell System Tech. J." 34, pp. 637.
Griffiths, J.W.R., Hudson, J.E.: 1977, in "Aspects of Signal Processing", pt 1. ed. G. Tacconi, Reidel.
Gull, S.F., Daniell, G.J.: April 1978, "Nature" 272, pp 686.
Jain, A.K.: September 1977, "Jour. Optim. Theory and Applic." 23, 1, pp. 65.
Jones, R.H.: August 1976, "Geophysics" 41, 4, pp. 771.
Kikuchi, R., Soffer, B.H.: December 1977, "J. Opt. Soc. Am." 67, 12, pp. 1656.
Lacoss, R.T.: August 1971, "Geophysics" 36, 4, pp. 661.
Lacoss, R.T.: 1977, in "Aspects of Signal Processing", pt 2, ed. G. Tacconi, Reidel.
Lewis, J.B., Schultheiss, P.M.: 1970, "Jour. Acoust. Soc. Am.", 49, 4 (part 1), pp. 1083.
Makhoul, J.: April 1975, Proc. IEEE, 63, 4.
Newman, W.I.: January 1977, IEEE Trans. IT-23, 1, pp. 89.
Newman, W.I.: 1977, "Astron. Astrophys." 54, pp. 369.
Owsley, N.L.: 1973, in "Signal Processing", op. cit.
Papadopoulos, G.D.: January 1975, IEEE Trans. AP-23, 1, pp. 45.
Papoulis, A.: 1965, Probability, Random Variables and Stochestic Proces-es, McGraw-Hill.
Parzen, E.: December 1974, IEEE Trans. AC-19, 6, pp. 723.
Ponsonby, J.E.B.: 1973, Mon. Not. R. Astr. Soc. 164, 0, pp. 1.
Schultheiss, P.M.: 1977, in "Aspects of Signal Processing", pt 1, ed. G. Tacconi, Reidel.
Shannon, C.E., Weaver, W.: 1949, The Mathematical Theory of Communicat-ion, Univ. of Illinois Press.
Taylor, T.T.: 1955, IRE Trans. AP-3, 16.
Ulrych, T.J., Bishop, T.N.: February 1975, Rev. Geophys. and Space Phys. 13, 1, pp. 183.
Wernecke, S.J., d'Addario, L.R.: April 1977, IEEE Trans. C-26, 4, pp.351.
Wernecke, S.J.: Sept.-Oct. 1977, Radio Science, 12, 5, pp. 831.
Woods, J.W.: March 1972, IEEE Trans. IT-18, 2, pp. 232.

DISCUSSION

Comment R.H.T. BATES
Do you not think that superresolution can work (using Gerchberg's algorithm, for instance) provided the resolution is only a few Shannon samples? See, for instance, McDonnell and Bates, Ap.J. 208, Sept. 1976, 443.

Reply C. van SCHOONEVELD
The only useful application of superresolution that I know of, has to do with very short antenna's or antenna arrays, i.e. arrays with a length of one or two Nyquist intervals Δ, or less. One can then extrapolate the coherence function over a distance of the order of Δ beyond the physical array limits, before the effect of internal noise becomes prohibitive. Unfortunately, this limits superresolution to antenna

systems where the resolved angle is an appreciable fraction of the field of view. In addition, some useful applications were reported in relation to short end-fire arrays.

Comment J.C. DAINTY
The application of MEM to power spectrum analysis leads to a "log B" maximization implying certain structural constraints on the random process. Astronomers sometimes use "B log B" -MEM- methods. What are the structural constraints in this case?
Reply C. van SCHOONEVELD
I wish I knew! This is an interesting area for further research. The variational problem involved is probably too complicated to lead to a closed-form solution, but it might give a hint about the structure of the solution. This might then lead to a guess about an underlying assumption concerning the process.

Comment L.R. D'ADDARIO
I think that the MLM is not useful for fourier synthesis in radio astronomy. It computes, for each map point, the most likely flux assuming the rest of the map is empty. This is often an unreasonable assumption. Also, the simple matrix inverse does not apply to rotation synthesis data; an instantaneous array with all elements cross-correlated is required.
Reply C. van SCHOONEVELD
I do not agree with your first statement. The MLM, or HRM, is fully aware of the flux in other map points and it deliberately minimizes the effect of these fluxes in the map point under consideration. Your second comment is completely to the point! Lack of time prevented me to mention this problem in my presentation; it is considered in the paper.

Comment J.E.B. PONSONBY
You said that one dimensional time series analysis and two dimensional imaging are analogous. You said that the Wiener-Khintchine theorem becomes the van Cittert-Zernike theorem, etc. You then discussed the time-series analysis of autoregressive processes with the implication that there is an analogy in 2 dimensions. Can you give examples of 2-dimensional autoregressive processes?
Reply C. van SCHOONEVELD
Yes and no. In my paper I give the example of a causal 2-dim. AR process, where the corner of a rectangular block of data can be predicted from the other points in the block. This is a direct analogy with the 1-dim. case. The trouble is, that one can imagine many other prediction schemes in 2 dimensions, most of them non-causal. Whether such processes do indeed exist, and whether they lead to modified MEM algorithms, is not yet completely clear.

THE MAXIMUM ENTROPY METHOD

(Invited paper)

S.F. Gull
Mullard Radio Astronomy Observatory
Cavendish Laboratory, Cambridge, United Kingdom

G.J. Daniell
Department of Physics
University of Southampton, United Kingdom

1. INTRODUCTION

We have heard so far at this meeting from the practical-minded men of radio astronomy. Faced with the problem: 'my map has sidelobes and noise on it', they promptly reach for the nearest available weapons - 'cleaners, polishers and kneaders'. All these use varying amounts of subjective human judgement, as for example in 'window cleaning'.

The ideas of the maximum entropy method are intriguing, since they suggest that one can in principle remove the effects of sidelobes from the map, reduce the apparent noise, and give extra resolution. But maximum entropy is not just another image reconstruction technique, it is a technique for the idealist; the man who cares what should be done with his data. It is more fundamental than other techniques and deals with problems in a very general way. What the maximum entropy method sets out to do is to seek the most likely answer to a particular question that can be obtained from a given body of scientific data.

2. BAYES' THEOREM

We declare immediately our scientific philosophy. We are followers of Jeffreys (1) and are unashamed Bayesians. We further believe, following Jaynes (2), that the various techniques and ideas of maximum entropy should be regarded as simplified ways of applying Bayes' Theorem in probability theory. Although the fundamental ideas are very general, we refer here to a radioastronomical context; the problem is to make a map of the radio sky.

All our knowledge of the sky can be expressed, we believe, as the relative probabilities of various possible maps of the sky. Bayes' Theorem is then the basic theorem of probability theory that tells us how to modify our knowledge or prejudice about the sky when some

C. van Schooneveld (ed.), Image Formation from Coherence Functions in Astronomy, 219–225.

additional data are acquired. It says:

$$\text{Posterior Prob.(sky} \mid \text{data)} \propto \text{Prior Prob.(sky)} \times \text{Prob.(data} \mid \text{sky).}$$

The constant of proportionality can be found by normalizing the posterior probability.

The second term on the right is easy to obtain: we can calculate what we ought to observe from any given sky, and hence the probability of this giving rise to our measured data. For the special case when our measurements have independent Gaussian errors this second term is $\exp(-\chi^2/2)$ where:

$$\chi^2 = \sum_{\text{data}} \frac{(\text{misfit to data})^2}{(\text{error in data})^2},$$

so that:

$$\text{Posterior Prob. of sky} \propto \text{Prior Prob. of sky} \times \exp.(-\chi^2/2).$$

When the number of sample points in the sky (pixels) is large, and certainly in the continuous limit, this probability distribution of maps of the sky is far too complicated to comprehend. One, albeit very inadequate, way of summarizing it and displaying a single, useful map is to pick the map for which the posterior probability is a maximum.

This maximum in the distribution may, or may not, be sharp; there may even be several local maxima. If the maximum is broad a single map is not very useful, and if there are several maxima it is downright misleading. The occurrence of these situations simply means that the data collected do not determine the sky properly; you must do a better experiment!

We emphasize that in this formulation the data enters into a probability distribution, prior prejudice or prior knowledge of the sky is modulated by a combination of the data values and their errors. Highly uncertain data simply changes one's opinion less than more accurate data.

The vital importance of the errors in the data is not always realized; for example in the Burg (3) algorithm for maximum entropy spectral analysis the data are estimates of the autocorrelation function of a time series. These estimates will have errors, nevertheless the spectrum is made to agree exactly with the estimates. We suggest that the deficiencies of maximum entropy spectral analysis, the line splitting and line shifts, are basically due to ignoring the errors of estimation.

There is danger in interpreting any map of the sky that fits the data exactly, it will almost certainly contain features arising from the noise in the data. The above analysis shows the importance of the statistic χ^2 in any problem where the data values have independent Gaussian errors.

3. THE PRIOR AND MAXIMUM ENTROPY

The fundamental limitation of Bayesian statistics lies in the selection of a suitable expression for the prior probability; in our example the prior probability of various maps of the radio sky before we are given any data. If we write this exp (S/λ), where S is a function of the map and λ a fixed constant, we see that to display the single map that maximizes the posterior probability we have to maximize $S - \lambda\chi^2$. This is the recipe used by Gull and Daniell (4) and by Wernecke and D'Addario (5) for constructing maximum entropy maps.

We can therefore identify S as the 'entropy of a map of the sky' and the result of Bayes' theorem is identical with the maximum entropy algorithm in which we maximize an entropy expression, S, subject to the constraint that the final map fits the data. The constant, λ, has become a Lagrange multiplier. It follows, in accordance with the philosophy of Jaynes (2), that the maximum entropy method is a simplified Bayesian approach and the various entropy expressions in use are really statements of a prior prejudice about the sky, given no data.

The problem of determining appropriate prior distributions corresponding to such ignorance has led to considerable controversy since the time of Laplace and there is, as yet, no entirely satisfactory way of assigning priors, even in the simplest situations. Therefore, at this stage, we suggest that it is most useful to understand the properties of the prior distributions that arise from the various entropy expressions and to collect arguments that lead to different priors, without attempting to decide which, if any, are correct.

4. THE ENTROPY EXPRESSION

There are two schools of thought concerning the entropy of a map of the sky. If we label a set of areas on a map by an index i and the flux of radio waves from the ith area is f_i, then we can express the results thus: The H_1 school says that $S \propto \sum_i \log f_i$, and the H_2 school says $S \propto -\sum_i f_i \log f_i$. We are members of this second school and we can give an outline derivation of the entropy expression using a model incorporating prior assumptions about the sky.

For ease of counting maps we assume the flux f_i from each cell of the sky is quantized. We then employ a secret weapon - the canonical team of monkeys. Each monkey is assumed to be given a large number N of flux quanta which he scatters randomly across the cells of the sky to make a trial map. Suppose n_i quanta land in the ith cell. Not all sets of n_i generated in this way by the monkeys are equally likely. The probability that a particular set occurs is proportional to $N!/\prod_i n_i!$. If N and all the n_i are large, we can use Stirling's approximation and obtain $S \propto \log(\text{probability}) \propto -\sum_i n_i \log n_i/N$. When the question of relating our artificial flux quanta to the continuous quantities f_i is considered we are led to the entropy expression:

$$S \propto -\sum_i \left(\frac{f_i}{\Sigma f}\right) \log \left(\frac{f_i}{\Sigma f}\right)$$

Suppose now we are given a body of data, for example samples of a spatial coherence function, having independent errors. The fit of a trial map to these data may be defined using χ^2 and the probability of the data given the map $\{f_i\}$ is then proportional to $\exp(-\chi^2/2)$. Maximizing $S - \lambda\chi^2$ yields:

$$f_i = \left[\exp\left\{\frac{\sum_j f_j \log f_j}{\sum_j f_j}\right\}\right] \exp\left(-\lambda \frac{\partial\chi^2}{\partial f_i}\right) .$$

The first term in brackets is just a constant and it ensures that the final map does not depend on the units in which the data are given. The value of λ is not determined by this argument and we have suggested (Gull and Daniell (4)) that λ is chosen so that χ^2 for the final map equals its statistically expected value, i.e. the number of data samples. In this way we get a safe map that is unlikely to contain artifacts arising from the noise in the data. Of course, if there are several maxima in the probability this map will still be misleading. However, for the problem of aperture synthesis, with phase information, or for deconvolution, this cannot happen. The prior based on the monkey mechanisms is strongly peaked about uniform maps so we can loosely say that we have constructed the most uniform map consistent with the data and their errors.

The above discussion shows in outline how a prior probability distribution or entropy expression follows from assumptions or prejudice about the sky. Other entropy expressions follow from other prejudices corresponding apparently to different types of 'complete ignorance' about the sky. The expression H_2 is a configurational entropy of the radio flux, although the above calculation is similar to those used in statistical mechanics. One cannot run a heat engine on an image!

5. OTHER ENTROPY EXPRESSIONS

We will not deal in such detail with the alternative expression H_1, but will give some idea of how it arose and the assumptions it makes. It has been used for many years in Burg's (3) algorithm and is based on

Shannon's information theory.

Burg is concerned with the estimation of the power spectrum of a Gaussian random process or time series. The relation of this to the autocorrelation function certainly involves a Fourier transform, and so indeed does aperture synthesis, but here the analogy ceases; there is no 'space series' in radio astronomy. The expression $H_1 = \int \log P(\nu)d\nu$ (Bartlett (6)) is the entropy rate of a noise source with power $P(\nu)$ at frequency ν. It is distinct from the configurational entropy of the spectrum itself. In more recent work on the Burg method the information theoretic aspects have been relatively neglected and interest has centred on the fact that the method is equivalent to fitting an autoregressive process to the time series. This has a natural interpretation for the estimation of power spectra of time series but it does not make sense to regard the signals from successive aerials in an interferometer as being derived by spatially filtering random noise. We conclude that Burg's and Shannon's arguments, whilst sound within their terms of reference for time series, are not appropriate to radio astronomy.

More recently Kikuchi and Soffer (7) have claimed that both H_1 and H_2 expressions are limiting cases of a single one, and the choice between them depends on a parameter n/z, n being the number of photons received and z their number of degrees of freedom. Their expression is based on a discussion of the entropy of the radiation field coming from the sky and the fact that photons obey Bose-Einstein statistics is central to their argument. They claim that if $n/z \gg 1$, which is true in radio astronomy, the H_1 expression is correct, whereas in optical astronomy where $n/z \ll 1$ the H_2 expression should be used. The trouble with this is that, in our language, they are saying that we should have a different prior prejudice about the shape of radio sources from the shape of optical sources, which is ludicrous. Surely we are interested in extracting information about the sky itself, not the radiation field. We cannot accept that the analysis of the shape of radio sources depends on the real quantized nature of light. We have used quanta to assist in the counting of states, as in elementary treatments of classical statistical mechanics, and have used this to derive H_2, but we then take the limit as the quanta become indefinitely small. The results of Kikuchi and Soffer seem to imply that, by studying the relative efficiency of the H_1 and H_2 algorithms on the restoration of a photograph, one could determine Planck's constant, or at least its order of magnitude.

6. PRACTICALITIES

So much for theory, can it be done in practice? Yes!

The H_1 algorithm has been used by Wernecke and D'Addario (5) and the H_2 by Gull and Daniell (4), on real astronomical data. In aperture synthesis the latter implementation takes 5 minutes on an IBM 370/165 for a 128 x 128 point image and a program for 256 x 256 points is working. For deconvolution there exists a working program for 128 x 128 pixels and the results are significantly better than the ART algorithm. Gull and Daniell have also successfully applied their method to aperture synthesis without phase information, and to VLBI data.

7. CONCLUSIONS

1. The maximum entropy method can do what its inventors hoped, it is a fundamentally sound, but different, way of looking at data, derived from Bayes' theorem and rooted in the foundations of scientific methodology.

2. It can be used for any problem for which one can predict the data that would be observed from a trial map.

3. It provides an objective, uniquely defined, procedure for analysing data and therefore runs counter to the trend for interactive data processing.

REFERENCES

1. Jeffreys, H. Theory of Probability, 3rd edition, Oxford 1961.
2. Jaynes, E.T. Probability Theory in Science and Engineering, Field Research Laboratory, Socony Mobil Oil Co., Inc. 1959.
3. Burg, J.P. Maximum Entropy Spectral Analysis, presented at the 37th meeting of the Society of Exploration Geophysicists, Oklahoma City 1967.
4. Gull, S.F., and Daniell, G.J. Nature, 272, 686 1978.
5. Wernecke, S.J., and D'Addario, L.R. IEEE Trans. on Computers, C26, 351 1977.
6. Bartlett, M.S. An Introduction to Stochastic Processes, 2nd Edition, Cambridge University Press 1966.
7. Kikuchi, R., and Soffer, B.H. J. Opt. Soc. Am., 67, 1656 1977.

DISCUSSION

Comment C. VAN SCHOONEVELD.
1) What is the difference between the results when the H_1- and H_2-definition are applied to the same data? 2) How can we expect a resolution improvement if we start from the a-priori assumption of a most uniform sky?
Reply S.F. GULL.
1) We have not made calculations using H_1, but with nearly complete and accurate data the choice of prior distributions has comparatively little effect and both H_1 and H_2 will give similar results. (See also comment to question by CRANE.) 2) For accurate data a modest improvement results (never more than a factor of 2). This arises from the positivity and

from the fact that ungraded Fourier data can be used without creating sidelobes.

Comment R. GORDON.
In what sense is Max.Entropy the spatially smoothest solution?
Reply S.F. GULL.
In no sense whatsoever. The Max.Entropy solution gives the statistically most uniform map, in the sense that it has the most uniform histogram of flux values.

Comment L.R. D'ADDARIO
The sense in which Max.Entropy maps are "smooth" is made precise by a theorem due to Wernecke (1977, Radio Science, 12, 831-844), which says that given the ME-map with measurement discrepancy D, any linear smoothing of the map results in a new map with a discrepancy $\geq$ D.

Comment Y.G. BIRAUD.
1) Analogous work was done by R. Herschel and B.R. Frieden. 2) What is the importance of the pixel size? 3) What is the gain in resolution versus S/N?
Reply S.F. GULL.
2) Decreasing the pixelsize and increasing their number improves the detail visible on the map until the limit set by the signal-to-noise ratio. After that, smaller pixels simply interpolate smoothly between their neighbours. 3) We made quantitative tests for the deconvolution case. Until a signal-to-noise ratio of 3:1 is reached (per pixel), the MEM gives a smoother result than conventional analysis. Beyond that ratio the method gives increasing resolution until a super-resolution of a factor 2 is reached at S/N $\approx$ 100. After that, virtually no improvement is possible.

Comment P.C. CRANE.
Two versions of Wernecke and D'Addario's algorithm operating at NRAO, one implementing H_1 and the other H_2, give, in one case, results identical to $\lesssim$ 10%.
Reply S.F. GULL.
I am not at all surprised. For the radio astronomical case the data are often extremely good. The posterior probability distribution is then largely determined by the data, not by the prior. It is when only a small amount of noisy data is available that the influence of the prior distribution will be seen.

Comment U.J. SCHWARZ.
How stable is the MEM solution against the choice of λ?
Reply S.F. GULL.
Very stable. The quantity χ^2 decreases as λ is increased but the map does not change much unless χ^2 is very different from its expected value.

THE USE OF BAYESIAN STATISTICS IN IMAGE ESTIMATION FROM INTERFEROMETER DATA

T. J. Cornwell
Nuffield Radio Astronomy Laboratories,
Jodrell Bank, Macclesfield, Cheshire.

1. INTRODUCTION

The problem of estimating radio sky brightness distributions from incomplete, and noisy, visibility measurements, such as those collected by a long baseline interferometer, has recently been attacked using non-linear data-adaptive techniques such as the Maximum Entropy method (Ables (1974), Wernecke (1976), Wernecke and D'Addario (1977), Gull and Daniell (1978)) and the Maximum Likelihood method (Papadopoulos (1975)).

In general, there are many brightness distributions consistent with the measured visibilities (Bracewell (1956)). The Maximum entropy method selects, as its estimate, the brightness distribution which has maximum entropy and is consistent with the data. It has been interpreted as the most objective way of assigning the missing values of the visibility (Ables (1973), Ponsonby (1973)), and as selecting the most probable image given the Bose-Einstein statistics of photons (Kikuchi and Soffer (1977)). The second interpretation has the advantage that the two entropy measures, in current use namely log f and $-$f log f where f is the brightness (Wernecke (1976)), can be derived as limiting cases of a more general expression (Kikuchi and Soffer (1977)).

The Maximum Entropy equations possess no known closed-form solution in two dimensions and hence they must be solved numerically using the techniques of constrained optimization (Wernecke and D'Addario (1977)). The value of the Lagrange multiplier used in this method can only be estimated and hence the solution is approximate.

A method of extending the statistical interpretation of the Maximum Entropy method advanced by Kikuchi and Soffer (1977) is presented here and it is shown that a definite value can be assigned to the Lagrange multiplier introduced by Wernecke and D'Addario (1976). This method involves the use of Bayes' rule connecting conditional probabilities (Hoel (1947)).

We discuss first the Maximum Likelihood method and the CLEAN algorithm.

C. van Schooneveld (ed.), Image Formation from Coherence Functions in Astronomy, 227–234.

2. THE MAXIMUM LIKELIHOOD METHOD

We wish to estimate the sky brightness f(x,y) within a region of sky described by orthogonal coordinates x,y. We have M samples of the measured visibility

$$m_i = m(u_i, v_i) \qquad i = 1,2,\ldots M \tag{1}$$

The measured visibility is related to the real visibility $\hat{m}_i$ by the addition of an error term

$$m_i = \hat{m}_i + \varepsilon_i \qquad i = 1,2,\ldots M \tag{2}$$

where

$$\hat{m}_i = \int_D f(x,y) \exp\{-j.\, 2\pi.\, (u_i x + v_i y)\}\, dxdy \tag{3}$$

We will approximate this integral by a sum over N pixels of area $\Delta A = \Delta x.\Delta y$

$$\hat{m}_i = \Delta A \sum_{k=1}^{N} f_k \exp\{-j.2\pi.(u_i x_k + v_i y_k)\} \tag{4}$$

The pixel area ΔA must be chosen to be less than the resolution of the interferometer in order that this approximation can fit the observed visibilities.

For convenience we will use vector notation for all variables.

$$\underline{x} = \{x_i : i=1,\ldots \text{ (no. of elements in } \underline{x})\} \tag{5}$$

We assume that the error terms ε_i are uncorrelated, zero mean, complex Gaussian distribution variables with σ_i^2 on the ith sample point. Therefore the probability of the vector of error terms $\underline{\varepsilon}$ lying between $\underline{\varepsilon}'$ and $\underline{\varepsilon} + \underline{d\varepsilon}'$ is

$$P(\underline{\varepsilon}')\underline{d\varepsilon}' = (2\pi)^{-M} \prod_{i=1}^{M} \sigma_i^{-2} \exp \frac{-1}{2} \sum_{i=1}^{M} \frac{|\varepsilon'_i|^2}{\sigma_i^2}\, d\varepsilon' \tag{6}$$

The function $P(\underline{\varepsilon}')$ is the probability density function of the error vector $\underline{\varepsilon}'$. from equation (2) we can see that

$$\underline{\varepsilon}' = \underline{m} - \hat{\underline{m}} \tag{7}$$

and therefore $P(\underline{\varepsilon})\ \underline{d\varepsilon}'$ tells us the probability of measuring the visibility between $\underline{m}$ and $\underline{m} + \underline{dm}$ given that the true visibility is $\hat{\underline{m}}$. Thus

$$P(\underline{\varepsilon}') = P(\underline{m}|\hat{\underline{m}}) \tag{8}$$

is the conditional probability density function of the observed visibility vector $\underline{m}$ given that the true visibility vector is $\hat{\underline{m}}$. Since the true visibility $\hat{\underline{m}}$ can be calculated from the true brightness $\hat{\underline{f}}$ we have that $P(\underline{m}|\hat{\underline{m}}) \equiv P(\underline{m}|\hat{\underline{f}})$.

One obvious estimate of the brightness distribution is that which gives maximum probability of measuring the observed visibility vector $\underline{m}$. Formally, we maximise the Likelihood function given by $L = P(\underline{m}|\hat{\underline{f}})$, in fact this means that the true visibility $\hat{m}$ is fitted to the observed visibility $\underline{m}$ in the weighted least squares sense. Schwarz (1978) has shown that the CLEAN algorithm performs such a fit. However, this Maximum Likelihood estimate may have negative components, though usually only at the noise level.

A simple physical constraint is that the brightness must be positive. We may incorporate this constraint into the Maximum Likelihood estimate by the use of barrier functions (Adby and Dempster (1974). A barrier function can be used in optimization to restrict the range of possible solutions to within a subspace of the original space. For example to maintain positivity we may add a logarithmic term to the function to be maximised; this discriminates against solutions near zero and forbids negative solutions. Using this approach we find the restricted Maximum Likelihood estimate by requiring that

$$J = \alpha \sum_{k=1}^{N} \log_e \hat{f}_k + \log_e L \qquad \alpha << 1 \tag{9}$$

be maximised. For convenience, we have used the logarithm of L in equation (9); this produces the same estimate since the logarithm is a monotonically increasing function of argument. From equation (6) we have that

$$\log_e L = -\tfrac{1}{2} \sum_{i=1}^{M} \frac{|m_i - \hat{m}_i|^2}{\sigma_i^2} + \text{constant} \tag{10}$$

This estimate may be found using the iterative Maximum Entropy algorithm developed by Wernecke and D'Addario (1977) with their Lagrange multiplier set to $\Delta A(2\alpha)^{-1}$.

Therefore using a technique of constrained optimization, namely the use of barrier functions we may introduce a physical constraint, the positivity of brightness distributions.

We will now describe another way of introducing physical constraints.

3. THE BAYESIAN APPROACH

The philosophy of Bayesian statistics (Silvey (1970)) is to introduce a priori knowledge about the parameters to be estimated. We introduce a weighting function $Q(\hat{\underline{f}})$ which describes our degree of belief in the vector $\underline{\hat{f}}$ prior to measurement. This prior knowledge is transformed to posterior knowledge by the sampling distribution $P(\underline{m}|\hat{\underline{f}})$. Bayes' rule gives the conditional probability density function of the brightness vector $\hat{\underline{f}}$, given the measurements $\underline{m}$, to be

$$R(\hat{\underline{f}}|\underline{m}) = \frac{Q(\hat{\underline{f}})\ P(\underline{m}|\hat{\underline{f}})}{N(\underline{m})} \tag{11}$$

where

$$N(\underline{m}) = \int_{\text{all } \hat{\underline{f}}} Q(\hat{\underline{f}})\ P(\underline{m}|\hat{\underline{f}})\ \underline{df} \tag{12}$$

The presence of $N(\underline{m})$ in the denominator means that $Q(\hat{\underline{f}})$ does not have to be normalizable (Silvey 1970), i.e.

$$\int_{\text{all } \hat{\underline{f}}} Q(\hat{\underline{f}})\ \underline{df} \neq \text{finite}$$

This leads to an extension of the Maximum Likelihood method, we choose the brightness vector which is most probable given that the measured visibility is $\underline{m}$.

Therefore if we can find a suitable form for $Q(\hat{\underline{f}})$, which we will call the prior, then we can find an estimate based on the measurements and on the knowledge contained in $Q(\hat{f})$.

Jaynes (1968) has discussed two "objective" methods of defining priors in cases where we have very little knowledge about the parameters to be estimated. One of these, now known as Jaynes' Principle involves the maximisation of the entropy of $Q(\hat{\underline{f}})$ subject to certain observations constraints. This principle forms the basis of the Maximum Entropy method (Ables 1974) which therefore should be viewed as an attempt to find a prior distribution suitable for use in Bayes' rule.

However we have some physical knowledge about the brightness distribution which may be used to select a prior.

4. INTRODUCTION OF PHYSICAL KNOWLEDGE

(i) As stated above, we know that the brightness distribution must be positive. Using this fact only we require

$$\begin{aligned} Q(\hat{\underline{f}}) &= \text{constant if all } f_i > 0 \\ &= 0 \text{ or undefined otherwise} \end{aligned} \tag{14}$$

Since we wish to find the maximum of $R(\hat{\underline{f}}|\underline{m})$ we require that the derivatives of $Q(\underline{f})$ be calculable. By analogy with the barrier function mentioned above we can easily select a suitable function

$$Q(\hat{\underline{f}}) = \prod_{i=1}^{N} f_i^{\alpha} \quad \alpha << 1 \tag{15}$$

This is an example of an unnormalizable function. This prior can be made arbitrarily close to unity for any $\hat{\underline{f}} > \underline{0}$ by selecting α small enough.

We now choose as our estimate the vector $\hat{\underline{f}}_B$ which maximises $P(\hat{\underline{f}}|\underline{m})$, or more conveniently its logarithm.

Therefore

$$\frac{\partial}{\partial \underline{f}}\left[\left(\alpha \sum_{i=1}^{N} \log \hat{f}_i - \tfrac{1}{2} \sum_{k=1}^{M} \frac{|m_k - \hat{m}_k|^2}{\sigma_{k^2}}\right)\right]_{\hat{\underline{f}}=\hat{\underline{f}}_B} = 0 \tag{16}$$

Again this may be solved using the algorithm of Wernecke and D'Addario (1977).

(ii) We should recognize that we observe the brightness distribution by photons and therefore we should use Bose-Einstein statistics to select a prior.

Kikuchi and Soffer (1977) use these statistics to derive a form for $Q(\hat{f})$ valid in the radio-astronomical regime.

$$\log_e Q(\hat{\underline{f}}) = z \sum_{k=1}^{N} \log_e \hat{f}_k + \text{constant} \tag{17}$$

where z is the number of degrees of freedom for each photon. Kikuchi and Soffer argue that z is due to a spatial uncertainty about the origin of the photon within the pixel and to a temporal uncertainty about the time of arrival of the photon. This temporal uncertainty clearly depends upon the noise level, bandwidth and integration time and thus, in our formulation, must be included in the sample distribution $P(m|\hat{f})$. Then z is given by the number of resolution elements in a pixel, therefore we have that $z = 1$ since z cannot be fractional and the pixels are, by definition, not resolved.

Note that the prior has the same form as that in (i) and thus, the estimate may also be solved by the algorithm of Wernecke and D'Addario (1977) with their Lagrange multiplier set to $\Delta A/2$.

We have used strictly physical knowledge in cases (i) and (ii) and have obtained the most probable estimates given that knowledge. However another application of Bayes' rule is possible and this we discuss now.

5. INTRODUCTION OF HUMAN JUDGEMENT

We have a set of noisy, incomplete measurements of the visibility which is compatible with many different brightness distributions. In the above formulation we have selected the most probable, in some sense, of these as our estimate.

Another approach would be to reject the selection of one of these distributions as an estimate but instead to try and characterise this set of possible brightness distributions in a smaller set.

Let us select a property, A, of brightness distributions and devise a measure M_A of that property. For example, Baker (1978) has proposed the measure

$$M_s = \sum_{k=1}^{N} \hat{f}_k^{\,2} \qquad (18)$$

for the sharpness or contrast of an image. Then we postulate a prior dependent upon this measure; for example

$$Q(\hat{f}) = \exp(\beta M_A) \qquad (19)$$

Using the above Bayesian formulation and this prior we can define a most probable image $\hat{\underline{f}}_A$ which obeys the condition

$$\frac{\partial}{\partial \hat{\underline{f}}}\left[(\beta M_A - \tfrac{1}{2}\sum_{i=1}^{M}\frac{|m_i - m_i|^2}{\sigma_i^2})\right]_{\hat{\underline{f}}=\hat{\underline{f}}_A} = 0$$

Then for β positive we select the distribution with property A dominant and for β negative we select the distribution with the opposite of property A dominant.

If we use sharpness as property A then we find both the sharpest and the least sharp brightness distributions consistent with the measurements.

We could now use these two brightness distributions as the subset characterising the full set. The presence of a feature in both estimates would be strong evidence that such a feature was contained in the true brightness distribution.

Under this philosophy an all purpose map such as that described by Gull and Daniell (1978) would not be satisfactory, instead we would produce a number of maps by different algorithms and study this set instead of one member.

6. CONCLUSIONS

We have used Bayesian statistics in two different modes.

In the first we have extended the interpretation of the Maximum Entropy method made by Kikuchi and Soffer (1978). We acknowledge that we observe the radio source by photons and therefore we can apply Bose-Einstein statistics to select, as the most probable, one of the set of distributions consistent with the data. This Bayesian formulation has the advantage that no arbitrary constraints, such as Lagrange multipliers, are involved.

In the second mode we choose not to select one map as the best in some sense but to use several different maps to characterise the set of brightness distributions consistent with the data.

Although the two modes are different philosophically the map produced by using Bose-Einstein statistics can be regarded as belonging to the characterising subset of the second mode. However this second mode has the advantage that the final judgement is left to the person processing the visibility measurement.

ACKNOWLEDGEMENTS

The author is especially indebted to Dr. R.E. Spencer who introduced him to data adaptive image processing and who shared both theoretical and practical insights into this subject. Many other people at Jodrell Bank discussed aspects of map-making with the author, in particular, Drs. J.E.B. Ponsonby, P.N. Wilkinson, R.G. Conway and R.J. Davis. Finally the author would like to thank Prof. J.G. Davies and Dr. J.E.B. Ponsonby for their supervision, the S.R.C. for financial support and Sir Bernard Lovell for the use of the facilities at Jodrell Bank.

REFERENCES

Ables J. G. 1974. Astron and Astrophys. Suppl. 15, 383.
Adby P.R. and Dempster M.A.H. 1974. "Introduction to Optimization Methods", Chapman and Hall, London.
Baker P.L. 1978. Astron. and Astrophys. to be published.
Bracewell R.N. 1956. Aust.J.Phys. 9, 297.
Gull S. F. and Daniell G.J. 1978. Nature 272, 686-690.
Hoel P.G. 1947. "Introduction to Mathematical Statistics", John Wiley and Sons, London.
Jaynes E.T. 1968. I.E.E.E. Trans. SSC-4 227-241.
Kikuchi R. and Soffer B.H. 1977, J.Opt.Soc.Am., 67, 1656-1657.
Papadopoulos G., 1975, Trans. I.E.E.E. AP-23, 45.
Ponsonby J.E.B., 1973. M.N.R.A.S., 163, 369.
Schwarz J., 1978, Astron. and Astrophys. 65, 345.

Silvey S. D. 1970. "Statistical Inference", Chapman Hall London (1970).
Wernecke S.J., 1976, Ph.D. Thesis, Stanford Univ.
Wernecke S.J., 1977, Rad.Sci., 12, 831.
Wernecke S.J. and D'Addario L.R., 1977, Trans. I.E.E.E., C-26, 351.

PROBLEMS WITH MAXIMUM ENTROPY

J.E.B. Ponsonby
University of Manchester
Jodrell Bank, United Kingdom.

ABSTRACT

Two rival expressions for entropy have been used in maximum entropy image formation. Attention is drawn to the fact that both have been shown to be limiting cases of a general expression for the entropy of the photon distribution swept up during an observation. All measures of entropy are based on some physical model of the radiation mechanism. It is argued that knowledge of its character should be fully included in the image formation process.

C. van Schooneveld (ed.), Image Formation from Coherence Functions in Astronomy, 235.

THE INTRODUCTION OF A PRIORI KNOWLEDGE IN CERTAIN PROCESSING ALGORITHMS

J.A. Högbom
Stockholm Observatory

A common problem in radio synthesis work is that of determining the brightness at N grid points in the map domain when there are only $n < N$ independent interferometer measurements available. The missing $(N-n)$ equations can in principle be replaced by an equivalent amount of information in the form of a priori knowledge about the brightness distribution. One way of doing this is to add an equation of the form

$$F(\text{map}) = \max \,, \tag{1}$$

i.e. some function of the brightness distribution is maximized subject to the condition that the map be compatible with the n measurements. This automatically gives $(N-n)$ new equations, leaving us in the pleasant situation of having as many equations as there are unknowns.

The expression (1) is very general and potentially able to absorb all kinds of a priori knowledge that we care to introduce. On the other hand, this generality is inconvenient when designing efficient processing algorithms. Several algorithms in current use are instead based on a simplified version of (1):

$$\Sigma f(x_i) = \max \,, \tag{2}$$

the sum to be taken over all the values x_i of the brightness at the N grid points of the map. The function $f(x)$ is to be specified in such a way that it becomes an expression of the a priori information. The left hand side of (2) is a function of the individual values x_i but not of their spatial distribution over the map. Hence, this simple formulation cannot be used to express a priori information about spatial relations between features on the map, but it does permit the use of information related to the statistics of the values x_i.

The argument has been advanced that the function $f(x)$ should relate to what may be termed the 'entropy' of the map (Ables 1974, Gull and Daniell 1978). The resulting maximum entropy map becomes, in a certain sense, that which contains the least amount of information -

C. van Schooneveld (ed.), Image Formation from Coherence Functions in Astronomy, 237-239.

and hence presumably the least amount of false information - among all possible maps that are compatible with the measurements. It has even been claimed at this symposium that, of all possible inversion procedures, the maximum entropy method is fundamental and unique in some deep philosophical sense.

There are two main competing views as to what should be the form of the function $f(x)$ in this context. They are:

$$f(x) = \log(x), \quad \text{and} \tag{3}$$
$$f(x) = -x \log(x) \tag{4}$$

At first sight these two functions appear to be very different (Figure 1). If one is fundamental and unique, then one would expect the other to perform very badly in comparison. However, when confronted with real data, both alternatives do quite well.

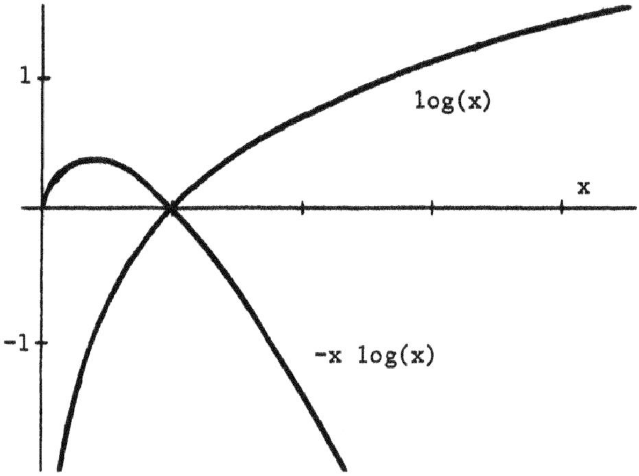

Figure 1

The functions log(x) and -x log(x)

A good expression for the entropy of a radio brightness distribution - if such a thing exists - is not necessarily identical to one that applies to idealized distributions encountered in physics or mathematics. However, there may well be similarities. Thus one should concentrate on the similarities between (3) and (4) in order to find out what makes these functions work, rather than debate which of the two is the one and only expression for the map entropy.

Both functions are real for $x > 0$ and their first derivatives go to infinity as x approaches zero. This represents a convenient way of introducing the positive constraint, the a priori knowledge that brightness distributions are exclusively positive. The progressive steepening as x approaches zero means that eq(2) will prevent map amplitudes x_i from becoming zero or negative.

The two functions are convex, their second derivatives are negative everywhere. This has the effect of biasing the map in favour of a uniform distribution. Any unnecessary dispersion of the amplitudes x_i about their mean value (assuming this to be specified in advance) does not satisfy eq(2) since it will lower the entropy of the map. Gull has pointed out that the form $f(x) = -x \log(x)$ will give maps over which the relative amplitude errors tend to be constant. This is a desirable property when one wants to find out what radio features are present in the mapped region of sky. It is less desirable if one wants to derive accurate structures and intensities of those features. Special precautions are needed in order to avoid overinterpretation of noisy data in low brightness areas.

As noted above, eq(2) cannot be used to introduce a priori knowledge about the spatial distribution of the values x_i over the map. Such information must be provided by the n measurements. In interferometry these specify certain spatial frequency components that are present in the brightness distribution. Some dispersion of the x_i values is required to accomodate these components but, since any unnecessary dispersion is discouraged by eq(2), large amplitudes will occur where they most effectively contribute to satisfying the measurements, i.e. where sources are likely to be situated.

The question of what is the 'best' map that is compatible with the measurements and the a priori knowledge has different answers depending on what one wants to do with the map. The maximum entropy approach is very valuable because it has demonstrated the potentialities of the simple approach represented by eq(2). The 'entropy' functions suggested so far may be regarded as examples of a class of functions $f(x)$ which allows the user some freedom in matching the general properties of the map to the type of information he wants to extract from his measurements.

REFERENCES

Ables, J.G.: 1974, "Astron. Astrophys. Suppl." 15, p. 383.
Gull, S.F., Daniell, G.J.: 1978, "Nature" 272, p. 686.

EXTENDING THE FOURIER TRANSFORM - THE POSITIVITY CONSTRAINT

M.M. Komesaroff and I. Lerche†
Division of Radiophysics, CSIRO, Sydney, Australia
†Dept. of Physics, University of Chicago, Chicago,
Il. 60637, U.S.A.

1. INTRODUCTION

In radio astronomy it is often necessary to estimate a brightness distribution from a limited number of samples of its Fourier transform. The manifest requirement that the brightness distribution be everywhere positive imposes definite constraints on its Fourier transform which yield information about unmeasured Fourier components. Here we discuss the question: given the first n+1 values, ρ_0, ρ_1 ... ρ_n, of a uniformly sampled Fourier transform of a real positive function, what can we say about Fourier terms of higher order?

We first show that ρ_{n+1} must lie within a circle in the complex plane, whose radius and centre can be calculated from ρ_0, ρ_1, ... ρ_n. For a value of ρ_{n+1} outside this circle there is no positive function whose Fourier transform passes through the values ρ_0, ρ_1, ... ρ_{n+1}.

The known values ρ_0, ρ_1, ... ρ_n constrain higher-order terms to successively larger, calculable, areas in the complex plane. If these areas are small enough, we may decide that ρ_{n+1}, ... ρ_{n+m} are determined to sufficient accuracy and it is therefore unnecessary to measure them.

Now in view of the residual uncertainty in each ρ_{n+m} ($m \geqslant 1$), how do we select a value for each within its "allowable" range? It is shown in Section 3 that <u>any</u> choice of ρ_{n+1} places additional constraints on all higher-order terms but if we estimate ρ_{n+1} by selecting the centre of its "circle of constraint" this imposes the minimal additional constraint on ρ_{n+m} ($m \geqslant 2$). The value of ρ_{n+2} then lies within a calculable circle, and if we select the centre of this circle as our estimate of ρ_{n+2}, once again the additional constraints on higher terms are minimized. The process can be iterated indefinitely. This prescription provides the greatest likelihood that a measurement of ρ_{n+m} (where m is an arbitrarily chosen integer $\geqslant 2$) will fit with the terms of lower order to form an acceptable auto-correlation sequence. The procedure is illustrated numerically in Section 4. In Section 5 it is shown that choosing the centre of the circle of constraint for each successive higher-order term leads to the maximum entropy solution.

C. van Schooneveld (ed.), Image Formation from Coherence Functions in Astronomy, 241-247.

2. THE FOURIER TRANSFORM OF A POSITIVE FUNCTION

If $F(x)$ is a real positive function we may write

$$F(x) = f(x) \cdot f^*(x) \equiv |f(x)|^2 .$$

Then the Fourier transform of F is the auto-correlation of g, the Fourier transform of f. If we are interested in $F(x)$ for only those values of x within the range $-x_0/2 \leq x \leq x_0/2$ the auto-correlation function need be specified only at discrete intervals yielding an auto-correlation sequence with mth term

$$\rho_m = \sum_{k=-\infty}^{\infty} g_k \, g^*_{k-m}$$

where g_ℓ is the discrete Fourier transform of $f(x)$. This expression can be regarded as the scalar product of two complex vectors. Consider a set of unit vectors such that the jth vector, V_j, has as its ℓth component $g_{j+\ell}$. It can be shown that the mth term of the auto-correlation sequence may be written as the scalar product of the jth and (j-m)th vectors

$$\rho_m = \bar{V}_j \cdot \bar{V}^*_{j-m} . \tag{1}$$

Since the vectors $\bar{V}_j$ are unit vectors, $\rho_0 = 1$. In fact it may be shown that the numbers ρ_0, ρ_1, ... ρ_n are successive members of a normalized auto-correlation sequence if and only if a set of n+1 unit vectors exists such that (1) is satisfied for all j and m.

Without loss of generality we may write

$$\bar{V}_j = \sum_{k=1}^{n} \alpha_{k,n} \bar{V}_{j-k} + \beta_n \bar{\varepsilon}_0 , \tag{2}$$

where β_n and the $\alpha_{k,n}$ are constants and the unit vector $\bar{\varepsilon}_0$ is such that

$$\bar{\varepsilon}_0 \cdot \bar{V}^*_{j-p} = 0 , \quad 1 \leq p \leq n . \tag{3}$$

It then follows that

$$\bar{V}_j \cdot \bar{V}^*_{j-p} = \sum_{k=1}^{n} \alpha_{k,n} \bar{V}_{j-k} \cdot \bar{V}^*_{j-p} + \beta_n^2 \delta_{p,0} , \quad 0 \leq p \leq n .$$

In view of equation (1) these may be written

$$\rho_p - \sum_{k=1}^{n} \alpha_{k,n} \rho_{p-k} = \beta_n^2 \delta_{p,0} , \quad 0 \leq p \leq n . \tag{4}$$

Knowing the values of $\rho_0(\equiv 1)$, ρ_1, ρ_2 ... ρ_n, we may solve the n+1 equations represented by (4) for the n+1 unknowns β_n^2, and $\alpha_{k,n}$, $(1 \leq k \leq n)$, yielding values which are independent of j. Accordingly, in place of equations (2) and (3), we may write

$$\bar{V}_{j+1} - \sum_{k=1}^{n} \alpha_{k,n} \bar{V}_{j+1-k} = \beta_n \bar{\varepsilon}'_1 , \tag{5}$$

where

$$\bar{\varepsilon}_1' \cdot \bar{V}^*_{j+1-p} = 0 \;, \quad 1 \leqslant p \leqslant n \; . \tag{6}$$

The unit vector $\bar{\varepsilon}_1'$ may be written quite generally

$$\bar{\varepsilon}_1' = \lambda_{1,1}\, \bar{\eta} + \lambda_{1,2}\, \bar{\varepsilon}_1 \;, \tag{7}$$

where the unit vector $\bar{\eta}$ is a linear combination of the unit vectors $\bar{V}_j$, $\bar{V}_{j-1}$ ... $\bar{V}_{j-n}$, such that

$$\bar{\eta} \cdot \bar{V}^*_{j+1-k} = 0 \;, \quad 1 \leqslant k \leqslant n \; .$$

This defines $\bar{\eta}$ uniquely with

$$\bar{\eta} = \left\{ \bar{V}_{j-n} - \sum_{k=1}^{n} \alpha^*_{k,n}\, \bar{V}_{j-n+k} \right\} \beta_n^{-1} \; . \tag{8}$$

The unit vector $\bar{\varepsilon}_1$ satisfies the relation

$$\bar{\varepsilon}_1 \cdot \bar{V}^*_{j+1-q} = 0 \;, \quad 1 \leqslant q \leqslant n+1 \; ;$$

and the numbers $\lambda_{1,1}$ and $\lambda_{1,2}$ satisfy $|\lambda_{1,1}|^2 + |\lambda_{1,2}|^2 = 1$. From equations (5), (7) and (8)

$$\bar{V}_{j+1} - \sum_{k=1}^{n} \alpha_{k,n}\, \bar{V}_{j+1-k} = \lambda_{1,1} \left\{ \bar{V}_{j-n} - \sum_{k=1}^{n} \alpha^*_{k,n}\, \bar{V}_{j-n+k} \right\} + \lambda_{1,2}\, \beta_n\, \bar{\varepsilon}_1 \; . \tag{9}$$

Taking the scalar product of (9) with $\bar{V}^*_{j-n}$, we obtain

$$\rho_{n+1} - \sum_{k=1}^{n} \alpha_{k,n}\, \rho_{n+1-k} = \lambda_{1,1}\, \beta_n^2 \; . \tag{10}$$

An essentially identical argument yields

$$\rho_{n+2} - \sum_{k=1}^{n} \alpha_{k,n}\, \rho_{n+2-k} - \lambda_{1,1} \left\{ \rho_1 - \sum_{k=1}^{n} \alpha^*_{k,n}\, \rho_k \right\}$$
$$= \lambda_{2,1}(1 - |\lambda_{1,1}|^2)\, \beta_n^2 \;, \quad |\lambda_{2,1}| \leqslant 1 \; . \tag{11}$$

Equations (10) and (11) define the constraints imposed on ρ_{n+1} and ρ_{n+2} by the known values of ρ_0, ρ_1 ... ρ_n. Any value of $\lambda_{1,1}$ (such that $|\lambda_{1,1}| \leqslant 1$) substituted into (10) yields an "acceptable" value of ρ_{n+1}, and any combination of $\lambda_{1,1}$ and $\lambda_{2,1}$ such that $|\lambda_{1,1}| \leqslant 1$ and $|\lambda_{2,1}| \leqslant 1$ yields an "acceptable" value of ρ_{n+2} when substituted into (11).

3. EXTRAPOLATING THE FOURIER TRANSFORM

The foregoing argument has shown that, given ρ_0, ρ_1 ... ρ_n, and hence $\alpha_{k,n}$ and β_n^2 from equation (4), we can write

$$\rho_{n+1} = \sum_{k=1}^{n} \alpha_{k,n}\, \rho_{n+1-k} + \lambda_{1,1}\, \beta_n^2 \tag{12}$$

$$\rho_{n+2} = \sum_{k=1}^{n} \alpha_{k,n}\, \rho_{n+2-k} + \lambda_{1,1}(\rho_1 - \sum_{k=1}^{n} \alpha^*_{k,n}\, \rho_{k+1})$$
$$+ \lambda_{2,1}(1-|\lambda_{1,1}|^2)\, \beta_n^2 \; ; \quad |\lambda_{1,1}| \text{ and } |\lambda_{2,1}| \leqslant 1 \, . \tag{13}$$

Equation (12) tells us that ρ_{n+1} lies somewhere within a circle in the complex plane, the centre of the circle being at $R_{n+1}(\equiv \sum_{k=1}^{n} \alpha_{k,n}\, \rho_{n+1-k})$ and its radius being β_n^2. The value of β_n^2 depends of course on the nature of the power spectrum, and it may be shown that β_n^2 either decreases or remains constant as n increases. Clearly the smaller β_n^2 the more accurately we know ρ_{n+1}.

In the absence of further information we have no reason to prefer any one point within the circle as representing the "true" value of ρ_{n+1}. However, consider the result of assigning to ρ_{n+1} its extreme value - i.e. consider the result of setting $|\lambda_{1,1}| = 1$, with the phase of $\lambda_{1,1}$ fixed. This constrains ρ_{n+1} to a particular, but arbitrary, value as can be seen from equation (12). Furthermore, from (13), ρ_{n+2} is then also determined as are all terms of higher order. If we had some independent way of determining the value of ρ_{n+10} (say) we have no reason to expect agreement with the value obtained by extrapolating from n with $|\lambda_{1,1}| = 1$.

On the other hand, if we set $\lambda_{1,1} = 0$, ρ_{n+2} is no longer completely constrained and ranges within values determined from equation (13). The range of values within which ρ_{n+2} can lie is largest when $\lambda_{1,1} = 0$. Similarly it can be shown that if $\lambda_{1,1} = 0$, and also $\lambda_{2,1} = 0$, the range of values within which ρ_{n+3} can lie is greatest. Thus: sequentially setting each term to the middle of its permitted range maximizes the possible range of all higher-order terms. If we had some independent method of determining ρ_{n+10} (say) this procedure provides our best chance that its value will lie within the range permitted by the lower terms.

Setting $\lambda_{1,1} = 0$ leads to the iterative algorithm

$$\rho_{n+m} = \sum_{k=1}^{n} \alpha_{k,n}\, \rho_{n+m-k} \, . \tag{14}$$

In Section 4 this procedure is illustrated.

In general, the constraints on ρ_{n+m} weaken as m increases. Thus it can be shown (Komesaroff and Lerche, 1978) that

$$\rho_{n+m} - \sum_{k=1}^{n} \alpha_{k,n}\, \rho_{n+m-k} = |\beta_n|^2\, S_{n+m} \, , \quad m \geqslant 1 \, , \tag{15a}$$

where S_{n+m} satisfies

$$S_{n+m} = \bar{\varepsilon}^*_n \cdot \varepsilon_m + \sum_{k=1}^{n} \alpha^*_{k,n}\, S_{n+m+k} \, , \quad m \geqslant 1 \, , \tag{15b}$$

where $\bar{\varepsilon}_n$ is an arbitrary complex unit vector. It can also be shown that

$$\rho_{n+m} - \sum_{k=1}^{n} \alpha^*_{k,n}\, \rho_{n+m+k} = |\beta_n|^2\, P_{n+m} \;, \quad m \geqslant 1 \;, \tag{16a}$$

where P_{n+m} satisfies

$$P_{n+m} = \bar{\varepsilon}^*_n \cdot \bar{\varepsilon}_m + \sum_{k=1}^{n} \alpha_{k,n}\, P_{n+m-k} \;, \quad m \geqslant 1 \;. \tag{16b}$$

Equations (15) and (16) can be solved (Komesaroff and Lerche, 1978) leading to expressions for S_{n+m} and P_{n+m} in terms of sums of the $\alpha_{k,n}$ and products of two ε factors. But see Komesaroff and Lerche (1978) for the general constraint conditions on ρ_{n+m}. Here we note that equations (15) and (16) can be used as the basis for recursive algorithms expressing the uncertainty in ρ_{n+m} in terms of the ρ_{n+m-k} $(n+m > k \geqslant 1)$. This point has already been anticipated by the illustration involving ρ_{n+1} and ρ_{n+2}.

4. A NUMERICAL EXAMPLE

An application of the procedure is illustrated in Figure 1. A non-negative function, sampled at 512 points, is illustrated in Figure 1(a). The Fourier transform has been calculated for $\rho_0 \dots \rho_{32}$ by the usual Fourier inversion technique and the values of $\rho_{33} \dots \rho_{512}$ estimated by the extrapolation method just outlined. Transforming the result yields Figure 1(b). In the case of Figure 1(c) the usual Fourier inversion technique has been used to calculate $\rho_0 \dots \rho_{32}$ and, in addition, $\rho_{64}, \rho_{96}, \rho_{128}, \rho_{160}, \rho_{192}, \rho_{224}, \rho_{256}$. The missing values have been estimated from the extrapolation method. For comparison, the effects of merely truncating the transform to the values indicated above are shown in Figures 1(d) and 1(e).

5. RELATION TO MAXIMUM ENTROPY

If we knew the ρ_m (all m) exactly then $Q(x) \left[\equiv \sum_{m=-\infty}^{+\infty} \rho_m \exp(2\pi i m x/x_0)\right]$ is, of course, just proportional to $F(x)$. But if we use relation (14) to estimate ρ_{m+n} $(m \geqslant 1)$, the resulting estimate for $Q(x)$ is equivalent to the maximum entropy spectral estimate. We show this as follows. Equation (14) is equivalent to demanding

$$\bar{\varepsilon}'_m \cdot \bar{\varepsilon}^*_0 = \delta_{m,0}\, \beta_n^2 \;, \tag{17}$$

where

$$\bar{\varepsilon}'_m = \bar{V}_{j+m} - \sum_{k=1}^{n} \alpha_{k,n}\, \bar{V}_{j+m-k} \;. \tag{18}$$

Introduce the following change in nomenclature

$$\gamma_k = \delta_{k,0} - \alpha_{k,n}(1-\delta_{k,0}), \quad 0 \leqslant k \leqslant n \;. \tag{19}$$

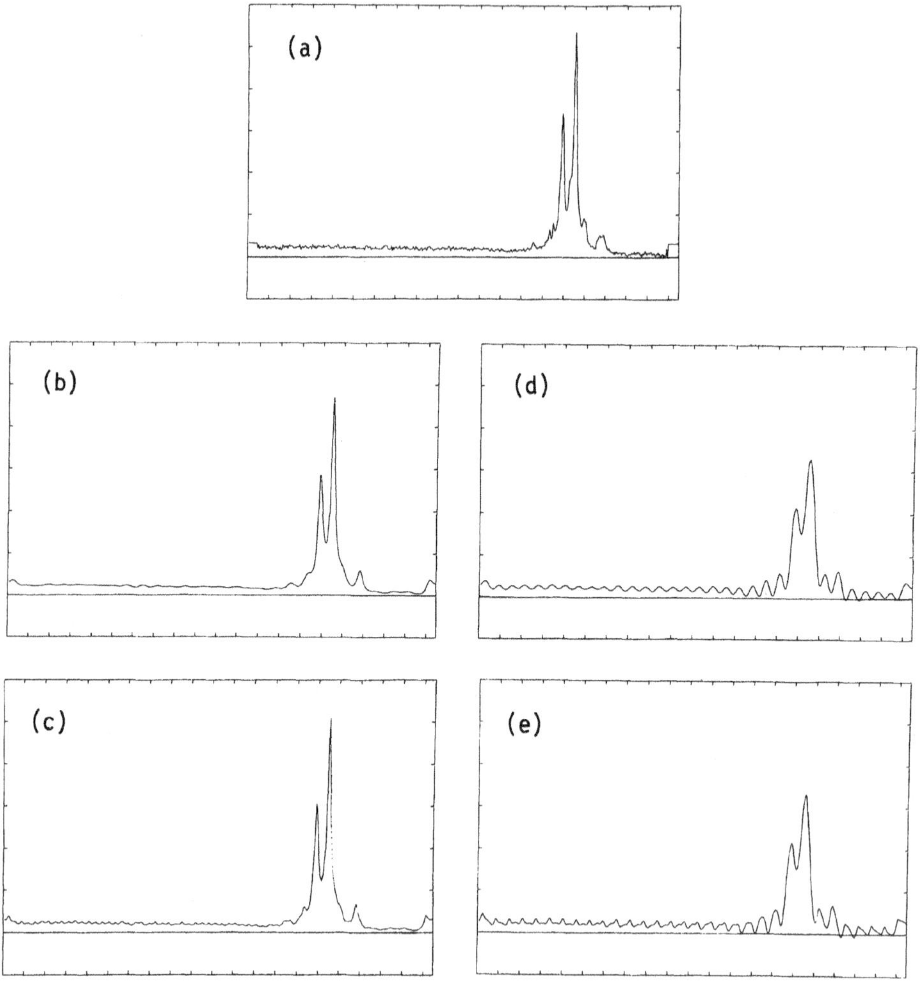

Fig. 1(a) - A 512-point H_2O spectrum measured at Parkes.

(b) - For the data of Figure 1(a) Fourier components $\rho_0 \ldots \rho_{32}$ were calculated by Fourier inversion, and the missing higher-order terms were estimated by the method outlined in the text. The result was then retransformed.

(c) - Fourier components $\rho_0 \ldots \rho_{32}$ and also ρ_{64}, ρ_{96}, ρ_{128}, ρ_{160}, ρ_{192}, ρ_{224}, ρ_{256} were calculated by Fourier inversion and the missing terms were estimated as in Figure 1(b).

(d) & (e) - Fourier transform truncated as for Figures 1(b) and (c) respectively and the missing terms set to zero.

Then

$$\sum_{k=0}^{n} \gamma_k \, \overline{V}_{j+m-k} = \beta_n \, \overline{\varepsilon}'_m \, , \qquad (20)$$

and thus from (17) and (18)

$$\sum_{m=-\infty}^{+\infty} \sum_{k=0}^{n} \gamma_k \sum_{\ell=0}^{n} \gamma^*_\ell \, \rho_{m-k+\ell} \exp(2\pi imx/x_0) = \beta_n^2 \, . \qquad (21)$$

With the replacement $M = m-k+\ell$ we have

$$\sum_{M=-\infty}^{+\infty} \rho_M \exp(2\pi iMx/x_0) = \beta_n^2 \left| \sum_{k=0}^{n} \gamma_k \exp(2\pi ikx/x_0) \right|^{-2} \qquad (22)$$

which result is formally identical with the maximum entropy algorithm.

Many authors have written on the subject of maximum entropy. In particular Burg (1967) and Van den Bos (1971) have discussed its connection with the positivity constraint. The papers of Ables (1974) and Newman (1977) are oriented towards its radio astronomy application.

Acknowledgments. Discussions with Dr. J.G. Ables and Dr. R.J.-M. Grognard have been extremely helpful. IL is grateful to Mr. H.C. Minnett, Acting Chief, and Dr. B.J. Robinson, Cosmic Group Leader, for the courtesies afforded him during the tenure of a Senior Visiting Scientist appointment at the Division of Radiophysics, CSIRO.

REFERENCES

Ables, J.: 1974, *Astron. Astrophys. Suppl.* 15, p. 383.

Burg, J.P.: 1967, *Maximum Entropy Spectral Analysis* - paper presented at 37th Annual International SEG Meeting, Oklahoma City.

Komesaroff, M.M., and Lerche, I.: 1978 - to be submitted to *J. Math. Phys.*

Newman, W.I.: 1977, *Astron. Astrophys.* 54, p. 369.

Van den Bos, A.: 1971, *IEEE Trans. Inf. Theory* 17, p. 493.

THE INFORMATION OF IMAGES AND ITS DEPENDENCE ON NOISE AND SPATIAL RESOLUTION

Peter M.W. Kalberla
Max-Planck-Institut für Radioastronomie, Bonn, F.R.G.

1. INTRODUCTION

Radioastronomical observations are often incomplete in the sense that either the angular resolution desired or the signal-to-noise ratio required are not adequate. Attempts to fill in the missing information by either smoothing or deconvolving the observed data date back to Bracewell (1956, 1958). More recently the need for more sophisticated restoration procedures has increased since observations with radioastronomical interferometers frequently suffer from incomplete coverage of the u - v plane. Procedures like CLEAN (Högbom 1974, Schwarz 1978) and MEM (Wernecke 1977, Gull and Daniell 1978) have been developed to interpolate the unknown data in the u - v plane.

All these methods do not explicitly take into account the change of the signal-to-noise ratio and the angular resolution of the observed map during the restoration process. In the present paper we point out that the information of any observed temperature distribution is critically dependent on these parameters. An analytical expression is presented which gives the information measure I of a map as a function of angular and temperature resolution. Further, the relevance of I for convolutions, deconvolutions and nonlinear restorations of interferometer maps is discussed. We find that the information measure I is a useful quantity to judge the quality of any of those restorations.

2. THE INFORMATION MEASURE I

Observing a temperature distribution T on the sky with a radiotelescope affects T in two ways. T is convolved with a beam function f and contamination by statistical errors which are characterized by a noise function n.

$$T_{obs} = f * T + n$$

For any interpretation of such a map only structure is of interest which is statistically significant if compared to the noise and angular reso-

C. van Schooneveld (ed.), Image Formation from Coherence Functions in Astronomy, 249-255.

lution. The information of a given map increases with signal-to-noise ratio, spatial resolution and observed area. Hence it is plausible that a measure for the information I contained in the map can be derived from the number of significant details. This number can be determined, if the observed solid angle A is measured in units of the beam solid angle

$$\Delta a = f_{max} / \int_{4\pi} f \, da \quad .$$

Then there are

$$m = A / \Delta a$$

locations in the map which may carry statistically independent details. Similarly, we measure the brightness temperature in units of the RMS noise level τ. Then the observed signal (assumed to be always positive) is decomposed into

$$l = \frac{\int_A T \, da}{\Delta a \quad \tau}$$

signal elements. The degrees of freedom m and l derived in this way can now be used to derive an expression for the information I contained in our map.

If P is the number of equally possible outcomes of an experiment, then the information I which is gained by observing one of these P cases is given by (Brillouin, 1963):

$$I = k \ln P \text{ (here we use } k = 1)$$

To derive P we count the total number of possible outcomes of the experiment by distributing the l undistinguishable signal elements over m degrees of freedom in position. From Bose - Einstein statistics we get

$$P = \frac{(m-1+l)!}{(m-1)!\,l!}$$

If m and l are large numbers we derive I by applying Stirlings formula

$$I = (m+l) \ln (m+l) - m \ln m - l \ln l \ .$$

After derivation of an analytical expression for the information content in a radioastronomical map we will study, how the information is changed by a manipulation of the data. For any reasonable transformation we require that the observed solid angle A and the total intensity $\int_A T_{obs} \, da$ remain unchanged. Thus the information I is merely influenced by changing the resolution limits Δa and τ. Both limits are affected in a complementary way: convolution increases Δa and decreases τ, deconvolution decreases Δa and increases τ. For linear transformations this complementarity is described by a relation $\Delta a \cdot \tau > \text{const}$ (Kalberla, 1978).

During any restoration procedure the information measure I changes in a well defined way. As an hypothesis we assume that I will become maximal for an optimal restoration. There is a simple example which shows that I indeed may be maximized. If the data are smoothed out completely by a convolution we obtain m = 1 and I ≑ 0. In the other extreme case of superresolution (Bracewell, 1958) the errors may be amplified such that they overcome the signal completely, thus 1 = 0 and I = 0 again.

3. COMPUTER TESTS.

To test, whether maximizing I eventually leads to an optimal restoration, some computer generated brightness temperature distributions have been degraded and restored. The fidelity of the restored distributions was measured by calculating the RMS deviation with respect to the model.

One dimensional test scans (delta functions, triangular, rectangular and gaussian distributions) have been convolved with a gaussian beam function, afterwards noise was added. In any case the scan length, the total intensity and the noise level have been adjusted in a way that for the corresponding degrees of freedom m >> 1. Then apparently a deconvolution is necessary to increase the information of the test scans. The restored scans T_{rest} have been calculated from (Kalberla, 1978)

$$T_{rest}(x) = q \cdot T_{obs}(x) - q \int_{scan} Q\ (x-x')\ T_{obs}(x')\ dx'$$

where

$$Q\ (x) = \lim_{n \to \infty} \sum_{i=1}^{n} (-qf)^{(i)}$$

is the resolving Kernel function. $f^{(i)}$ denotes i-fold convolution of f. The parameter q determines the strength of the restoration. The width Δa of the beam function decreases with increasing q, the amplification of the errors is proportional to q.

For any degraded model distribution a set of restored scans with different restoration parameters q has been calculated. For any of these restored scans the information I and the RMS deviation with respect to the model has been determined. Figure 1 shows two typical cases with different signal-to-noise ratios. The maximum of the information I is well correlated with the minimum of the RMS deviation between model and restored scan. Thus the information measure I is a useful quantity to determine the optimal restoration parameter q.

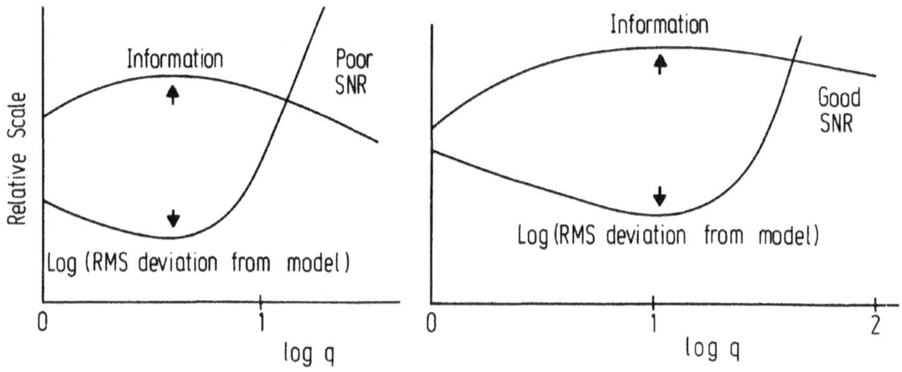

Fig. 1
Information I and RMS deviation between restored scan and model are plotted as a function of the restoration parameter q. The extrema of both curves are closely related. Two typical cases for poor and good signal-to-noise ratio are shown.

In a second set of computer tests aperture synthesis observations have been simulated. Figure 2 shows an example: A model for the brightness temperature distribution (A) was degraded by noise (B). After Fourier transformation into the u - v plane a subset of u - v data was erased. Figure 2 C shows the direct Fourier transform after eliminating 40% of the significant data. To obtain the restored map (D), the missing data have been replaced starting from the origin of the u - v plane proceeding to higher spatial frequencies. For this purpose a modified Gerchberg - Saxton algorithm (Gassmann, 1977 other references herein) was used. For any cycle of the iterative procedure the dirty map was convolved by a gaussian function. Then the remaining negative intensities have been forced to zero. From cycle to cycle the width of the smoothing function was decreased.

During this restoration procedure the RMS deviation between model and restored map was calculated. Simultaneously the information I was determined from the width of the smoothing function and the uncertainties which are introduced by the interpolation of the missing data points in the u - v plane. As for the one-dimensional test cases mentioned before maxima in I were found to correspond to minima in the RMS deviation between model and restored map. For some test models no extrema could be obtained but then both, information as well as RMS deviation, were found to converge.

These two cases can be distinguished by different m/1 ratios. A maximum in I was obtained if m/1 >> 1 which corresponds to typical aperture synthesis observations (e.g. for Westerbork maps

$m = \frac{A}{\Delta a} \approx 8000$ and 1 of the order of 1000).

Convergence was obtained for $m/1 \lesssim 1$ which in general means very high signal-to-noise ratio.

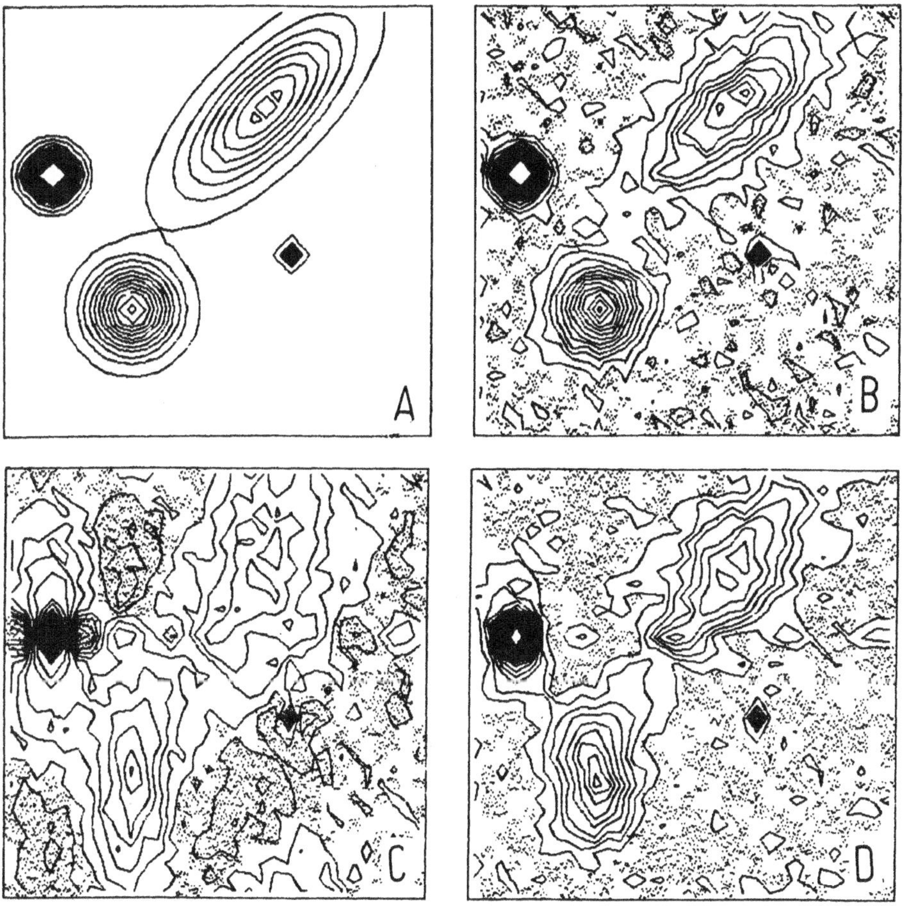

Fig. 2
Example for the restoration of a model distribution:
A) model B) model with noise, C) dirty map derived from an incomplete covered u - v plane, D) restored map.
Negative regions are shown shaded. The contour interval is constant.

The purpose of the computer tests was to study whether or not there is any relation between the maximum in the RMS deviation between model and restored map. The results obtained so far suggest that such a relation exists for our restoration procedure. Although no tests with other restoration procedures have been made we propose that such a relation should hold in general. Otherwise the relevance of the information measure I has to be reconsidered

For a practical use of the information measure I in restoring observations it should be possible to predict those restoration parameters for which I becomes maximal. In order to do this we have to know how the restoration procedure changes the spatial resolution Δa and the temperature uncertainties τ (including the errors which are introduced by any interpolation process). While the change in resolution Δa for our procedure can be derived from the smoothing function which is applied to the map we cannot calculate yet the amplification of the errors τ. This problem needs further investigations.

4. DISCUSSION

It was shown that the information content I of a radioastronomical map is critically dependent on the resolution limits Δa and τ. During an restoration procedure both, spatial resolution Δa as well as temperature resolution τ are changed. Consequently the information I of a map is affected, too.

To test the hypothesis that maximizing I leads to an optimal restoration, computer tests have been performed. The restoration used here is a special one in the sense that it allows to calculate the resolution limits Δa and τ at any stage of the procedure. This is necessary for a simultaneous calculation of I. Whether or not other restoration procedures, e.g. CLEAN or MEM - methods, allow also to calculate I during the iterations has not been investigated.

To calculate the information measure I we only need the total intensity within the observed solid angle A and the resolution limits Δa and τ. Then maximizing I is possible without introducing further constraints to the data. Thus the information measure I can be considered as a new independent criterion to judge the quality of a map. It would probably be worth comparing the restoration methods mentioned above by applying them to the same set of observations and calculating the resulting information measure I for the restored maps.

REFERENCES

Bracewell, R.N: 1956, Aust. J. Phys. 9, pp. 297

Bracewell, R.N: 1958, Proc. IRE 46, pp. 106

Brillouin, L.: 1963, "Science and Information Theory", Academic Press Inc., New York.

Gassmann, J.: 1977, Optik 48, pp. 347
Gull, S.T. and Daniell, G.J.: 1978, Nature 272, pp. 686
Högbom, J.A.: 1974, Astron. Astrophys. Suppl. 15, pp. 417
Kalberla, P.M.W.: 1978, Thesis, Bonn.
Schwarz, U.J.: 1978, Astron. Astrophys. 65, pp. 345.
Wernecke, S.J.: 1977, Radio Science 12, pp. 831

MAXIMUM ENTROPY SPECTRAL ANALYSIS - SOME COMMENTS

Rajendra Bhandari
Raman Research Institute
Bangalore, 560 006, India.

ABSTRACT

Study of a model data analysis situation with the help of computer experiments reveals that the Maximum Entropy Method of Spectral Analysis owes its popularity to a peak-sharpening property which is found to be a strong function of the level of background white noise present in the spectrum. In one limit, misleading results may be obtained by this technique. Some suggestions for a more fruitful use of the technique are made.

C. van Schooneveld (ed.), Image Formation from Coherence Functions in Astronomy, 257.

PART VI

OTHER IMAGE IMPROVEMENT METHODS

THE METHOD 'CLEAN' - USE, MISUSE AND VARIATIONS

(Invited paper)

U.J. Schwarz
Kapteyn Astronomical Institute
University of Groningen

SUMMARY

The mathematical background of the method CLEAN, as given by Schwarz (1978) is summarized. A method to derive source parameters (flux, position, extent) as a direct application of the theory is described. The ambiguity of the CLEAN method is discussed from a theoretical and practical point of view. Empirical results on the influence of the loop-gain are given. A new approach for the processing of maps with limited phase information is presented and a novel application of CLEAN for beam-switching observations (as an example for the case of a non-symmetric beam) is given.

1. INTRODUCTION

It is probably not necessary to give an outline of the algorithm of the method CLEAN, after its introduction by Högbom (Högbom, 1974) and first pioneering work by Rogstad (Rogstad and Shostak, 1971) eight years ago. The fact that it is still so widely used, speaks for the method or against the users.

In the first part I give a short summary of the mathematics of the method, as recently published (Schwarz, 1978, hereafter called Paper I), in order to introduce those concepts, on which the further discussion will be based. In the second part I give some direct applications of the theory. The third part is more empirically oriented, where some ambiguities and pitfalls are mentioned. In the final section some variations and other applications of the method are presented.

1.1 Algebraic Description and Convergence of CLEAN

Although the method CLEAN is conceived entirely in the map-domain, it is of great use to analyse the working of the method in the plane which is the Fourier transform of the map, the UV-plane. In order to do this one must first describe the method mathematically in the map-plane

C. van Schooneveld (ed.), Image Formation from Coherence Functions in Astronomy, 261-275.

and then as a second step interpret it in the UV-plane.

i) CLEAN in the map-plane: We can write the result of the CLEAN process (actually the first phase, namely the decomposition) after some iterations by

$$\underline{d} - B\,\underline{t}' = \underline{r} \tag{1}$$

where $\underline{d}$ is the original map (the 'dirty' map), $\underline{t}'$ the ensemble of components, $\underline{r}$ the residuals and B a matrix containing the values of the synthesized beam, $\underline{b}$, (the 'dirty' beam) as elements, such that the convolution of the components $\underline{t}'$ with the dirty beam can be written in this simple way. This notation in vector form does not mean that one is restricted to the one-dimensional case. The multi-dimensional case can be put exactly in this form as well.

The CLEAN process of searching for the maximum of the residuals and subtraction of g times (g = 'loop-gain') this value convolved with the dirty beam, $\underline{b}$, is nothing else than a well known iterative method from the pre-computer time to solve a system of linear equations

$$B\,\underline{t} = \underline{d}\,. \tag{2}$$

For our purpose this method is attractive since we can assume that the sky is almost empty; this aspect will be examined in more detail later. When discussing the convergence of the CLEAN process it is useful to introduce a norm Q. One can prove that the following norm,

$$Q = \underline{\Delta t}\; B\; \underline{\Delta t} \quad \text{with} \quad \underline{\Delta t} = \underline{t}' - \underline{t}\,, \tag{3}$$

diminishes with each iteration and hence converges to zero, if $0<g<2$ (g = loop-gain), B is symmetric and Q cannot become negative. This last condition can be fulfilled easily, as will be shown below. $\underline{t}$ is the solution of equation 2, therefore:

$$\underline{t}' \to \underline{t} = \underline{t}_o + \underline{z}_o\,. \tag{4}$$

The final solution is the 'true' map, $\underline{t}_o$, if eq. 2 has a unique solution. Normally in our applications this is not the case and any zero-eigenvector $\underline{z}_o$ ('ghost') can be added.

Since one does not know $\underline{t}$ in advance one can calculate another quantity H which is related to Q by

$$Q = Q_o - H \quad \text{with} \quad H = \underline{t}'(\underline{d} + \underline{r}), \quad Q_o = \underline{t}\;\underline{d}\,. \tag{5}$$

H increases with each iteration by

$$\Delta H = g\,(2 - g)\, r_m^2 \tag{6}$$

where g is the loop-gain and r_m the maximum residual. H is a measure of the quality of $\underline{t}'$; the starting value is 0 and can reach a maximum of Q_o, when $\underline{r} = 0$.

ii) Interpretation of CLEAN in the UV-plane: We define the visibility, V, as the FT of the 'true' map. This visibility is observed at some sample points and is given a weight. This sampling and weighting function is w. The FT of the weighting function is the dirty beam, b. The observed visibility which includes noise we call V_O. The FT of wV_O is the dirty map, d.

Starting the iteration process with a subtraction of the convolution of the 'dirty' beam with the first component, corresponds in the UV-plane to the subtraction of a sine-wave multiplied by w. After some iterations, the FT of the residuals r is left. This can be expressed as $w\Delta V = w(V_O - V')$, where V' is the FT of the set of components, t', and $\Delta V = V_O - V'$. Since $r \to 0$, $w\Delta V$ will converge to zero. This means that V' approaches V_O. This is illustrated in Figure 1, where the first iterations of a model example are shown; since the visibility is a complex quantity, the cosine- and sine-components are shown separately with subscripts c and s respectively.

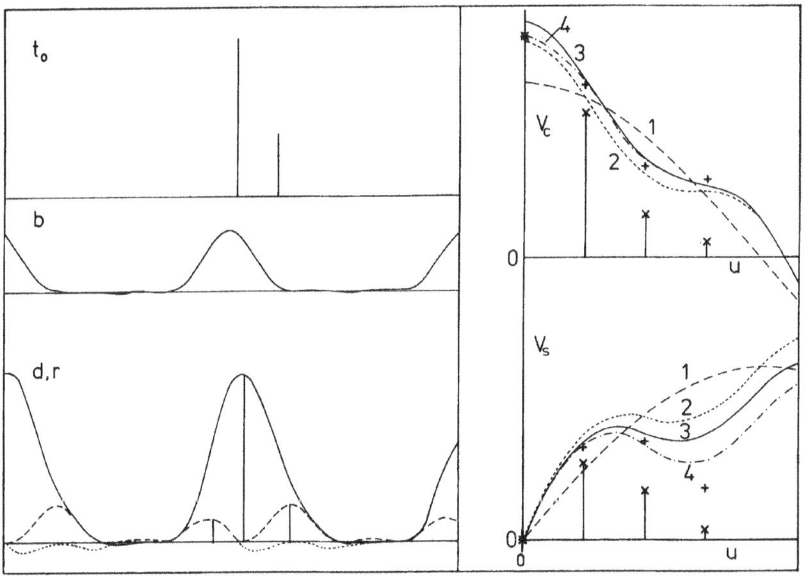

Figure 1. One-dimensional CLEAN of a double source. The first few iterations are displayed. The loop-gain is 1. Left is the map-plane, right the U-plane. Left, from top to bottom: t_O = the true map, b = the dirty beam, d,r = dirty map and residuals; the vertical bars indicate the first three components. Right at top and bottom the real and imaginary parts respectively of the following complex quantities: curves 1 to 4 = FT of the components, for the 4 first iterations. +: the visibility of the true map, at the sample points, V_O. ×: the weighted visibility wV_O. *: + and × coincide. The visibility of the components approaches closer and closer the visibility of the true map (but not the weighted ones) with increasing number of components.

Note that V' does not approach the weighted visibility $\underline{w}V_o$, but V_o, regardless of the amplitude of the applied weighting function $\underline{w}$. This shows that the decomposition in components is in fact a deconvolution.

The convergence of ΔV to zero can be described mathematically more precisely by using the norm Q introduced above. It can be shown that

$$Q \;=\; \Sigma w_k \;\Delta V_k^2 \;/\; \Sigma w_k \tag{7}$$

is the normalized sum of squares of the difference between the FT of the components V' and the observed visibility V_o. We see immediately that one of the criteria of convergence, $Q \geq 0$, is fulfilled if $w_k \geq 0$ for all k. Since V_o is affected by random noise, Q converges towards a minimum in a proper least square sense, if the weights are 'natural' weights, i.e. inversely proportional to the μ_k^{-2}, where μ_k is the rms-error of an observation.

We can also calculate Q_o in a simple way, which is the starting value of the norm Q or the maximum value of H (cf. eq. 5):

$$Q_o \;=\; \Sigma w_k \;\; V_{o_k}^2 \;/\; \Sigma w_k \tag{8}$$

We can use Q as a criterion to stop the iteration process, namely, when Q reaches the value which would be expected from random noise only. For natural weighting we get from eq. 7

$$\langle Q\rangle \;=\; \Sigma w_k \;\mu_k^2 \;/\; \Sigma w_k \;=\; \mu^2\; R/\Sigma w_k \;=\; R\sigma^2 \tag{9}$$

where μ is the noise of an observation per unit weight, and σ is the rms noise of a point in the map.

1.2 CLEAN as an Iterative Harmonic Analysis

Suppose that the true map consists of one point-like source. In harmonic analysis if one wanted to fit a sine-wave to the observed visibility one would FT the visibility and look for the largest harmonic, i.e. the maximum in the dirty map. This is idential to the CLEAN method. Therefore one could interpret the CLEAN method as an iterative harmonic analysis.

1.3 CLEAN as a Least Squares Fit

Normally the iteration process of CLEAN is stopped if the residuals become smaller than a given threshold. But in order to get the advantages of a proper least squares fitting, one has to deviate from the above scheme slightly and to distinguish between two phases:

i) The search phase: This is the usual iteration process, where the maximum residual is searched in the whole map or part of it.

ii) Optimization: The CLEAN process goes on, but the search of the maximum residuals is done only within the positions, which have already a component, say in W positions. In this way the residuals within this 'window' approach zero and one solves a subsystem of the eq. 2.

Outside the window the residuals are not necessarily zero. The optimization insures that one has fitted the FT of W components to the observed visibility in a proper least squares fit. Naturally the number of positions W may not exceed the number of observations R (R/2 complex quantities), otherwise the matrix B_W will be singular.

1.4 Error Calculation

Now we have a standard problem for a least squares solution. It involves the inversion of a matrix, which for the case of uniform taper is just B_W^{-1}, the inverse matrix of the submatrix of order W. One easily can calculate the errors of the components or of their FT. An example is shown in Figure 4 of Paper I. There one sees, that the uncertainty of the high spatial frequencies become very large. Thus an increase in resolution requires high S/N ratios! In order to cut down the uncertain high spatial frequencies, one normally convolves the components with a 'clean' beam.

In this summary, many aspects and problems of CLEAN are not mentioned, such as the question of the clean beam, adding the residuals to the clean map, etc. Those and details of the points above can be found in Paper I.

2. APPLICATIONS OF THE THEORY

2.1 Determination of Parameters of Point-like or Slightly Resolved Sources

There are several methods to determine the parameters of discrete sources. Most use a fitting procedure in some way. Clearly the best method is to fit some model function directly to the observed visibility, but this requires quite a large computational effort, at least for the two-dimensional case. Therefore the fitting is mostly done in the 'sky'-plane; but then one must correct the parameters found for the convolution effect due to the 'dirty beam'. An alternative approach is to 'clean' the sources and to determine the source parameters using the components, $\underline{t}$. These components are implicitly based on a least squares fit to the visibility as has been shown above; they are therefore free of the effects of the convolution by the dirty beam.

A simple straightforward method to determine the source parameters is to use the <u>moments</u> with respect to the x-y coordinates in the sky. The moments up to second order are:

$$\begin{aligned} M_o &= \Sigma_i t_i \\ M_x &= \bar{x} = \Sigma_i t_i x_i / M_o \\ M_y &= \bar{y} = \Sigma_i t_i y_i / M_o \\ M_{xx} &= \Sigma_i t_i (x_i - \bar{x})^2 / M_o \end{aligned} \tag{9}$$

$$M_{xy} = \Sigma_i t_i (x_i - \bar{x})(y_i - \bar{y})/M_o$$
$$M_{yy} = \Sigma_i t_i (y_i - \bar{y})^2/M_o \qquad (9)$$

The zeroth order moment gives the flux, the first moments give the position and the second moments allow one to calculate parameters of models of extended sources, such as a two-dimensional Gaussian or an equal point double. An important fact to note is that the second order moments need not to be positive; they can be zero or even negative. For a point-like source they are on the average zero but deviate from zero due to statistical fluctuations.

In order to calculate the mean errors of the parameters, derived from the above moments, one needs the derivatives with respect to the components, $\partial M/\partial t_i$. These derivatives are (van Dijk, 1977, priv. comm.):

$$\partial M_o/\partial t_i = 1 \qquad (10)$$
$$\partial M_x/\partial t_i = (x_k - \bar{x})M_o \qquad \partial M_y/\partial t_i = (y_k - \bar{y})/M_o$$
$$\partial M_{xx}/\partial t_i = ((x_k - \bar{x})^2 - M_{xx})/M_o \qquad \partial M_{yy}/\partial t_i = ((y_k - \bar{y})^2 - M_{yy})/M_o$$
$$\partial M_{xy}/\partial t_i = ((x_k - \bar{x})(y_k - \bar{y}) - M_{xy})/M_o$$

For instance the error of one of the second order moments becomes then

$$\mu^2(M_{xx}) = \sigma^2\Sigma(\partial M_{xx}/\partial t_i)(\partial M_{xx}/\partial t_k)B^{-1}_{Wik}. \qquad (11)$$

Error estimates for diameters, position angles, etc. can be calculated based on these errors in the moments. Extensive Monte Carlo tests have been made to check these errors and showed good agreement with prediction.

The method was tested with artificial sources and with real sources in synthesis maps made from Westerbork data. It was found, using typically 5 to 9 components, that for accurate positions a small area around the source must be chosen (approx. 3/2 beam diameters), but for accurate fluxes and the second moments the area should be larger (about twice as large). A compromise was found by applying a Gaussian weighting function (size about twice the beam width) to the components in order to calculate the positions. This yields typically flux accuracies of the order of 20% of the flux and in positions of 8% of the beamwidth for sources with a flux of about 10σ.

A different approach to overcome the difficulty of this 'trade-off' in accuracy between various parameters would be to use constraints on the minimization of the norm Q. The constraints can be that the second order and higher moments must agree with a model to be determined. For instance if one likes to get the flux and position of a source, assuming it to be a point-like source, then one would require that the second and higher moments are zero. Then one would get from CLEAN, components which are subject to these constraints. I have not worked out an algorithm or error analysis in CLEAN to solve this problem.

2.2 The Ambiguity of the Solution

It is obvious that if the sampling in the UV-plane is not complete there can be components of the brightness-distribution which we can never reconstruct (with any method!). The most simple example is a uniform background which will be missed completely if we do not measure the zero-spacing. The 'dirty' map contains no trace of it and in the UV-plane this contribution is contained in a single point, therefore no interpolation will help. But such a distribution contradicts our condition that the true brightness distribution is zero in most parts. We can ask the general question: Does some brightness distribution exist, which is confined in the map-plane, and whose FT is also confined in the UV-plane? This is not possible, since any confined distribution in the map can be multiplied by a box-function. In the UV-plane this corresponds to the convolution of the visibility with the FT of the box-function (in one dimension a sin x/x-function), which extends to infinity. Therefore the visibility of such a confined source must necessarily be spread over the whole UV-plane, and therefore it cannot be missed completely by the observations. This proof is not completely tight, since some brightness-distribution with periodities matched to periodities in the sampling function could escape unnoticed. These would be distributions of the type of grating responses and one can argue whether these do or do not satisfy the criterion of emptiness of the sky.

Apart from this case, it is then possible in principle to extrapolate the non-observed parts of the visibility assuming noise-free data. Can CLEAN achieve such an extrapolation? In general this will not be the case, at least in this form of CLEAN presented here, where the components are restricted to positions on a grid. However if the true map is restricted to grid-points too, then it is shown in Paper I that the solution is unique if $Q = 0$ and if the solution contains less than R/2 components (where R/2 is the number of measured complex visibilities).

The above arguments are all based on noise-free observations and hence are only of theoretical value. In presence of noise it is meaningless to make $Q = 0$, since the expectation value of Q for noise is non-zero, see eq. 9. From eq. 3 we know that if $Q > 0$ then $\underline{\Delta t}$, the difference between components and the true distribution, is not zero either. Can we make some estimates about the size of $\underline{\Delta t}$?

A simple approach is the following. How large can the total flux ΔF be of a distribution $\underline{\Delta t}$, which would produce a constant change in zero-level of the dirty map? Let us relate this zero-shift to the noise in the map, σ, say $\eta\sigma$, where η is a number smaller than one, then we get from eqs. 1 and 3

$$|B_W \underline{\Delta t}| = \eta\sigma \qquad (B_W \text{ is a submatrix of B}) \qquad (12)$$

and therefore

$$|\Delta F| = |\Sigma \Delta t_i| = \eta \sigma \Sigma B^{-1}_{W_{ij}} \ . \qquad (13)$$

For dirty beams with small sidelobe level,

$$\Sigma B^{-1}_{W_{ij}} \simeq W,$$

but for beams with larger sidelobes this sum can become considerably larger. B_W^{-1} is usually not calculated explicitly, but is required if an error analysis is made.

3. PROBLEMS OF THE PRACTICE

3.1 Influence of Loop-Gain

In the many years that the method CLEAN has been used I believe that mostly a loop-gain of less than unity has been found to be optimum. In fact a very small loop-gain can be practical. It is even possible to use an infinitesimal small loop-gain by an algorithm closely related to the theory of CLEAN, being equivalent of solving a system of equations.

A systematic analysis of the problem of loop-gain has been done by Ron Harten (yet unpublished) by a purely empirical method. He cleaned a noise free point-source lying exactly in between two gridpoints, using a different number of gridpoints per halfwidth of the dirty beam, and for various loop-gains (0.9, 0.5 and 0.25). He looked at speed of convergence accuracy of flux determination and area needed, for the case of a free running CLEAN, without optimization. I would like to discuss a few aspects of this work:

i) Area required: The smaller the loop-gain the smaller the area needed.

ii) Convergence: The closer the loop-gain to unity, the higher is the frequency of oscillation. The amplitudes are in the test cases about equal. In the beginning, after only a few iterations large deviations from the true flux do occur.

iii) Influence of number of points per halfwidth of dirty beam: The more points, the smaller the amplitude of oscillation.

It is quite plausible that a small loop-gain needs a smaller area. This we see more clearly for a very small loop-gain, then the dirty map is steadily cut-down from the top, making more and more residuals equal. If one has one dominant source, CLEAN with small loop-gain will start with the highest point and steadily use new points thereby going further from the centre of the source in both directions. This is not only an advantage for extended sources but also for point-like sources giving a more accurate final result.

Interesting experiments were made by J.M. van der Hulst (1977) in connection with the observations of neutral hydrogen on the 'antenna'-pair of galaxies, NGC 4038/39, which has a negative declination and can therefore be observed at Westerbork only for a limited range in hour-angle. This missing information results in strong negative sidelobes (34% at a distance of 1.25 the beamwidth) in the antenna pattern in the EW-direction. From the optical picture (and theoretical dynamical models of an interacting pair) one would expect radiation in a filament running partly E-W. Due to the characteristics of this particular beam, the amplitude of any elongated E-W structure is strongly reduced. He showed that CLEAN can only partly reconstruct the missing information, using a realistic model with superimposed noise.

Somewhat disappointed by this result, I started to make similar experiments, using the same data, but a more simple model, namely a thin bar lying exactly in an E-W direction which gives, from the point of view of CLEAN, the strongest effects. I systematically varied the amount of noise superimposed and the loop gain. The results from these experiments can be summarized as follows:

i) The ability of CLEAN to reconstruct the true distribution is strongly affected by the S/N ratio. There is a rather sharp transition, where CLEAN starts to fail.

ii) The S/N ratio at which the true distribution can be reconstructed showed in some cases to be dependent of the loop-gain. For a bar of constant amplitude (thus with sharp edges) a high loop-gain, up to 1.5, was more successful than a small loop-gain. On the other hand a Gaussian amplitude (smooth edges) distribution along the bar showed no great differences in the results for various loop-gains.

The conclusion is that a small loop-gain is in general advisable, especially if one has a dirty beam with significant positive sidelobes. Ambiguities in CLEAN, due to different loop-gains, number of iterations, etc. can occur at too low a S/N ratio.

3.2 Increase of Resolution

It is tempting to use (or misuse) the method CLEAN to achieve higher angular resolution than one would obtain by the usual techniques. I would like to distinguish two categories:

i) First category: One would like to extrapolate the visibility to higher spatial frequencies than are contained in the observation. In Paper I application of the error theory shows clearly, that this is only possible for very high signal-to-noise ratios. But even in such cases one should keep in mind, that non-statistical errors of the observations then can play an important role.

ii) Second category: In Fourier transforming the observations, one usually applies a weighting or taper function, in order to reduce the side-lobe level of the dirty beam, resulting in prettier looking maps. As already mentioned in Section 1.1-(ii) CLEAN does recover these artificially degraded high spatial frequencies. Ron Harten used CLEAN to increase the resolution of maps, where such a taper function has been applied. He could demonstrate that the undoing of the taper function is possible, but any further increase in resolution leads to unrealistic maps. - But this method looks rather a cumbersome way to begin with; this is however not true, since the cleaned map is restored with a clean beam, therefore one gets higher resolution and a pretty looking map.

3.3 Some Pitfalls

a) CLEAN of smoothed maps: There is no objection using smoothed maps, if the dirty beam is smoothed as well, and if the FT of the smoothing function is non-negative. This condition can sometimes be violated however, if the smoothing is done in the map (e.g. in a small region of interest to save computing time), since the truncation of the smoothing function can in fact introduce negative weights in the UV-plane, which will lead to divergence during the CLEAN process.

b) CLEAN of extended sources: A basic assumption we made in earlier sections was that the sky is essentially empty. Suppose we sampled the visibility including a zero-spacing, which picks up a general background. We could use CLEAN in a restricted area, keeping the number of points W in the window smaller than the number of observations, R. In this case we obviously cannot expect to reconstruct the true distribution correctly. On the other hand if we were to use the whole map for CLEAN (thus violating the condition $W < R$) strange things can happen. Figure 2 shows a case where an artificial map of constant amplitude has been cleaned with loop-gain $g = 1$. The result is a sinusoidal looking wave, a perfect artifact.

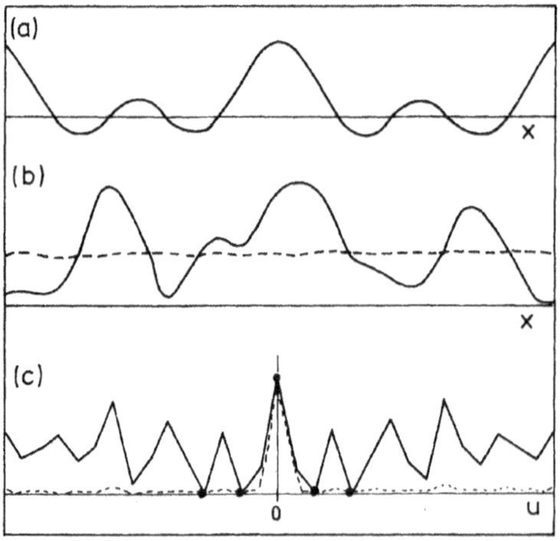

Figure 2. Experiments with a continuous brightness distribution in one dimension. (a) dirty beam, (b) clean map (the clean beam is equal to the dirty beam to the first zero), (c) FT of components. In (b) and (c): full lines, loop-gain = 1; dashed lines, loop-gain = 0.01, 1000 iterations. The dots give the assumed 'observation'. One sees that CLEAN with g = 1 gives rise to artifacts.

In the UV-domain, one sees that CLEAN has done its job perfectly well, the FT of the components - although wavy - equals exactly the 'observed' visibility at the positions of sampling. Another experiment was made, using a loop-gain g = 0.01, for 1000 iterations. The final residuals are at the 10% level of the dirty map. In Fig. 2 the components are shown by the broken line; one sees they are almost uniform. This demonstrates that a small loop-gain does not turn a smooth background into pronounced artifacts.

The above example gives a possible explanation of some features in a controversial map of the galactic centre made by Balick et al. (1974), based on observations with the NRAO interferometer. This map showed lobes which were not present in a more completely sampled map by Ekers et al. (1975). The odd orientation of the baseline of the NRAO interferometer, combined with the effect of low declination, results in projected short spacings, which are sensitive to the extended source. If one applies CLEAN to the map (what has been done by Balick et al.) in a restricted area, one can expect artifacts similar to those in above example, especially if a high enough loop-gain is used. Using the original data it can be shown that these lobes can be avoided with a small loop-gain.

The main question is how to avoid this effect? The most safe method is to clean an area with less points than the number of observations (W < R) and if an extended source may be present, which exceeds the size of the area to be cleaned, the corresponding short spacings should be excluded. The use of a small loop-gain helps to avoid difficulties as above experiments and also those of other users of CLEAN have shown.

4. VARIATIONS AND OTHER APPLICATIONS OF CLEAN

4.1 Limited Phase Information

In VLBI observations the phase cannot be measured absolutely, only relative phases. In recent years a new reduction technique has been developed, using the concept of closure phase. In simultaneous multi-interferometer observations phases must add up in some simple way to zero. Assuming some phases one can calculate the remaining phases from these conditions. The resulting map is then processed using CLEAN in an iterative way (Readhead et al., 1978). I do not intend to describe this method in detail, but I would like to make a remark related to the previous section. Due to the erroneous phase, the calculated 'true' source may extend over a large part of the map. If CLEAN is used in a restricted area one can also introduce artifacts by using a high loop-gain and by continuing the iteration-process too long.

I have tried a different approach to a similar problem, by using the CLEAN concept in its original form. Suppose one has observations with partial phase information. Putting the unknown phases to zero, one gets by Fourier transforming the data a dirty map. If we search this map for the maximum and subtract a beam at this position, we have to keep in mind, that due to the missing phase, the beam cannot simply be shifted to this position, but has also to be calculated for this particular position, setting the phase to zero at those sample points where no phase information is available, because in this case the shape of the beam depends on the position in the map. This is the only difference from the usual CLEAN. The analogy with the normal CLEAN is evident, because one also puts non-measured samples of the visibility to zero (amplitude and phase), whereas the described method puts only the phase to zero, since the amplitude is observed. As in the normal CLEAN the missing information is interpolated.

4.2 Other Applications of CLEAN

To my knowledge CLEAN has not been used much outside the field of radio synthesis observations. An example of another possible application in astronomy I would like to explain concerns the interpretation of beam-switching observations, used in radio-astronomy, but also extensively in a.o. infra-red observations. Two beams observe the brightness distribution and the differential output is recorded. This is equivalent to the observation with a beam of the form 1, -1. This results in a differentiated brightness distribution. A standard method of analysis is to integrate the signal. This method has the disadvantage that it relies on the model beam to give a really differentiated distribution and integration can only be done in small pieces, otherwise the statistical fluctuation leads to large excursions of the baseline. Could one apply CLEAN? The conditions of convergence of CLEAN are that the dirty beam has to be symmetric and its FT non-negative. In both senses the beam-switching beam is not conform to these requirements. Already Temple (1938) mentions the possibility in order to be able to apply his relaxation method (essentially CLEAN) to convolve the data with the asymmetric beam and to use the self-convolution of the beam as the 'dirty' beam. Obviously one then gets a symmetric dirty beam, whose FT is non-negative (being squared). Now it is possible to apply CLEAN, in the idealized case with a dirty beam $-\frac{1}{2}$, 1, $-\frac{1}{2}$. The matrix B will be positive definite, the resulting components are a linear combination of the convolved data (the coefficients being the elements of the inverse matrix, B^{-1}). This makes an error analysis possible; note that it will be different from the case of FT data.

I tentatively tried this method with real IR data of a high altitude flight, published by M. de Muizon, 1978. The dirty beam is based on the observations of Saturn, see Figure 3, and the data is sampled at a 2' interval. In Figure 4 the result of the integration by Muizon of the source M17 is shown, together with the results of a tentative CLEAN with 15 components, which are convolved with a 4'.5 Gaussian.

There are other potential fields of application of CLEAN, such as seismography, tomography etc. But I think the method is only then successful, if one can assume that the 'true' map is almost empty.

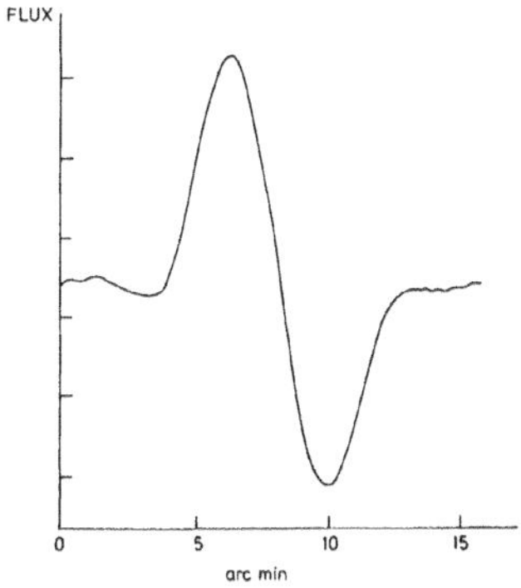

Figure 3. Profile of the beam-switching IR observations of Saturn (Muizon, 1978). The self-convolution gives the dirty beam, which is used in CLEAN shown in Figure 4.

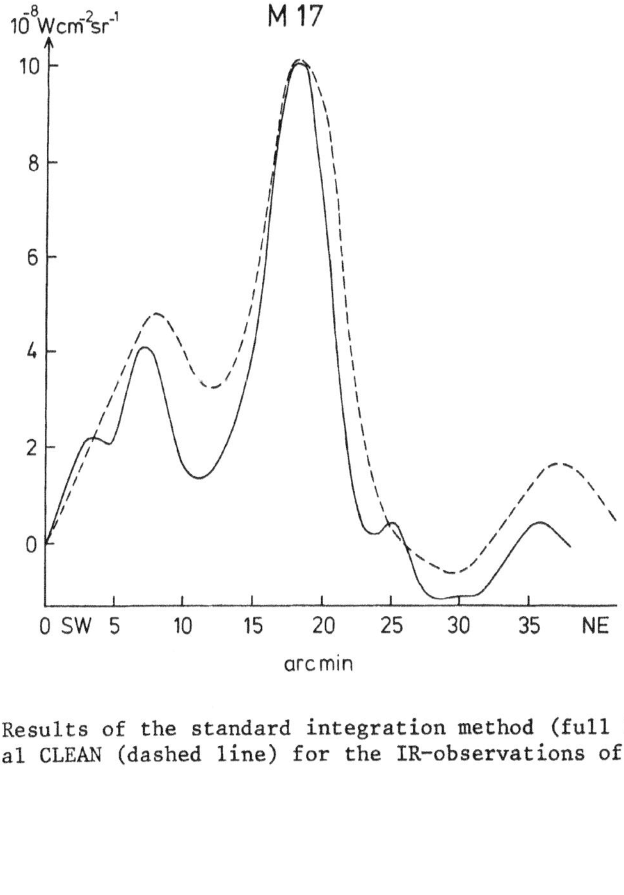

Figure 4. Results of the standard integration method (full line) and experimental CLEAN (dashed line) for the IR-observations of M17 (Muizon, 1978).

ACKNOWLEDGEMENTS

I like to thank Dr. R.H. Harten for making available to me the results of experiments discussed in Section 3.1 and for the critical reading of the manuscript, Mr. B. van Dijk for his contributions for Section 2.1 and Dr. R.D. Ekers for many discussions.

REFERENCES

Balick, B., Sanders, R.H.: 1974, "Astrophys. J." 192, p.325.
Ekers, R.D., Goss, W.M., Schwarz, U.J., Downes, D., Rogstad, D.H.: 1975, "Astron. Astrophys." 43, pp.159-166.
Högbom, J.A.: 1974, "Astron. Astrophys. Suppl." 15, p.417.
Hulst, J.M. van der: 1977, Thesis, University of Groningen.
Muizon, M. de: 1978, Thesis, l'Université de Paris.
Readhead, A.C.S., Wilkinson, P.N.: 1978, "Astrophys. J." 223, pp.25-36.
Rogstad, D.H., Shostak, G.S.: 1971, "Astron. Astrophys." 13, p.99.
Schwarz, U.J.: 1978, "Astron. Astrophys." 65, p.345.
Temple, G.: 1938, "Proc. Roy. Soc. A." 169, p.476.

DISCUSSION

Comment E.B. FOMALONT
Is it always better to use a small loop gain in cleaning a map?
Reply U.J. SCHWARZ
Mostly yes, but not always, see reply by Ron Harten.
Reply R.H. HARTEN
The high gain factor can be of use for sources with high negative sidelobes. Generally only the first few components need to be cleaned with a high gain factor.

Comment J.P. HAMAKER
In section 3.1 you discuss the mapping of a structure containing a bar. The only reason why you prefer the solution including the bar is that you have a priori information making it likely for something to be there. The u-v area that contains the pertinent information is not accessible in this case. Without your a priori knowledge you would most probably have accepted the two-component picture without a bridge as a reasonable one. The preference for the outcome using a high loop gain in this case is purely fortuitous.
Reply U.J. SCHWARZ
This is not completely true. There is no confined brightness distribution which is completely unaccessible by any sampling. But if one has noisy data, such a bar-like structure as above can stay undetected. My point is to show that for sufficient S/N one can recover the true distribution, and the quality measure can give an indication of the reliability of the reconstruction.

Comment L.R. D'ADDARIO
You said very little about the clean beam. Would you comment on the fact that the final map, after convolution with the clean beam, disagrees significantly with the available visibilities?
Reply U.J. SCHWARZ
As shown in my paper, CLEAN fits the FT of the components to the unweighted visibility of the true distribution. The components contain the most information, which can be used directly to calculate for instance moments. But in order to suppress the highly uncertain large spatial components, which can be disturbing in some applications, one can convolve the components with a clean beam. If one doesn't like to degrade the good reliable part of the fitted visibility but still likes to depress the large spatial components one can use the FT of a box function (method used by Rogstad and Shostak, A&A, 1973, 13, 99). But in general one uses a clean beam similar to the FT of the taper or weighting function, having the same motivation as in making maps by the simple FT.

FITTING A GAUSSIAN MODEL TO APERTURE SYNTHESIS DATA BY AKAIKE'S INFORMATION CRITERION (AIC)

MASATO ISHIGURO
The Research Institute of Atmospherics, Nagoya University
Toyokawa, 442 Japan

MAKIO ISHIGURO
The Institute of Statistical Mathematics, Tokyo, Japan

ABSTRACT

The problem of determining the optimum number of components in fitting a Gaussian model to aperture synthesis data is considered. As a measure of the badness of the fitted model, we propose the use of Akaike's Information Criterion (AIC).

1. INTRODUCTION

The output of an aperture synthesis telescope is a set of sampled values of the visibility function which is the Fourier transform of the sky brightness distribution. If the visibility function is completely sampled along the coordinate of spatial frequency under the circumstance of high signal-to-noise ratio, the true sky brightness distribution is obtained by a simple Fourier inversion process. In many practical cases, however, measurement of the visibility function is limited by a small number of antennas and by the measurement errors.

When the measured visibilities are incomplete and noisy, the synthesis map estimated through the direct Fourier inversion is disturbed by undesirable sidelobe effects, which make the physical interpretation of the structure very difficult. Various reconstruction techniques to cope with the problem of estimating a reliable brightness distribution from a limited set of visibility data have been developed. Some of these are model fitting (Fomalont, 1968), CLEAN (Högbom, 1974), and Maximum Entropy Method (Gull and Daniell, 1978) and they have been used successfully for some class of brightness distribution suitable for each reconstruction technique.

We have taken a new look at the problem of model fitting. A model brightness distribution is usually represented by a set of simple-shaped

C. van Schooneveld (ed.), Image Formation from Coherence Functions in Astronomy, 277-286.

components. We use a Gaussian model because it has the non-negative property and it is convenient for the Fourier transform. For mathematical simplicity, we considered the one-dimensional case. So far, the goodness of fit has been evaluated in a least square sense, and the number of components has been determined rather arbitrarily and/or empirically. If the number of degrees of freedom is too large compared with that of the independent observations, the results of fitting are satisfactory but are very sensitive to the noise, and *vice versa.*

As a measure of the badness of the fitted model, we propose the use of Akaike's Information Criterion (AIC). We can select the optimum model on the basis of the Minimum AIC Estimate (MAICE). In the following sections, the basic principle of the method is described and some numerical examples are illustrated.

2. STATISTICAL NATURE OF THE OBSERVATION

Let $T^*(x)$ be the true brightness distribution. If the observation noise is additive, N observations of the visibility function are expressed by

$$z_j = \tilde{T}^*(u_j) + n_j \qquad (j=1,2,\ldots,N), \tag{1}$$

where $\tilde{T}^*(u)$ is the Fourier transform of $T^*(x)$, and $\{u_j\}$ are the spatial frequencies at which the $\{z_j\}$ are measured. The $\{n_j\}$ are the noises. If $n^N = (n_1,n_2,\ldots,n_N)$ is a random vector whose probability density function (PDF) is $g^*(n^N)$, the PDF of the random vector $z^N=(z_1,z_2,\ldots,z_N)$ is given by

$$f^*(z^N) = g^*(y^N) \tag{2}$$

where y^N is a N-vector whose j-th element is

$$y_j = z_j - \tilde{T}^*(u_j) \qquad (j=1,2,\ldots,N).$$

3. MODELING

As an estimate of $T^*(x)$, we adopt the Gaussian model:

$$T_M(x|\theta^M) = \sum_{m=1}^{M} w_m \frac{1}{\sqrt{2\pi}\,\sigma_m} \exp\{-\frac{(x-\mu_m)^2}{2\sigma_m^2}\}, \tag{3}$$

where

$$\theta^M = (w_1,\sigma_1,\mu_1,w_2,\sigma_2,\mu_2,\ldots,w_M,\sigma_M,\mu_M)$$

is the parameter vector of the model to be adjusted. Taking the Fourier transform of Equation (3), the visibility function is modeled as

$$\tilde{T}_M(u|\theta^M) = \sum_{m=1}^{M} w_m \exp\{-\frac{1}{2}(2\pi u\sigma_m)^2 + i2\pi u\mu_m\}. \quad (4)$$

We also assume that the noise n^N has the PDF:

$$g_N(n^N|\sigma) = \prod_{j=1}^{N} \frac{1}{2\pi\sigma^2} \exp\{-\frac{|n_j|^2}{2\sigma^2}\}. \quad (5)$$

Equation (5) is derived from the assumption that the real and the imaginary parts of $\{n_j\}$ are statistically independent samples from the same normal distribution with zero mean and standard deviation σ. From (2) and (5) we obtain the expression for the joint PDF of z^N as

$$f_M(z^N|\theta^M,\sigma) = \prod_{j=1}^{N} \frac{1}{2\pi\sigma^2} \exp\{-\frac{|z_j - \tilde{T}_M(u_j|\theta^M)|^2}{2\sigma^2}\}. \quad (6)$$

4. MEASURE OF THE BADNESS OF THE FITTED MODEL

If the number of Gaussian components M of the model (3) is specified, the parameters of (6) can be estimated by the maximum likelihood method. Let $(\hat{\theta}^M, {}_M\hat{\sigma})$ be the maximum likelihood estimate (MLE). Then estimates of $T^*(x)$ and $f^*(z^N)$ are given by $T_M(x \mid \hat{\theta}^M)$ and $f_M(z^N \mid \hat{\theta}^M, {}_M\hat{\sigma})$ respectively.

As a measure of the badness of the fitted model, we propose the use of Akaike's Information Criterion (AIC), which is defined by

$$\text{AIC} = -2 \times \log(\text{maximum likelihood of a model}) + 2 \times (\text{number of parameters to be adjusted}). \quad (7)$$

This criterion is introduced by Akaike (1973,1977) as an estimate of minus twice the entropy of the true PDF with respect to the MLE of PDF. In our case, the entropy and AIC are given by

$$B\{f^*(.);f(.\,|\hat{\theta}^M, {}_M\hat{\sigma})\} = \int f^*(z^N) \log \frac{f(z^N \mid \hat{\theta}^M, {}_M\hat{\sigma})}{f^*(z^N)} dz^N, \quad (8)$$

and

$$\text{AIC}(M) = -2\log\{f_M(z^N \mid \hat{\theta}^M, {}_M\hat{\sigma}) + 2(3M+1), \quad (9)$$

respectively. Although it is impossible to maximize (8) without knowing the true PDF $f^*(.)$, (9) can be minimized by a suitable choice of M. Thus AIC provides a criterion for the optimum choice of the number of Gaussian components. The estimated model which gives the minimum of AIC is called the Minimum AIC Estimate (MAICE).

5. NUMERICAL PROCEDURE

To maximize (6) for given z^N, it is sufficient to minimize

$$-\log\{f_M(z^N|\theta^M,\sigma)\} = N\log 2\pi + N\log\sigma^2 + \frac{1}{2\sigma^2}\sum_{j=1}^{N}|z_j-\tilde{T}_M(u_j|\theta^M)|^2. \quad (10)$$

If the function

$$\phi(\theta^M) = \sum_{j=1}^{N}|z_j-\tilde{T}_M(u_j|\theta^M)|^2 \quad (11)$$

is minimized by $\hat{\theta}^M$, the minimum value of (10) is given by

$$-\log\{f_M(z^N|\hat{\theta}^M,{}_M\hat{\sigma})\} = N\log 2\pi + N\log{}_M\hat{\sigma}^2 + N, \quad (12)$$

where ${}_M\hat{\sigma}^2$ is given by

$${}_M\hat{\sigma}^2 = \frac{1}{2N}\,\phi(\hat{\theta}^M). \quad (13)$$

Now the problem is reduced to the minimization of the non-negative function $\phi(\theta^M)$. Since the minimization of (11) leads to a problem of solving a set of nonlinear equations, it must be solved iteratively, starting with a suitable initial guess.

It is easy to evaluate the function $\phi(\theta^M)$ and its gradient for given θ^M. It is also not difficult to obtain a rough estimate of the Hessian matrix of second partial derivatives. Considering the situation, we propose the use of Davidon's Variance Algorithm for minimization (Davidon, 1968, 1969), which is essentially a modified Newton-Raphson algorithm, where the direct computation of variance, the inverse of the Hessian, is replaced by successive correction of the estimated variance.

6. NUMERICAL EXAMPLES

As a true brightness distribution, three Gaussian components are used. Simulated noisy data are generated by adding zero mean normal random numbers with a standard deviation of 0.1, to the real and imaginary parts of the Fourier transform of the true brightness distribution. In this example, the number of complex visibility data is 50.

We start with equally spaced and identical Gaussians as an initial estimate, the number of which is the same as that of the assumed model. In Table 1., AIC(M), its constituent terms, and $_M\hat{\sigma}$, for one- to six-component models are shown. Constant terms in (12) are omitted.

M	AIC(M)	$-2\log\{f_M(z^N \mid \hat{\theta}^M, {}_M\hat{\sigma})\}$	2(3M+1)	$_M\hat{\sigma}$
1	-359.29	-367.29	8	0.156
2	-393.86	-407.86	14	0.127
3	-421.01*	-441.01	20	0.108
4	-417.23	-443.23	26	0.107
5†	-411.23	-443.23	32	0.107
6	-418.39	-456.39	38	0.100

*MAICE †weight for one of the Gaussians is reduced to zero.

Table 1.

Figure 1. shows the result of MAICE together with that of direct Fourier inversion. The effectiveness of the model fitting by MAICE over the Fourier inversion is very clear. Fitted models for M=2,3,4 and 6 are shown in Figure 2.

We dropped some visibility data to see the applicability of the MAICE procedure when visibility data are incompletely sampled. The results are shown in Figure 3. Numbers of complex visibility data are 38 and 19 for Figures 3.a and 3.b, respectively. The main features of the true brightness distribution are still well reconstructed.

The convergence of the iterative procedure is quadratic, and the computing time for the three component model is less than 6 seconds.

7. CONCLUSION

We have shown by numerical examples that the MAICE procedure is very useful in fitting a Gaussian model to aperture synthesis data. The shapes of the components need not necessarily be Gaussian; other shapes can be used in specific cases. In any case, it is preferable that the component shape has the properties of non-negativity and smoothness. The first and the second terms in the AIC expression represent "badness of fit"

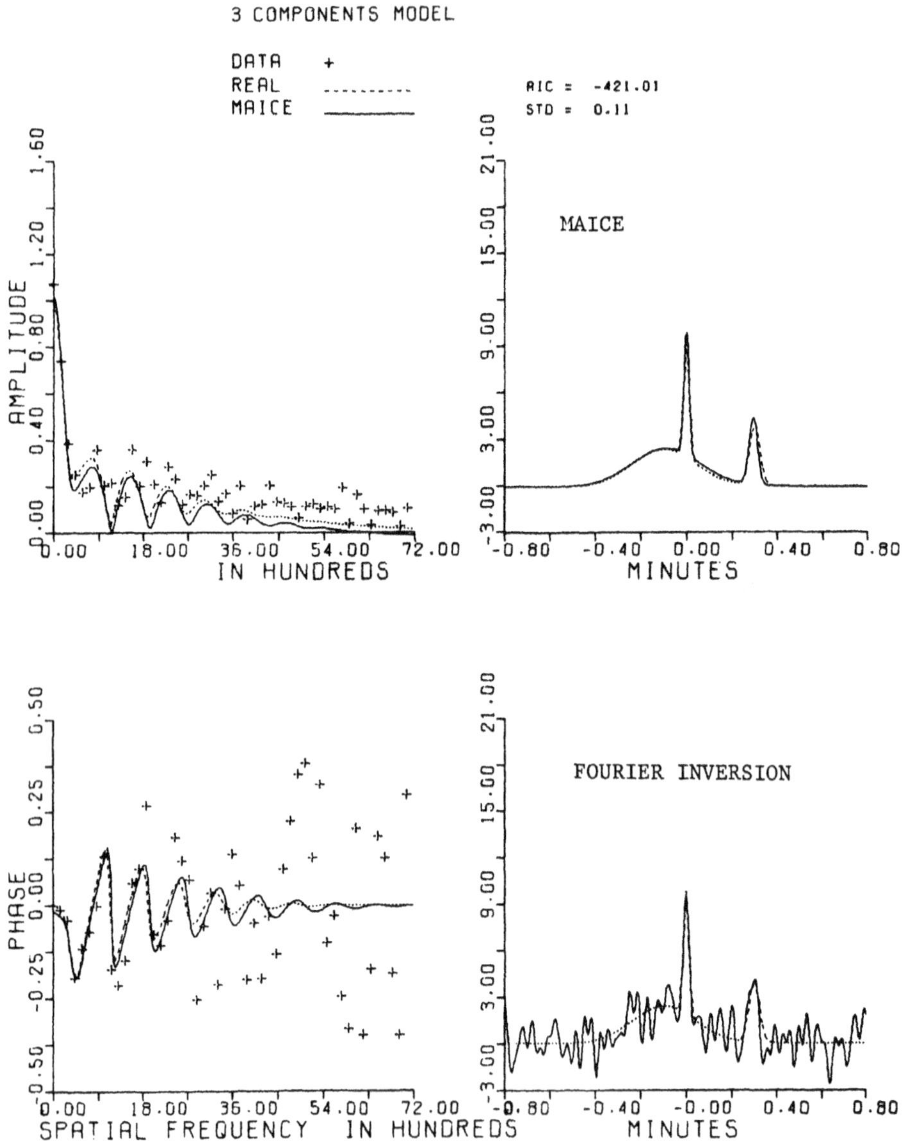

Figure 1. Selected model by MAICE (N=50).

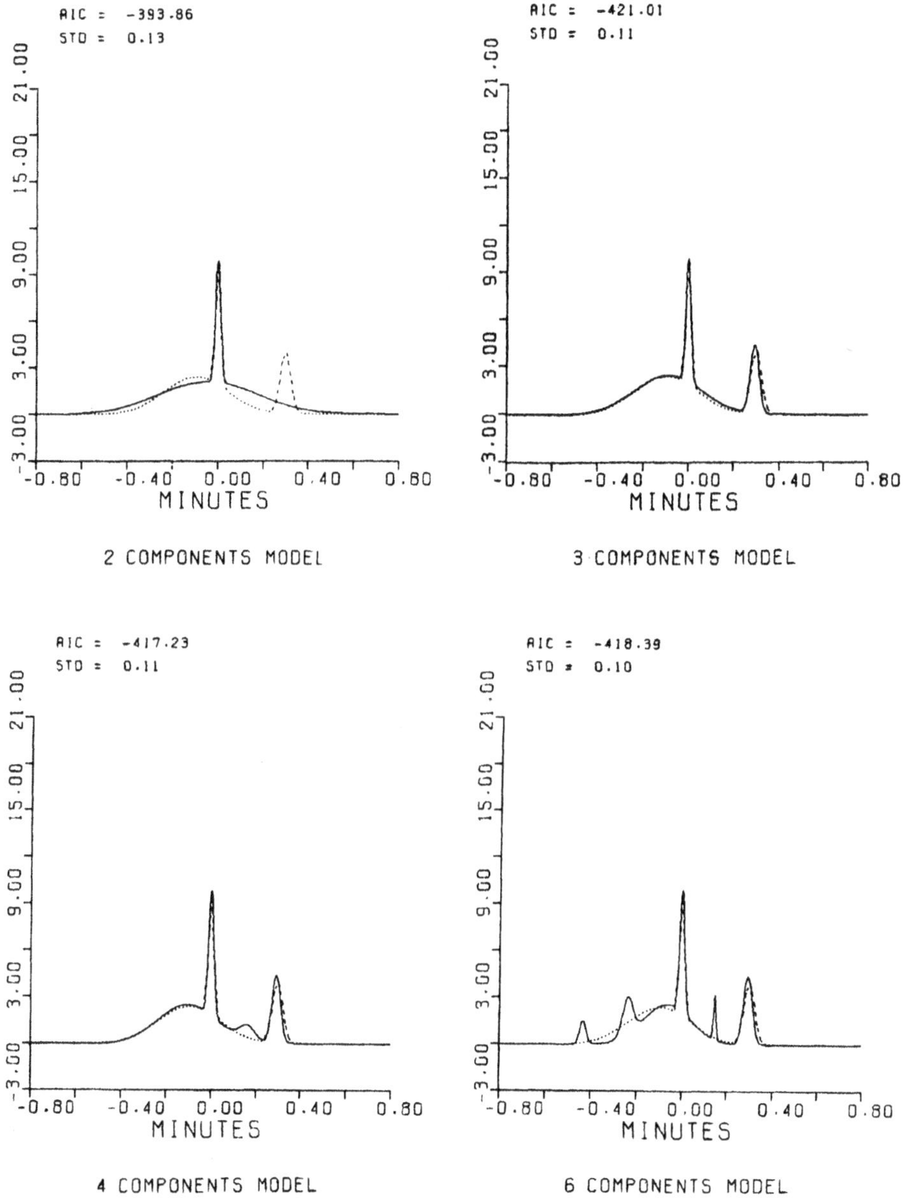

Figure 2. Fitted models for M = 2, 3, 4 and 6. AIC is minimum for M=3.

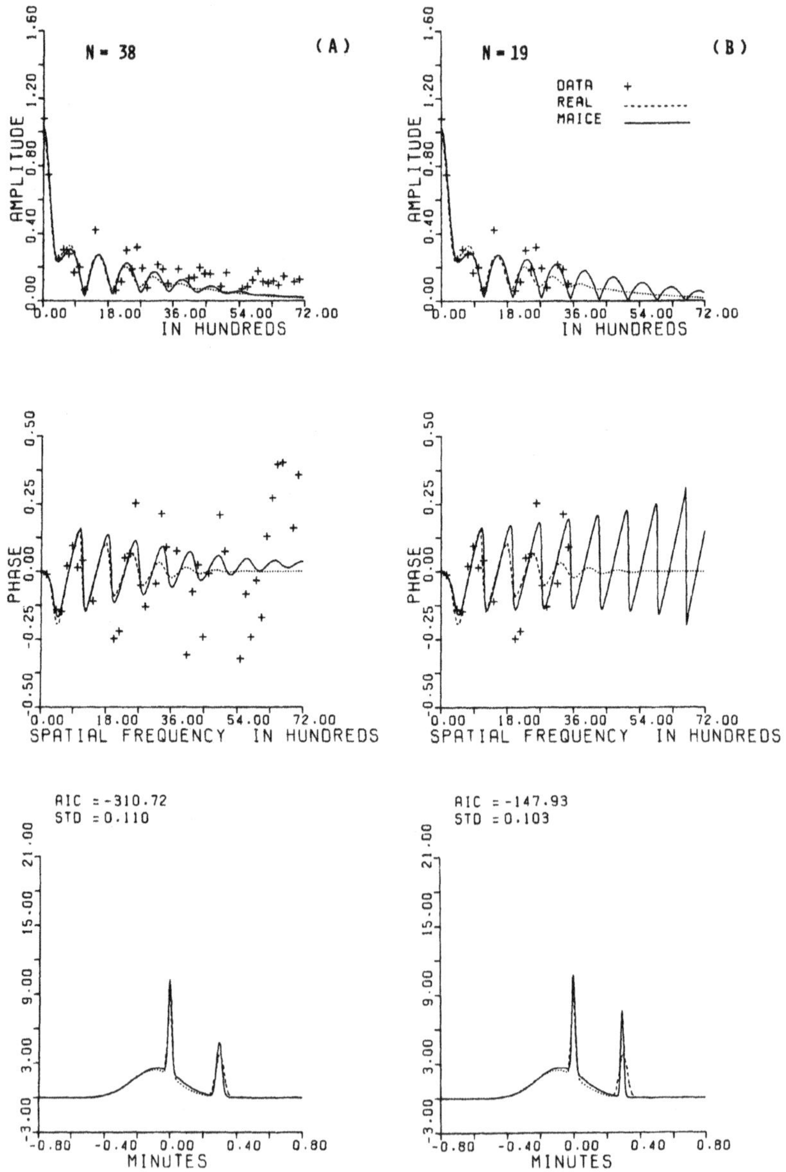

Figure 3. Results of MAICE for incompletely sampled visibility data. Numbers of complex visibilities are; (A) N=38, and (B) N=19.

and "unreliability of model" respectively. The model selected by MAICE is the most parsimonious one. If we have *a priori* information on the brightness distribution, visibility function, and noise statistics, it should be included in the model to reduce the number of degrees of freedom as far as possible. Although we have concentrated our discussion on the model fitting in aperture synthesis, application of the MAICE procedure to other imaging systems is straightforward.

ACKNOWLEDGEMENTS

We would like to thank Dr. H. Akaike of the Institute of Statistical Mathematics for his valuable suggestions and discussions. We are also grateful to Miss. F. Tada for her help in programming.

REFERENCES

Akaike, H.: 1973, *Information Theory and an Extension of the Maximum Likelihood Principle, in 2nd International Symposium on Information Theory* (B.N. Petrov and F. Csaki, eds) pp.207-281, Akademiai Kiado, Budapest.
Akaike, H.: 1977, *On Entropy Maximization Principle, in APPLICATION OF STATISTICS* (P.R. Krishnaiah, ed.) pp.27-41, North-Holland Publishing Company.
Davidon, W.C.: 1968, *Computer J.*, vol.10, pp.406-410.
Davidon, W.C.: 1969, *Optimization* (Fletcher, R. ed.) pp.13-20, Academic Press, London.
Fomalont, E.B.: 1968, *Astrophys. J. Suppl.* 15, pp.203-273.
Högbom, J.A.: 1974, *Astron. & Astrophys. Suppl.*15, pp.417-426.
Gull, S.F. and Daniell, G.J.: 1978, *Nature*, vol.272, pp.686-690.

DISCUSSION

Comment Y. BIRAUD
How to determine the number of degrees of freedom - or the number of Gaussian components - knowing the noise characteristics?
Reply M. ISHIGURO
We did not do experiments for a wide range of signal to noise ratio. If we know the noise characteristics, the parameters of the noise distribution may be fixed and need not be estimated.

Comment R.H.T. BATES
Is it not dangerous to model the "true" brightness distribution as a sum of Gaussians?

Reply M. ISHIGURO
It is dangerous to use smooth functions to fit true brightness distributions which have sharp boundaries. It is recommended to test several different classes of models. Selection of the optimum model can be done by MAICE.

Comment J.J. WITTELS
It is particularly self-deceptive to create a circular-Gaussian test model for an inversion program that models the data by superposing circular Gaussians. Very few astronomical sources display such a high degree of cylindrical symmetry.
Reply M. ISHIGURO
For a two dimensional example, we used elliptical Gaussians. So the number of parameters is (7M + 1).

Comment J.J. WITTELS
Would you comment on the impression that although, in your example, the model appeared to agree with the calculated curve for the test source, it did not appear to agree very well with the data. This situation is particularly disturbing since agreement with the data(that is both the minimum point to point scatter as well as the absence of systematic disagreement) is the primary criterion for selecting an acceptable model. In this vein, have you tried blind tests of your method?
Reply M. ISHIGURO
The brightness distribution can be modeled by many other suitable functions. We can select most objective ones from a set of possible models by the MAICE procedure.

Comment U.J. SCHWARZ
In relation to the question by Wittels, I would like to point out that the discrepancy between observations and model seems to me - at least for low intensities of the model - to be only apparent. In a display of the modulus of the complex visibility the noise will always give a positive contribution, whereas the model correctly has small or zero amplitude.

AN OPTIMUM DECONVOLUTION METHOD

C.R. Subrahmanya
Tata Institute of Fundamental Research,
Homi Bhabha Road, BOMBAY 400005, INDIA.

An optimum solution to a deconvolution problem has to fulfil three general criteria: (a) an explicit recognition of the smoothing nature of convolution; (b) a statistical treatment of noise, e.g., using the least-squares criterion; and (c) requiring the solution to conform to all our prior knowledge about it. In the usual least-squares method, one minimises a variance of 'residuals', or the departures of the observed data from the values expected according to the recovered solution. However, this condition does not lead to a stable solution in the case of deconvolution, since the only stable solutions are those conforming to a criterion of 'regularisation' or smoothness (see, e.g., Tikhonov and Arsenin 1977). In our method, the stability is achieved by minimising the variance of the second-differences of the solution simultaneously with the fulfilment of the least-squares criterion. Such a procedure was first used by Phillips(1962). However, the solution thus obtained is still unsatisfactory since it usually does not conform to our *a priori* information. When we seek the brightness distribution of an object, the most frequent violation of our prior knowledge is that of positiveness. This motivated us to develop an Optimum Deconvolution Method (ODM) which constrains the solution to satisfy prior knowledge while retaining the features of least-squares and smoothness criteria.

If prior knowledge is in the form of equalities, e.g., when the total intensity (area under the solution) is specified, it can be easily incorporated by using the Lagrange multiplier method of constrained minimisation. However, this is not applicable for positiveness or any specification of bounds on the solution since this leads to constraints in the form of inequalities. Such constraints are imposed in ODM by a new iterative algorithm, which minimises a weighted sum of squares of the departures of the solution from the specified bounds, simultaneously with the fulfilment of the criteria of least-squares and smoothness. The weights chosen depend on the degree of such departures in the previous iterations. The algorithm is found to converge rapidly, within 5 iterations in most of the several hundred applications tried by us for the Fresnel-diffraction curves produced by the lunar occultations of radio sources.

C. van Schooneveld (ed.), Image Formation from Coherence Functions in Astronomy, 287–290.

The method was extensively tested by using both the computer-simulated occultations and those of actual sources observed with the Ooty radio telescope. Restored profiles from these data obtained by using ODM were compared with those obtained by using the conventional method which was suggested by Scheuer(1962). The tests reveal that ODM often leads to an improvement in the attainable resolution by about a factor of two over the conventional method. In addition, it also gives a 'clean' solution masking the effects of noise on the recovered solution and hence leads to a more objective interpretation. A detailed description of the method and its application to lunar occultations has been given elsewhere (Subrahmanya 1979).

The usual purpose of seeking a nonclassical method like ODM is to improve our estimate of the size of a certain component of a source. For this, it is necessary to know the effective resolution with which the solution has been recovered. In a linear method, this effective resolution is always known independently of the solution to be recovered, e.g., as a property of the restoring function used. However, this is not the case in a nonlinear method like ODM or Maximum Entropy Method (MEM) which uses known properties of the solution before actually deriving it. In such a case, it becomes necessary to introduce an a posteriori definition of resolution, say, based on the statistical properties of the residuals. For this, we suggest the use of the mean zero-crossing interval of the residuals. For instance, if one examines the noise resulting from an observation with a Gaussian beam, the mean zero-crossing interval is about 1.9 times the half-power-width of the Gaussian beam.

We have seen that ODM can provide super-resolution and also a 'clean' output as compared to a classical method. The solution is optimum since it agrees statistically with the observed data and also conforms to all our prior knowledge about it. The main source of improvement over the conventional method is the incorporation of prior knowledge, just as in MEM. A few general remarks can be made by way of comparison of the two methods. Both are nonlinear methods. Both are able to use incomplete data like those with the absence of a few samples or the ignorance of phases of some Fourier components. They use the data as given and do not make any assumption on the unavailable data. The characteristics of 'super-resolution' and 'clean output' are also shared by both the methods. Since both are nonlinear methods, it may be worthwhile to compare their computational efficiencies. In our experience, ODM is particularly fast for a nonlinear method in lunar occultations, since the iterations generally converge within 3 to 5 iterations. On the contrary, an available figure for MEM is 8 to 40 iterations for the one-dimensional Fraunhofer diffraction problem studied by Frieden (1975). However, this cannot be used to make a categorical statement on the relative computational efficiency of the two methods, since they have only been used in different contexts.

Thus any fundamental difference between the two methods is to be expected only if the 'entropy' as used in MEM has a special sanctity apart from satisfying prior knowledge about the solution. Serious doubts have already been cast on this aspect and it is not clear at present whether the expressions being used for 'entropy' have the same significance as in information theory or thermodynamics(cf. Kikuchi and Soffer 1977). An essential difference in the usage is the definition based on *a posteriori* probability in MEM as opposed to that based on *a priori* probability in information theory. It is also possible that the existing schemes of MEM can be alternately viewed as merely numerical algorithms for incorporating positiveness. In this context, it may be noted that there is a numerical scheme for constrained minimisation called the 'logarithmic penalty-function method' (Fiacco and McCormick 1970) which is strikingly similar to MEM as being used by one class of workers (e.g. Ables 1974). Further work is needed to understand whether a concept like entropy is indeed pertinent to the problem or whether an arbitrary numerical algorithm which can efficiently incorporate prior knowledge is all that is relevant.

REFERENCES

Ables, J.G.: 1974, "Astron. Astrophys. Suppl.", 15, p.383

Fiacco, A.V. and McCormick, G.P.: 1970, "Nonlinear Programming", (New York: Wiley)

Frieden, B.R.: 1975, in "Picture-processing and Digital Filtering", (Ed.) Huang, T.S. (Berlin: Springer-Verlag)

Kikuchi, R. and Soffer, B.H.: 1977, "J. Opt. Soc. Amer.",67, p.1656

Phillips, D.L.: 1962, "J. Assoc. Comp. Machy.", 9, p.84

Scheuer, P.A.G.: 1962, "Aust. J. Phys.", 15, p.333

Subrahmanya, C.R.: 1979, submitted to Astron. Astrophys.

Tikhonov, A.N. and Arsenin, V.Y.: 1977, "Solution of Ill-posed Problems" (Washington: Wisconsin)

DISCUSSION

Comment U.J. SCHWARZ

Is it really desirable to impose positiveness? This is not a question concerning this method only, but other methods too. The point is that statistically the estimate is not asymptotically equal to the true distribution, since the noise always gives a positive contribution via - essentially - a detection process.

Reply C.R. SUBRAHMANYA

This is one of the reasons why I believe that it is desirable to incorporate prior knowledge explicity, with a provision to define the degree to which we know it. In ODM since an estimate of the uncertainty (σ) in total intensity (area) is known, the iterations for positiveness are terminated when the total contribution to the intensity from the negative points becomes $\sim 0.5\ \sigma$.

APPLICATION OF THE SIMPLEX ALGORITHM TO IMAGE RESTORATION

P.L. Baker
Max Planck Institut für Radioastronomie
Bonn, Fed. Rep. Germany

ABSTRACT

The data measured by an interferometer define the original image through linear equations of constraint containing the Fourier coefficients. These constraints and the requirement that all image points show a non-negative flux define a problem frequently encountered in operations research and numerically tractable through the Simplex Algorithm. When the data are undersampled, a continuum of original images is compatible with the measured points. The advantage of the operations research method in this situation is that the continuum of restorations is generated rather than a single example. As a practical matter, it appears most useful to select from this continuum the maximum sharpness restoration - which like CLEAN uses reasonable a priori information to chose one of the possibilities - and to select for comparison the maximum entropy solution - which embodies diametrically opposed a priori information. The first may be considered the best restoration for an astronomical source, while the second indicates the uncertainty due to undersampling. When our astrophysical deductions depend strongly on the restored structure of a source, a comparison of alternative solutions is necessary to acquire confidence in the deductions. The maximum sharpness solutions also have the advantage that small phase errors in the measured complex points can be compensated. In principle, large phase errors can be compensated, but uniqueness may be lost, and existing numerical procedures for this case are inefficient.

C. van Schooneveld (ed.), Image Formation from Coherence Functions in Astronomy, 291.

OBSERVATIONS OF EXTENDED RADIO SOURCES USING A MULTIPLE-BEAM TECHNIQUE

D.T. Emerson and U. Klein
Max-Planck-Institut für Radioastronomie, Bonn, F.R.G.

1. INTRODUCTION

The limiting sensitivity for conventional single-dish radio continuum observations of extended objects at wavelengths of a few cm is set mainly by weather; for much of the time a varying component of atmospheric attenuation, and hence thermal emission, causes deviations in the response of the receiver far in excess of any deviations due to receiver noise or receiver instabilities.

A dual-beam technique, where only the difference in power received by the two beams pointing at slightly separated points in the sky is recorded, has been used for many years for observations of small sources; atmospheric emission from within the near-field zone affects both beams equally, and so cancels. It has often however been stated that this technique is only suitable where the angular size of the radio source is less than the separation of the two beams.

This paper presents a dual-beam technique suitable for observations of sources many times greater in angular extent than the separation of the two beams. A consideration of the spatial frequency response of a dual-beam system and the effect of noise on the observations shows that such sources may be mapped and restored to the equivalent perfect-weather single-beam observation without degradation in signal-to-noise ratio, and with a complete sampling of spatial frequencies.

2. DUAL-BEAM OBSERVATIONS

An observation made with a dual-beam system is equivalent to a single-beam observation, but convolved additionally with the 2-beam function shown in Fig. 1(a). An example of such an observation is given in Fig. 2, which shows the source 3C 10 (diameter $\sim$ 9') observed at λ2.8 cm with the 100-m Effelsberg telescope using an angular separation of 5!5 between the two beams, each of which has a full-width to half-power of 70". The effects of bad weather have been removed by the dual-beam system — compare Fig. 2 with the single-beam map observed at

C. van Schooneveld (ed.), Image Formation from Coherence Functions in Astronomy, 293–298.

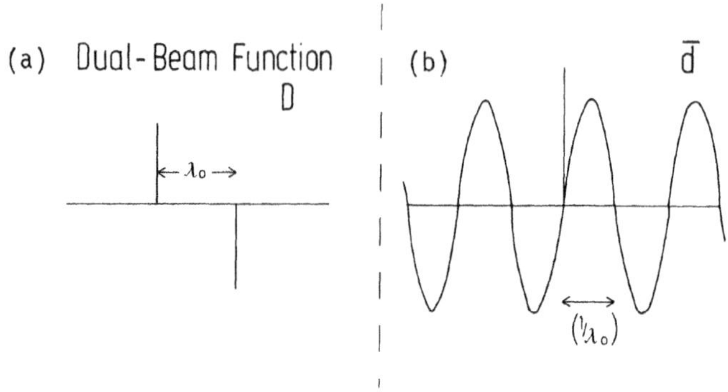

Fig. 1. (a) The dual-beam function, and (b) a representation of its Fourier transform. λ_o is the angular separation of the 2 far-field beams.

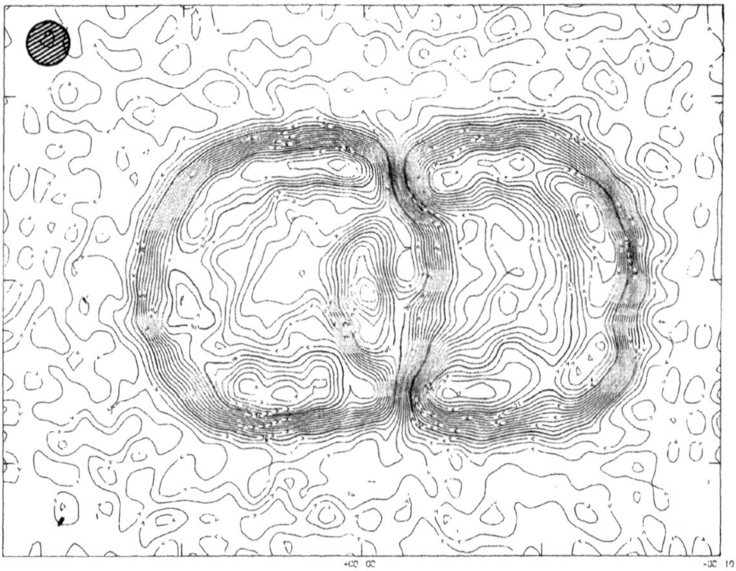

Fig. 2. A dual-beam observation of 3C 10 made at λ 2.8 cm with the 100-m Effelsberg telescope.

the same time, which is shown in Fig. 3 — but in the centre region of the map emission is detected by both beams simultaneously, giving a confused representation of the true emission.

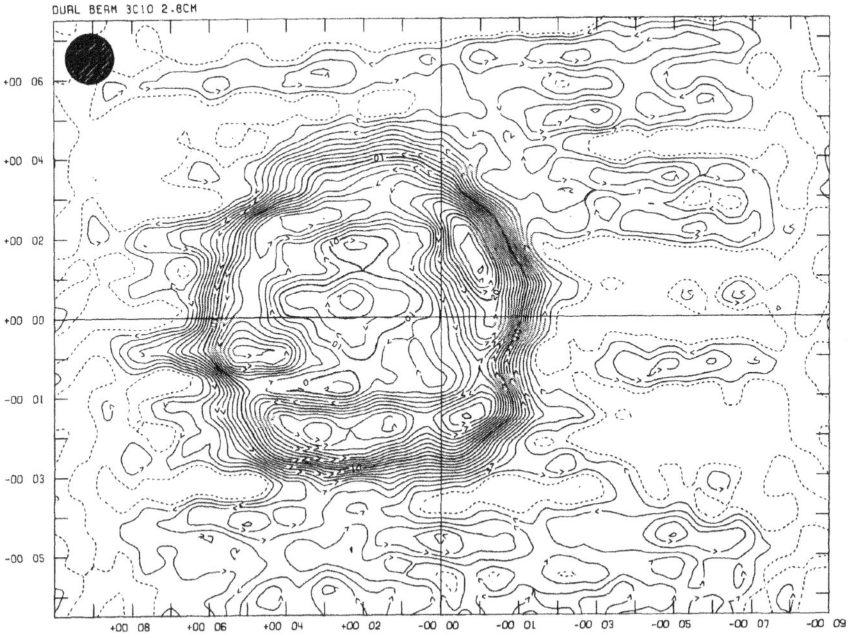

Fig. 3. A single-beam map of 3C 10, observed at the same time as Fig. 2.

The convolution with the dual-beam function of Fig. 1(a) is equivalent to multiplying the spatial frequency ($1/\lambda$) distribution of the data with the sine wave sin ($\lambda_0/\lambda \cdot \pi$) illustrated in Fig. 1(b). In principle it might be possible to reweight spatial frequencies by cosec ($\lambda_0/\lambda \cdot \pi$), but clearly spatial frequencies close to the zero-crossing points are undefined. However for a source of finite extent x it is only necessary to sample at finite intervals $1/x$ in spatial frequency, and these sample points can be arranged to avoid the undefined spatial frequencies. This is illustrated in Fig. 4; for a map of extent $3\lambda_0$, where λ_0 is the angular separation of the 2 beams, no spatial frequency needs to be sampled which has been attenuated by more than a factor of 2 by the convolution with the dual-beam function.

It may be computationally convenient to perform the reweighting and sampling of the spatial frequencies present in the original data by a 1-dimensional convolution with a function having the required spatial frequency response. This function is illustrated in Fig. 5. The result of applying the procedure to the data of Fig. 2 is shown in Fig. 6, which should be compared with the single-beam data recorded directly at the same time, shown in Fig. 3.

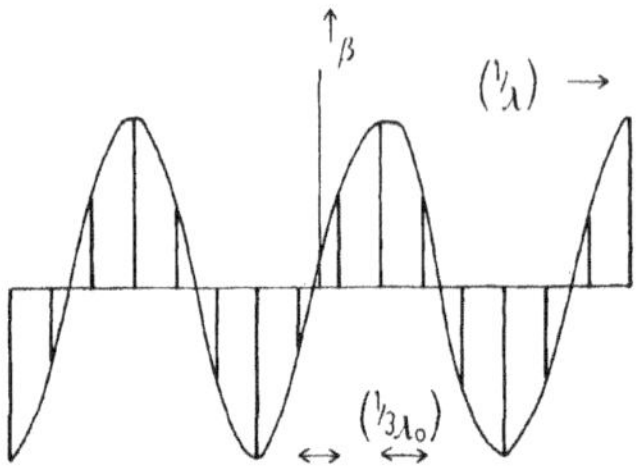

Fig. 4. The required spatial frequency sample points for a source of extent $3\lambda_o$ are illustrated.

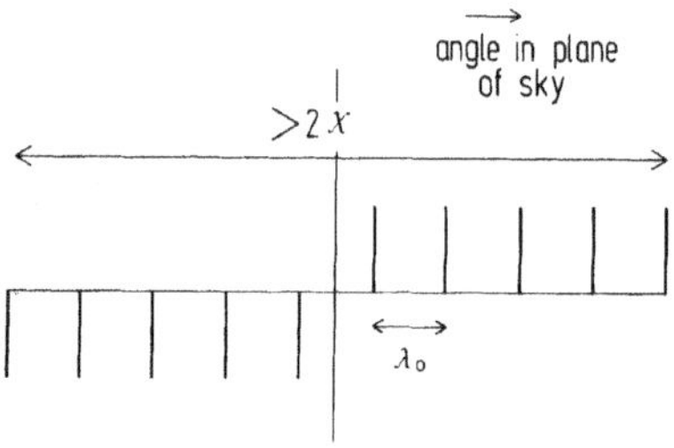

Fig. 5. The convolution function with the required spatial frequency response to restore a dual-beam observation to the equivalent single-beam observation.

3. AN EXTENSION OF THE TECHNIQUE TO LARGER SOURCES

After restoration to the equivalent single-beam observation, the noise level dT of a dual-beam observation of a field n units of the beam separation λ_o in extent, is given by:

$$dT = \frac{\sqrt{n} \cdot T_{sys}}{\sqrt{B \cdot \tau}}$$

This may be derived from a consideration of the noise of each spatial frequency sample after reweighting, and indicates that a higher noise level is expected on dual-beam maps of very extended regions. Compare this with the conventional Dicke-switching receiver, whose noise dT is given by:

$$dT = \frac{2 \cdot T_{sys}}{\sqrt{B \cdot \tau}}$$

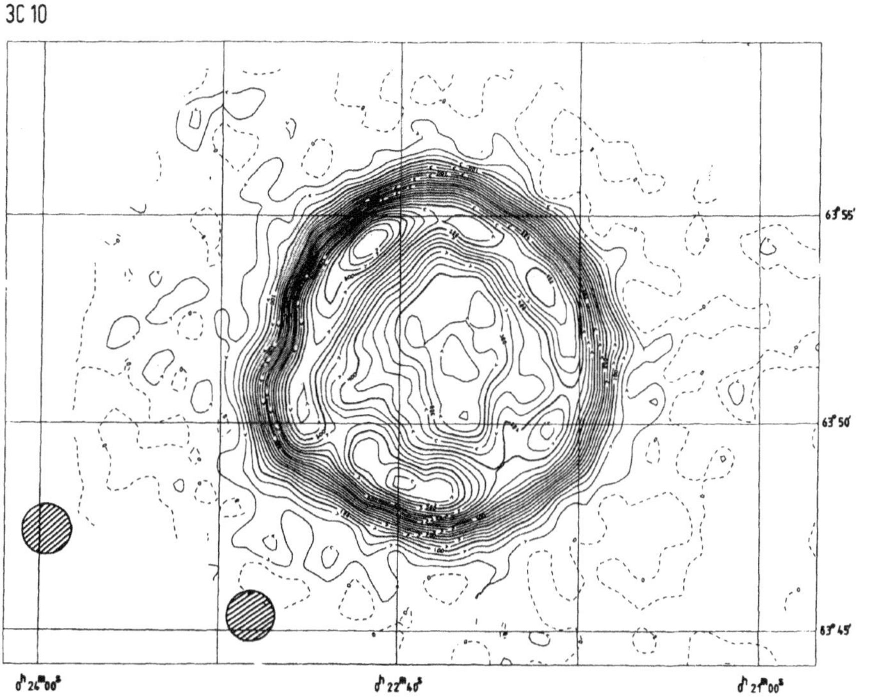

Fig. 6. The same data as Fig. 2, restored to the equivalent single-beam observation.

If additional dual-beam observations are made with a different beam separation λ_l, such that nulls in the respective spatial frequency distributions do not overlap, then all spatial frequencies except those close to zero will be well defined. This is illustrated in Fig. 7. All spatial frequencies except zero are now measured although the noise level of the very low frequency terms become worse closer to zero. The 2 sets of data are simply combined by weighting spatial frequencies from each dual-beam combination according to relative signal-to-noise ratio. With such a multiple-beam system, using for example 3 separated feeds, regions extending over many units of the larger far-field beam spacing may be mapped unambiguously without difficulty. Major continuum observing programs using this system at Effelsberg will begin towards the end of 1978.

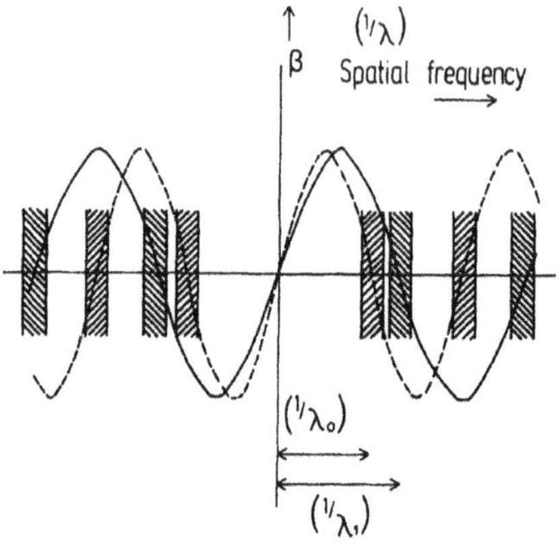

Fig. 7. Illustration of the complete spatial frequency coverage obtained by combining observations using 2 different beam spacings λ_o and λ_1.

DISCUSSION.

Comment T.W. COLE
1) How much blank sky do you need at the edge? 2) What is the noise degradation?
Reply D.T. EMERSON
1) It is necessary to scan far enough that both beams (& their sidelobes) are clear of the emission from the source at each end of the scan. A large "guard-ring" is not required. 2) For a source of ∿ 3 units of beam spacing in extent, this technique gives superior signal/noise ratio, even in the absence of atmospheric perturbations, compared to a Dicke-switching system. This is because a Dicke-switching system only sees the sky for ½ the observing time.

Comment U.J. SCHWARZ
One also could use different restoration techniques, such as CLEAN. Since the 'dirty beam' is non symmetric, one has to symmetrize it by a self-convolution and convolution of the data.
Reply D.T. EMERSON
Yes, one could use a version of CLEAN, but there is no need to. The technique I have presented gives a unique analytic solution, requiring no iteration and using negligible computing time.

FORMATION OF ACOUSTIC IMAGES FROM SPATIAL COHERENCE FUNCTIONS: PROCESSING BY SUPERRESOLUTION TECHNIQUES

Y. Biraud	B. Escudié	A. Hellion	J. Munier
Observatoire	ICPI	Sté Metravib	CEPHAG University
Meudon	Lyon	Ecully	Grenoble
France	France	France	France

Radioastronomy deals with radiowaves emitted by different kinds of sources. These sources are very far from the receiving antenna ; the distance R is very large compared with the wavelength. In acoustic imagery of noise sources the conditions are quite different.

The acoustic waves are scalar waves (pressure waves). The wave vector is very like the POYNTING vector related to transverse electromagnetic waves. The wavelength $\lambda = c/\nu$ is, in most cases, within 10 cm or a few meters, (frequency ν). The distance R from the sources to the antenna lies between :

$$100 \leq R \leq 3\,500 \text{ m}$$

The main problem is related to the mean power distribution of the incoming waves as a function of frequency and angle as shown by figure 1. An important feature of the interferometric receiver is that the signals are wideband ones [1] . The relative bandwidth may be of the order of 1/3 to 1/2 and 1 octave. In a few cases the bandwith may reach 2 or 3 octaves. The derivation of the receiver's performances never uses the assumption of narrow bandwidth. The array and the sensors have properties very like those of wideband antennas [2] [3] .

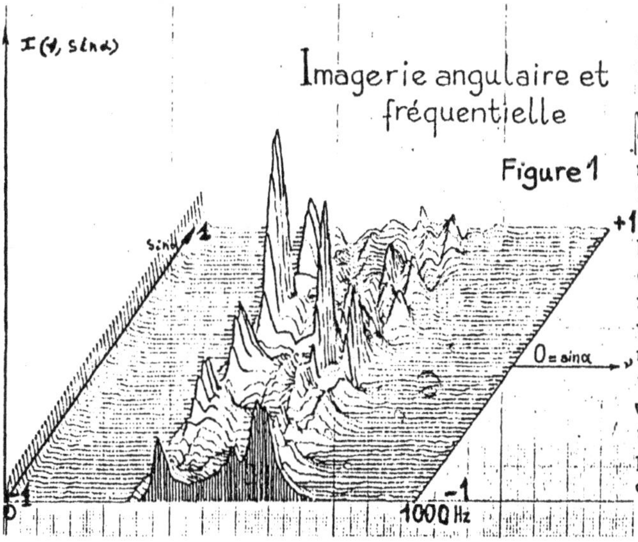

C. van Schooneveld (ed.), Image Formation from Coherence Functions in Astronomy, 299–313.

1. RANDOM PROPAGATION OF SOUND IN FREE TURBULENT ATMOSPHERE

Propagation of sound waves in free atmosphere is a random propagation. The fluctuations of air temperature and density generate turbulences. These turbulences modulate, in all places, the speed of sound. The local turbulent speed and the fluctuations of speed of sound affect the properties of the transmitted wave. The wave vector $\vec{k}$:

$$|\vec{k}| = \frac{2\pi}{\lambda} = \frac{2\pi\nu}{c_0}$$

is distorded by two effects :

- rotation
- modulus variation

Figure 2 displays these two effects. For a long range path ($R \simeq 1000\,m$) the effects of transmission in a turbulent medium transform a pure sine wave, the frequency of which is ν_0, into a narrow band random gaussian wave :

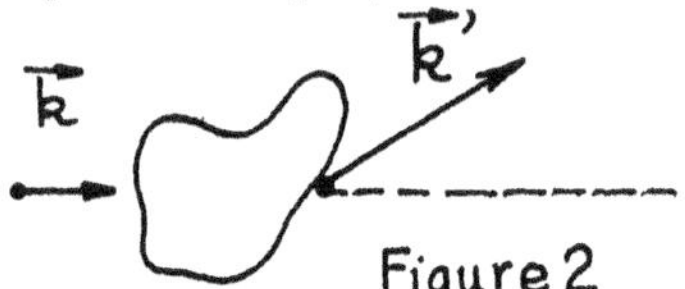

Figure 2

$$A_0 \sin(2\pi\nu_0 t + \varphi_0) \longrightarrow A(t)\sin(2\pi\nu_0 t + \Phi(t)) = X$$

X(t) is a stationnary gaussian process. The covariance function $\Gamma_X(t, \tau)$ is given by :

$$\Gamma_X(t,\tau) = \Gamma_X(\tau) = \sigma_A^2 \, e^{-\frac{|\tau|}{\tau_0}} \cos 2\pi\nu_0\tau$$

The constant τ_0 is such that : $\tau_0 \nu_0^q = cst$, $1 < q \leq 2$

The random medium may be represented by a "random parameter filter" whose transfer function $h(t, \nu)$ is such that :

$$h(t,\nu) = Re\,h(t,\nu) + i\,Im\,h(t,\nu)\ , \quad \Gamma_{Reh}(\tau,\nu_0) = K\,e^{-\frac{|\tau|}{\tau_0}}$$

The mean power spectrum density $\gamma_{Reh}(\nu, \nu_0)$ exhibits the spectral widening effect :

$$\Gamma_{Reh}(\tau,\nu_0) \underset{\tau}{\overset{\nu}{\rightleftharpoons}} \gamma_{Reh} = \frac{K}{2\tau_0}\,\frac{1}{1+4\pi^2\nu^2\tau_0^2}$$

as shown by figure 3.

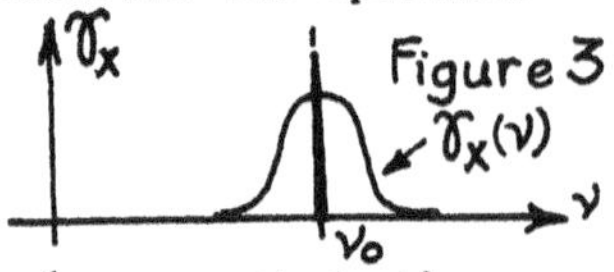

Figure 3

All along the propagation path turbulences act on the wave as a "space time" filter. Each turbulence, the correlation radius of which L is larger than λ, produces a "diffraction" effect on the wave. If we consider two paths R1 and R2 (see figure 4) the effects due to the turbulences will be the same if the two sensors, or microphones, are very near each other :

$$(k - j)\,\Delta < L$$

and these effects are independent or uncorrelated ones if the two sensors are spaced such that :

$$(k - j)\,\Delta \gg L$$

Many experiments done during the past three years have tried to measure the coherence function

$$C_{jk}(\nu, x)\ ,\ (k-j)\Delta = x$$

between the two received signals on two spaced sensors as shown in figure 4 :

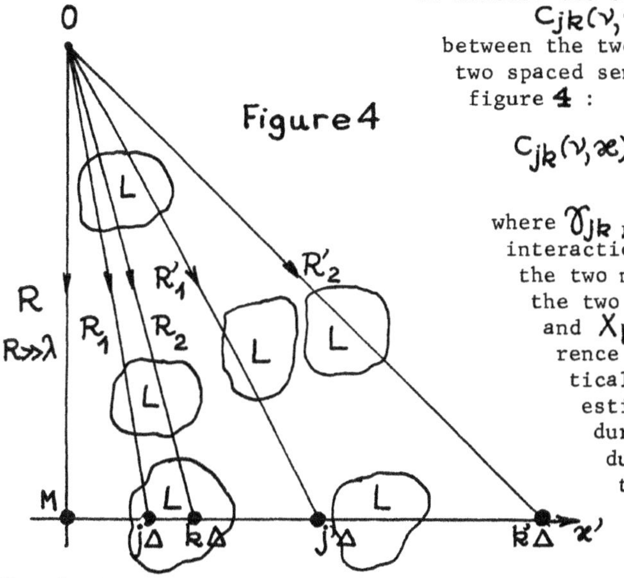

Figure 4

$$C_{jk}(\nu, x) = \frac{|\gamma_{jk}(\nu,x)|}{\sqrt{\gamma_j(\nu)\gamma_k(\nu)}}$$

where $\gamma_{jk}, \gamma_j, \gamma_k$ are the mean interaction power density, and the two mean power densities of the two received signals $X_j(t)$ and $X_k(t)$. Such a coherence function is a statistical moment, which was estimated on a long duration of time. This duration was determined to avoid bias and give a good statistical accuracy for the estimate [5].

For instance :

The conditions : T observation time $T = 1500\,s$

$\Delta\nu$ spectral broadening $\Delta\nu \simeq 1\,Hz$ bandwith

$\hat{C}_{jk} = 0{,}4$ correspond to : $\varepsilon_{\hat{c}} = \frac{\sigma_{\hat{c}}}{E\{c\}} \simeq 5.10^{-2}$

$\hat{c}$ estimate of $c(x,\nu)$ and $\sigma^2_{\hat{c}}$ variance of the estimate. [5]

All the experiments done on many different places and with many meteorological conditions lead to the result :

$$C(\nu, x) \simeq e^{-\frac{|x|}{n_\perp \lambda}}$$

$8 \leqslant n_\perp \leqslant 25$ $\qquad$ $400_{Hz} \leqslant \nu \leqslant 2\,400$ Hz

$n_\perp$ is a peculiar constant related to the turbulent effects of the medium on the transverse coherence. Such a result is an important one; we have to take it into account when we are deriving the angular and frequency response of the interferometric receiver.

2. ACOUSTIC IMAGERY BY VARIABLE BASELINE INTERFEROMETRY [5] [6]

M sources are radiating sound waves which are independant random ones. The wavelength λ is such that :

$$\lambda_m \leqslant \lambda \leqslant \lambda_M$$

The array or antenna is built with N sensors or microphones, which are "point sensors" compared with the values of λ . The spacing between two sensors Δ is : $\Delta \leq \frac{\lambda_m}{2}$

in agreement with the sampling theorem. Cross correlation between all the sensors in the array gives : $\Gamma_{jk}(x,\tau) = \sum_{\ell=1}^{M} \Gamma_\ell(\tau - \frac{x}{c_0}\sin\alpha_\ell) = \Gamma(\tau,x)$

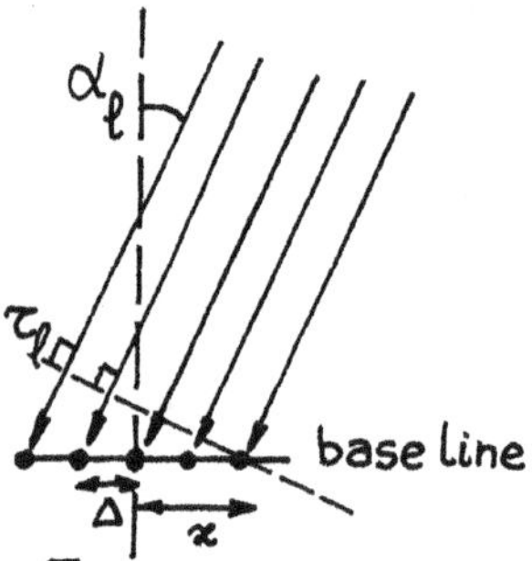

Figure5

Γ_ℓ is the correlation function of the random wave coming from the ℓth source in the direction α_ℓ as shown by figure 5.

The FOURIER transform of $\Gamma(\tau,x)$ is $\gamma(\nu,f)$ such that :

$$\gamma(\nu,f) = \int_{R^2} \Gamma(\tau,x)\, e^{-2i\pi(\tau\nu + fx)}\, d\tau\, dx$$

We may write it in the following form by a very easy derivation :

$$\gamma(\nu,f) = I(\nu,a) = \sum_\ell \gamma_\ell(\nu)\, \delta(f - \frac{\nu}{c}\sin\alpha_\ell)$$

$\gamma_\ell(\nu)$ is the power density spectrum of the ℓth random wave coming from α_ℓ direction and $\delta(\,)$ is the so called "DIRAC function" or DIRAC distribution. This result implies that, as figure 6 shows, in the (ν,f) plane the surface $I(\nu,f)$ have main traces such :

$$f = \frac{\nu}{c_0}\sin\alpha_\ell = \nu a$$

In fact, this result has to be modified to take into account two effects. The first one is the weighting effect of time window $F(\tau)$ multiplying the output of the correlator, and the second is the "coherence loss" effect expressed by the function $C'_\perp(x,\nu) = C_\perp(x,\lambda) = G(x)$ [5].

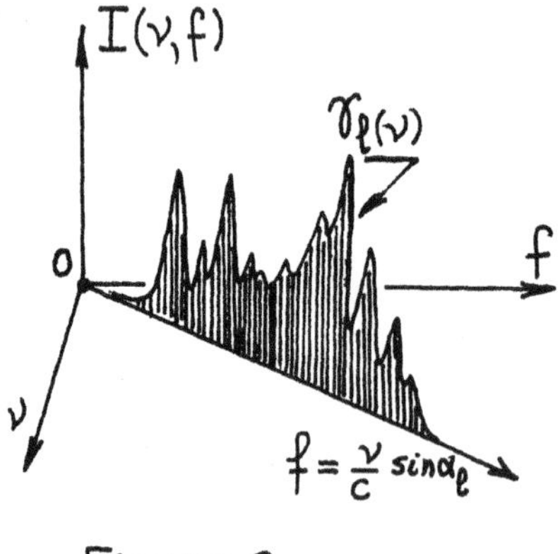

Figure 6

$$I(\nu,a) = \int_R \sum_{\ell=1}^{M} \gamma_\ell(n)\, w(\nu-n)\, c(na_\ell - \nu a)\, dn$$

$$f(\nu) \rightleftharpoons W(\tau),\quad G(x) \rightleftharpoons c(\nu),\quad a = \frac{\sin\alpha}{c_0}$$

a is the result of division between a wave number (m^{-1}) by a frequency ($Hz \sim s^{-1}$). c(f) is the coherence weighting function :

$$c(f) \overset{f,x}{\rightleftharpoons} G(x)$$

In most cases the maximum value of the delay $|\tau_M| \leq \frac{N\Delta}{c_0}$ is such that : $f(\nu)$ is a "pulse" function compared with $g(\nu a)$ which is the weighting function related to C(f), as shown by figure 6. We may express $I(\nu,a)$ in an easier way :

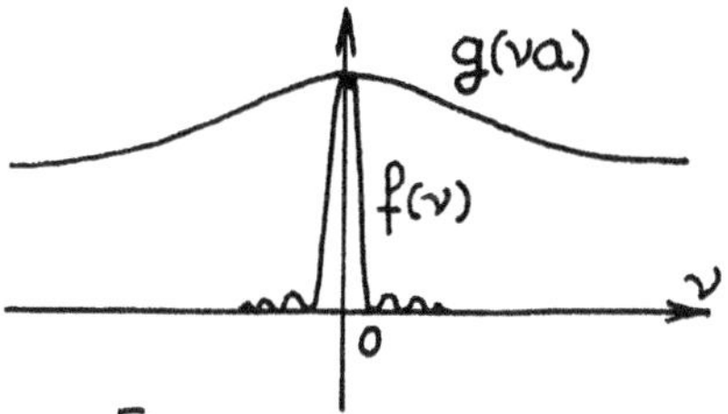

Figure 7

$$I(\nu,a)=\sum_{\ell=1}^{M} g(\nu(a_\ell - a))\,(\gamma_\ell * f)_{(\nu)}$$

under the previous assumption that f is very near a Dirac distribution.

This is the most frequent situation in noise imagery. [5]. The total length of the array is:

$$2N\Delta = 2N\frac{\lambda_m}{2} = 2D_M$$

The usual conditions are such that: $|x| \leqslant x_M$ becomes

$$x_M = p\frac{\lambda_m}{2}\ ,\quad \varsigma(x,\nu) \rightarrow G(x) = \Pi_{2x_M}(x)\,.\,e^{-\frac{|x|}{n_\perp \lambda}}$$

Π is the "rectangle" or "gate function" such that :

$$\Pi(x) = 1\ ,\quad 0 \leqslant |x| \leqslant \tfrac{1}{2}$$
$$\Pi(x) = 0\ ,\quad |x| \notin \,]0, \tfrac{1}{2}[$$

The FOURIER transform of G(x) is :

$$g(-\nu a) \overset{-\nu a \quad x}{\rightleftarrows} \Pi_{2x_M}\varsigma(x,\lambda)$$

$$g(-\nu a) = \frac{2n_\perp\lambda\,(1 - e^{-P/2n_\perp}[\cos(\pi p \sin\alpha) - 2n_\perp \sin(\pi p\sin\alpha)]}{1 + 4\pi^2 n_\perp^2 \sin^2\alpha}$$

$a = \frac{\sin\alpha}{c_0}$

We may define the angular resolution α' by:

$$\tfrac{1}{2}g(0) = g\left(-\nu\frac{\sin\alpha'}{c_0}\right)$$

We point out two different situations :

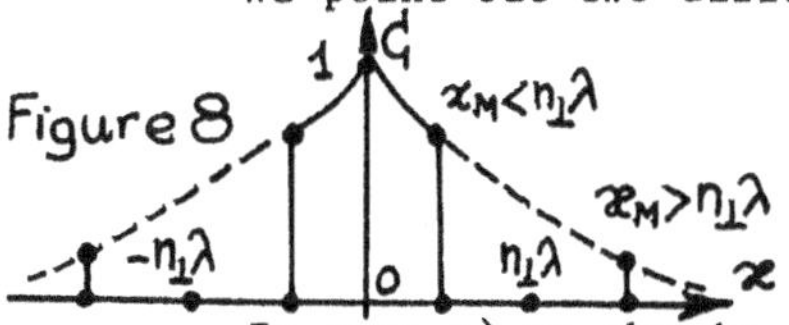

a) $x_M = p\frac{\lambda_m}{2} < n_\perp\lambda_m$

b) $n_\perp\lambda_m < x_M = p\frac{\lambda_m}{2}$

In case a) we obtain : $e^{-\frac{x_M}{n_\perp\lambda}} \simeq 0\ ,\ x_M \simeq 4n_\perp\lambda_m$

and in case b) : $e^{-\frac{x_M}{n_\perp\lambda}} \simeq 1\ ,\ x_M \simeq \frac{n_\perp\lambda_m}{3}$

as shown by figure 8. In the last case we may assume that :

$$\varsigma(x)\Pi_{2x_M}(x) \simeq \Pi_{2x_M}(x) \rightarrow g(-\nu a) \simeq \frac{\sin 2\pi\nu a x_M}{\pi\nu a}$$

α'_c is such that:

$$g\left(-\nu \frac{\sin \alpha'_c}{c_0}\right) = \frac{1}{2} g(0)$$ as a function of $n_\perp$ and p:
$$p \leqslant 15 \quad , \quad n_\perp \simeq 8 \text{ to } 25$$

The main result is that α'_c is a decreasing function with p and the different values of $n_\perp$ have very low effects on the function $\alpha'_c(p, n_\perp)$.

Practical constraints limit the usual values of p. If we deal with sources emitting random sine waves inside the "63 Hz octave" the wavelength λ lies between :

$$3m = \lambda_m \leqslant \lambda \leqslant \lambda_M = 6m$$
$$\Delta = \frac{\lambda_m}{2} = 1{,}5m \quad , p = 10 \rightarrow x_M = 15m \,!$$

The same computation in the case of "31 Hz OCTAVE" implies that : $x_M = 30$ m, $p = 10$ (i.e. large values for x_M and p).

Such constraints on the array have drastic limitations on x_M and p. In many cases : $$x_M = p \frac{\lambda_m}{2} \leqslant 5 \lambda_m$$

The array is designed as a "gracious array" very near the optimum solutions derived by F.BIRAUD and J.C. RIBES [8].

3. ANGULAR RESPONSE OF THE RECEIVER, COHERENCE LOSS AND ANGULAR RESOLUTION

The angular response of the receiver may be derived as a "Green function" under the assumption that a unique point source emits a pure tone wave and is located in the direction α_ℓ. This angular response is :

$$g(\nu a_\ell - \nu a) \quad , \quad G(x) = \Pi_{2x_M}(x)\, \zeta(x)$$

We note that : $$\zeta(x_M) \simeq 1, \quad x_M \leqslant 5 \lambda_m$$

This result may be explained by the properties of the instantaneous phase $\Phi_{12}(x, t)$ between two signals S_1 and S_2 received on two sensors located in the positions as shown by figure 10. The experimental set described on figure 10 provides an estimate of $\hat{\Phi}_{12}(x,t)$, S_0 the emitted wave being a sine wave the frequency of which ν_0 : $\nu_0 = 1$ KHz. $\hat{\Phi}_{12}(x, t)$ is a randomly sampled function, the mean sampling frequency is ν_0 , and :

$$\sigma_{\nu_S} \ll \nu_0 \qquad \nu_S \text{ sampling frequency}$$

A rather easy derivation shows that a low pass filter allows a very accurate restoration of $\Phi_{12}(x, t)$ [6]. The statistical properties of $\Phi_{12}(x, t)$ are such that :

$$\Gamma_{\Phi_{12}}(\tau) = E\left\{\Phi_{12}(x,t)\,\Phi^*_{12}(x,t-\tau)\right\} \rightleftharpoons \gamma_{\Phi}(\nu)$$

$\gamma_\Phi(\nu)$ is a "low pass" spectrum density, the cutoff frequency of which is very near 1 Hz. If $x \leqslant 5\lambda_m$ we obtain :

$$E\{\Phi(x,t)\} \equiv 0, \quad \sigma_\Phi \leqslant 0{,}2^\circ, \quad x \leqslant 5\lambda_m$$

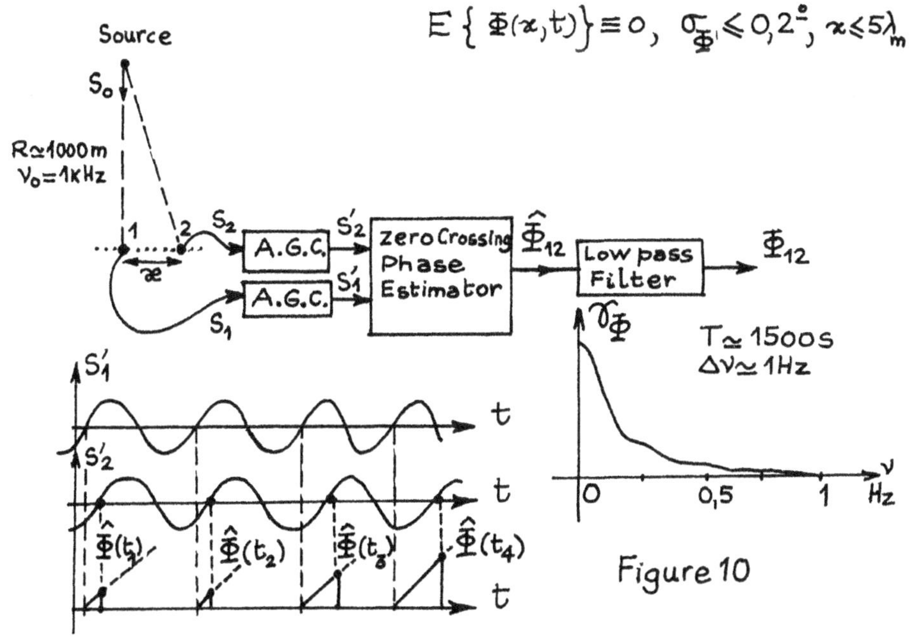

Figure 10

The coherence function describes the random modulation effects due to turbulences. These effects are modulating the group delay of the medium and modulate the path delay or the time of arrival of each wave. All these effects appear in the instantaneous phase of two signals received on two sensors distance x in the array. [6]

The coherence loss C(x) limits drastically the angular resolution of the interferometric receiver. A solution of the finite angle resolution is the technique of super-resolution and mainly "deconvolution." If we try to use the deconvolution technique we need a main information which is the angular response :

$$g(\nu(a_\ell - a))$$

We pointed out, in the last section, that we are facing two cases :

a) $n_\perp \lambda_m < x_M \quad , 8 \leqslant n_\perp \leqslant 20$

b) $x_M < n_\perp \lambda_m$

In the case a) the angular response is $g(\nu a) \rightleftharpoons G(x) = C\Pi_{2x_M}$ C(x) is the coherence function. We only deal with estimates of $C(x)$

and in every situation the condition described as case a) implies we have to estimate C(x). This is a very drastic limitation and may be a drawback in many cases.

On the other hand in case b) the assumption is :

$$x_M = p\frac{\lambda_m}{2} < n_\perp \lambda_m$$

and :

$$\Pi_{2x_M}(x)\,C(x) \simeq \Pi_{2x_M}(x)\ , \quad x_M \simeq \frac{n_\perp}{3}\lambda_m$$

In such a case the deconvolution technique may be used with the realistic assumption that :

$$g(-\nu a) \simeq \frac{\sin 2\pi\nu a x_M}{\pi\nu a} \qquad \begin{array}{c} x_M \leqslant 5\lambda_m \\ 10 \leqslant n_\perp \leqslant 25 \end{array}$$

4 . ACOUSTIC IMAGE PROCESSING BY SUPER-RESOLUTION TECHNIQUES

a - practical constraints and a priori information on acoustic images:

The processing of the interferometric image needs a resolution criterion. This criterion tends to define and use some "a priori" informations about the object or the sources. The image, or object's observation, or estimate, is such that :

$$I(\nu,a) = \int_R O(\nu,a')\,g(\nu a - \nu a')\,da' + b(\nu,a)$$

O is the object and $b(\nu,a)$ the estimation noise due to the random modulations in the medium and the finite integration time. In many cases the signal to noise ratio may be very high. O is a positive variable or density function and we get :

$$O(\nu,a) \geq 0\ , \quad \hat{I}(\nu,a) = \int_R O(\nu,a')\,g(\nu(a-a'))\,da' + \overbrace{b(\nu,a)}^{\text{NOISE}}$$

This is an important a priori information. We have to point out that a is such that :

$$-\frac{1}{c_0} \leqslant a \leqslant \frac{1}{c_0}$$

$$-\frac{\pi}{2} \leqslant \alpha \leqslant \frac{\pi}{2}$$

The usual acoustic interferometric image deals with angular information in the horizontal plane. $O(\nu,a)$ is a "finite length" object.

In many cases the noise sources are point sources emitting narrow band signals or pure sine waves. All these a priori informations are taken into account to get an efficient image processing.

b - estimation of intensity and location of point sources emitting pure tone waves:

In many cases the radiated field is generated by a few point sources, which may be considered as uncorelated sources. In such a case under this assumption, the estimation deals with a finite number of source parameters.

The data about the source location may be extracted from the cross spectrum matrix which describes the interaction spectra of all the signals received by the array's sensors. The number of point sources which may be identified is directly related to the number of sensors in the array. This problem was first described by H. MERMOZ, who uses the whole data to improve the propagation model under the assumption that the number of sources M is such that [9] : $M \leq N-1$.

If M point sources are radiating stationnary uncorrelated random waves, and if the antenna is built with N sensors, we may describe the kth output of the array as the output of a "linear filter" processing the signal emitted by the mth source. The transfer function is : $\xi_{km}(\nu)$

The cross spectrum density between the kth and the lth output $\gamma_{k\ell}$ is such that :

$$\gamma_{k\ell}(\nu) = \sum_{m=1}^{M} \gamma_m(\nu) . \xi_{km}(\nu) \xi^*_{\ell m}(\nu)$$

If the propagation medium is an isotropic, stable and non dispersive one, $\xi_{k\ell}(\nu)$ may be expressed as :

$$\xi_{k\ell}(\nu) = e^{-2i\pi\nu\tau_{k\ell}} h_k(\nu)$$

$\tau_{k\ell}$ is the propagation delay related to the mth signal received by the kth sensor, $h_k(\nu)$ is the complex transfer function of this sensor.

One may define "directional vectors" $\xi_m(\nu)$, which are k dimensional ones, related to one source, given the N sensors located in the array : [10] [11]

$$\xi_m(\nu) = \begin{bmatrix} \xi_{1m} \\ \vdots \\ \xi_{Nm} \end{bmatrix}$$

The cross spectrum matrix $\Gamma(\nu)$ is easily derived :

$$\Gamma(\nu) = \sum_{m=1}^{M} \gamma_m(\nu)\, \xi_m(\nu)\, \xi_m^+(\nu)$$

ξ_m^+ is the conjugate transposed vector.

If : $M < N$

the cross spectrum matrix $\Gamma(\nu)$ is a M dimension square matrix which has (N - M) non zero eigenvalues. Computing these eigenvalues leads to the number of sources.

One may define, in the same way, a "directional matrix" $X(\nu)$ which is a M x N dimension one, such that : $X(\nu) = [\xi_{km}(\nu)]$

The cross spectrum matrix is a function of $\mathbf{X}$:

$$\boldsymbol{\Gamma} = \mathbf{X}\,\mathbf{S}\,\mathbf{X}^{+}$$

with : $\mathbf{S} = [\gamma_m]$ which is a diagonal matrix.

H. MERMOZ defined the "source vectors" $y_m(\nu)$ by : [9]

$$y_m(\nu) = \sqrt{\gamma_m}\,[\xi_{km}] = \sqrt{\gamma_m}\,\xi_m$$

and a "source matrix" such that :

$$\mathbf{Y}_{(\nu)} = [\,y_1(\nu), \ldots, y_k, \ldots y_M(\nu)\,]$$ which is an N x P order matrix

$$\boldsymbol{\Gamma} = \mathbf{Y}\mathbf{Y}^{+}$$

To identify the M sources, we have to compute the mean power spectrum densities and the directional vectors from the eigenvalues and the eigenvectors of the $\boldsymbol{\Gamma}$ matrix. The solution of such problems requires in most cases a rather cumbersome algorithmic solution [12] .

When dealing with a rectilinear array and with remote sources, the solution may be easily derived. The source parameter to be estimated is the angle α_m and the mean power density $\gamma_m(\nu)$. The cross spectrum density related to the kth and l^{th} sensor is :

$$\gamma_{k\ell}(\nu) = \sum_{m=1}^{M} \gamma_m(\nu)\, e^{i(k-\ell)\,2\pi\frac{\Delta}{\lambda}\sin\alpha_m}$$

The cross spectrum matrix $\boldsymbol{\Gamma}$ is now a TOEPLITZ one. Estimation of the quantities $\gamma_{1\ell}(\nu)$ is a sufficient estimation to derive the whole matrix. In such a case a possible estimation of the source parameters uses the PISARENKO method [13] .

A peculiar case of interest is, when dealing with two sources located in very close directions (such as double sources in radioastronomy). For a rectilinear antenna built of N equally spaced sensors, the lower bound (as defined by CRAMER RAO) of the angular resolution is given by [14] [15]:

$$\delta\alpha = \frac{\lambda}{\pi\Delta}\left[\frac{45}{N(N^2-1)(2N-1)(8N-11)\mathcal{R}}\right]^{1/4}$$

$\delta\alpha = \alpha_1 - \alpha_2$ difference between the angular parameters, such that $\delta\alpha^2$ equals the variance of the estimate. This result deals with two sources the powers of which are the same. $\mathcal{R}$ is defined as :

$$\mathcal{R} = \frac{s^2}{\sigma^2}$$

$\mathcal{S}$ is the mean power of the source, σ is the standard deviation of the crossspectrum estimates, which are supposed to be uncorrelated with the same variance σ^2 .

A super-resolution gain (defined by reference to RAYLEIGH's criterion), about 20, needs for a three sensor antenna (M = 3) an accurate estimation of the cross spectrum densities such that :

$$\mathcal{R} \simeq 760 \ , \ \frac{\sigma}{\mathcal{S}} \simeq 3{,}5\ 10^{-2}$$

This solution requires that there are two and only two sources to be estimated from the data given by the three sensors. Under this assumption the solution describes the mean power density of each source by estimation of $\gamma_m(\nu)$, $\nu_{min} \leqslant \nu \leqslant \nu_{max}$. Even if the three sensors are such that their transfer functions have different modulus the solution may be derived under the assumption that the phases of these transfer functions are the same, by computation of the complex coherence matrix :

$$\mathbf{C} = \left[\frac{\gamma_{k\ell}(\nu)}{\sqrt{\gamma_k \gamma_\ell}} \right]$$

This C matrix is a TOEPLITZ one and the diagonal values are all equal to 1 ; the computation deals only with the two complex data C_{12} and C_{13} in order to estimate the four parameters : $\alpha_1, \alpha_2, \gamma_1, \gamma_2(\nu)$ $P_1 = \mathcal{S} = P_2$, (M = 2, N = 3). A straight forward solution is proved to exist [16] . Under the same assumptions, defined above, the angular resolution $\delta\alpha$ is :

$$\delta\alpha = \frac{\lambda}{\pi\Delta}\left(18\,\mathcal{R}\right)^{-1/4}$$

Such an estimation is not a true efficient one as defined by the CRAMER-RAO bound. Nevertheless this result is very near the optimum one :

$$\delta\alpha_{min} = \frac{\lambda}{\pi\Delta}\left(\frac{104}{3}\,\mathcal{R}\right)^{-1/4}$$

The RAYLEIGH criterion defines the angular resolution as : $\lambda/2\Delta$ then the super resolution gain is : $\eta = \frac{\pi}{2}\left(18\,\mathcal{R}\right)^{1/4} = 3{,}24\ \mathcal{R}^{1/4}$

If $\mathcal{R} = 760 \ , \ \frac{\sigma}{\mathcal{S}} \simeq 3{,}5\ 10^{-2} : \ \eta = 17$

This proves that the superresolution gain needs a very accurate estimation of the cross spectrum densities or of the coherence function.

c - positive deconvolution of acoustic images related to "finite length" objects [17] :

In many cases the assumption of a few uncorrelated point sources cannot be taken into account. We are faced with the deconvolution problem. As explained before we deal with an image

$$I(\nu, a) = (g * O)_{(\nu, a)} \quad , \quad O(\nu, a) \geq 0$$

$$I(\nu, a) = \int_R O(\nu, a')\, g(\nu a - \nu a')\, da'$$

$a = \frac{\sin\alpha}{c}$, $0 \leq |\sin\alpha| \leq 1$ The "object" is a finite length one, represented by a function which is a non zero one only if :

$$-\frac{\pi}{2} \leq \alpha \leq \frac{\pi}{2}$$

Y. BIRAUD has shown that a possible positive deconvolution method may use an iterative restoration whose first order solution is the so called "BRACEWELL'S principal solution". The iterative computation tries to get new informations about the FOURIER transform of $O(\nu,a)$ outside of a given bandwidth $(\mathcal{B})$, under the constraint : $O(\nu,a) \geq 0$

$$J \rightleftharpoons j \, , \qquad O(\nu, a) = |J(\nu, a)|^2 \overset{\nu a \,\, x}{\rightleftharpoons} \mathcal{O}(x) = (j * j)(x)$$

The iterative process deals with a function $j_n(x)$ such that :

$$\varepsilon_n^2 = \int_{(\mathcal{B})} (j_n^{*2} - j_o)^2 dx \longrightarrow (\varepsilon_n^2)_m \quad \text{minimum of } \varepsilon_n^2$$

where $j_o(x)$ is the BRACEWELL'S SOLUTION, j_n being an hermitian function, $J \in R \rightarrow j = j^*(-x)$. An energetic constraint ensures that :

$$j_n(0) = j_o(0)$$

The next step $j_{n+1} = j_n(x) + \omega_n(x)$ where $\omega_n(x)$ is an hermitian "perturbation pulse function". This pulse function is located at :

$x = k x_s$, where x_s is the sampling period of $j_n(x)$. The solution has to compute ω_n such the two constraints which were defined previously are fullfilled. The iterative process goes on till the error ε_n^2 is such that :

$$\varepsilon_n^2 \simeq \sigma_N^2$$

where σ_N^2 is the variance of the "noise" process added to the signal.

To compute the BRACEWELL'S solution we have to define the equivalent bandwidth $(\mathcal{B})$ and the sampling rate related to the impulse response :

$$g(\nu a) \overset{\nu a \,\, x}{\rightleftharpoons} G(x)$$

Under the assumptions previously defined we may assert that :

$$g(\nu a) = \frac{\sin 2\pi \nu a x_M}{\pi \nu a}$$

or we may choose a weighting function such that :

$$g(\nu a) = \left(\frac{\sin 2\pi\nu a x_M}{\pi\nu a}\right)^2 \rightleftharpoons \left(\Pi_{2x_M} * \Pi_{2x_M}\right)(x)$$

Some experiments with such a response show that a deconvolution solution may be found even if the Fourier transform of $g(\nu a)$ equals a "square" function [17].

The signal to noise ratio cannot be defined in an easy way. The effects of "turbulent fluctuations" are modulation effects. These effects are very near the so called "scintillation" effects. The previous results on "random propagation" in turbulent atmosphere show that they are very low frequency effects. A way to describe easily the sources is to estimate a first image, then to have a "deconvolution" solution. An other estimate computed on an uncorrelated sample, a few minutes after, may be processed to have a new solution, and so on. We are now trying to do systematic experiments with different sources and environmental conditions.

5. CONCLUSION

Incoherent imagery deals with uncorrelated sources, the statistical properties of which are only second order moments. In such a case the space and frequency coherence function provides an easy way to describe the properties of the far field. A rectilinear array with N sensors receives the incoming field.

The cross spectrum matrix may be processed to estimate the directions and the power spectra of the M sources, $M<N$. The length of the receiving array $2x_M = 2p\lambda/2$ has to be lower than $2n_\perp\lambda$, where $n_\perp$ is a peculiar parameter which describes the coherence loss due to the random propagation medium. Such an assumption $x_M < n_\perp\lambda$ allows to describe the space and frequency response of an interferometric image processor as the FOURIER transform of a constant function inside the "total length" $x_M = p\lambda/2$. The angular resolution is limited by the random propagation effect and mainly by the coherence loss. A solution to improve the frequency and angular resolution is to process the image by super resolution techniques such as source estimation and deconvolution. A priori informations such positiveness, finite length, finite number of sources may provide constraints leading to super resolution effects under the assumption that the estimation accuracy is a convenient one.

REFERENCES

1 Bard, C.: 1976, Eng. Doct. Thesis, Grenoble University, BP 15, 38040 Grenoble-Cedex.
2 Foissac, Y.: 1973, Proc. of the 4th GRETSI Symposium, T1, 247-275.
3 Larcher, J., et al.: 1975, Proc. of the 5th GRETSI Symposium, 211-218.
4 Beroujon, M., et al.: 1975, Proc. of the 5th GRETSI Symposium, 521-529.
5 Hellion, A., et al.: 1977, Proc. of the 6th GRETSI Symposium, 41/1-41/8.
6 Escudié, B., et al.: 1977, "Revue d'Acoustique" 40, 85-90.
7 Escudié, B.: 1976, MQV Contract No. 7409, ICPI Internal Report TS7506.
8a Biraud, F., et al.: 1974, IEEE Trans.Ant.Prop., 22, No. 1, 108-109.
8b Blum, E.J., et al.: 1975, "Astron. Astrophys." 41, 409-411.
9 Mermoz, H.: 1976, "Annales Télécommunications" 31, 17-36.
10 Bienvenu, G.: 1975, Proc. of the 5th GRETSI Symposium, 474-479.
11 Bienvenu, G.: 1977, Proc. of the 6th GRETSI Symposium, 17/1-17/9.
12 Liggett, W.S.: 1973, Proc. of the Loughborough NATO-ASI Symposium, J.W.R. Griffiths, P.L. Stocklin and C. van Schooneveld Ed., Academic Press, 327-345.
13 Pisarenko, V.F.: 1973, "Geophys. Jour. R. Astr. Soc." 33, 347-366.
14 Munier, J.: 1977, Proc. of the 6th GRETSI Symposium, 20/1-20/8.
15 Munier, J.: 1979, to be published in Signal Processing no. 1.
16 Munier, J.: 1977, Proc. of the EUROMECH 94 Meeting.
17 Biraud, Y.G., 1969, Proc. of the 2nd GRETSI Symposium, T2, 853-878.

References 2, 3, 4, 5, 10, 11, 14 and 17 are edited by B. Derouet, Gretsi Secrétariat, B.P 93 06802 Cagnes-sur-Mer, France.
Reference 7 may be obtained at ICPI, Laboratoire Traitement du Signal, 25 Rue du Plat, 69288 Lyon-Cedex 1.
Reference 16 is edited by Dr. Filippi, 31 Chemin J. Aiguier, 13274 Marseille.

DISCUSSION

Comment D.B. SHAFFER
Are you trying to understand the atmosphere or do you study acoustic sources in the atmosphere, like thunder or airplanes?
Reply B. ESCUDIE
We try to identify the properties of the random channel (atmosphere). These random properties are related to effects such as "coherence loss" on the array. Taking these data into account we make an image (estimate the brightness) of noise sources such as powerplants, factories, etc., and in the future of vehicles such as cars, trains and maybe aircraft.

Comment T.A. CLARK
In addition to the other non-linear effects, do you also have to consider

dispersion in your medium?
Reply B. ESCUDIE
Theoretically speaking the medium is a dispersive one. In fact the random modulation effects are the main ones, as far as we know now. These modulation effects are frequency dependent. The spectrum width is a function of frequency ν:

$$\Delta\nu = k\,\nu^q \; ; \; 1 \leq q \leq 2$$

This may also be called a frequency dependent effect or a dispersive effect.

Comment C. VAN SCHOONEVELD
Can the situation be summarized in the following way? In a frozen, but granulated medium, a point source can always be identified as such. If sufficiently rapid fluctuations occur, this is no longer possible and a point source can not be distinguished from an extended one, unless a-priori knowledge can be used in a deconvolution.
Reply B. ESCUDIE
This is perfectly correct! We have to use the a-priori information to do the "deconvolution" processing. The random modulation effects may be interpreted as a variable medium, the impulse response of which is a random one. The impulse response of the interferometric receiver is a highly medium dependent response. If we estimate the image of a "true point source" with the interferometric receiver, the estimated image will tend to a "pulse function" only if the integration time T_i is very high compared with the inverse of spectrum width ΔV. This pulse function is not very "narrow" one.

PART VII

RELEVANT INPUTS FROM OTHER FIELDS

RECONSTRUCTION FROM PROJECTIONS IN MEDICINE AND ASTRONOMY

(Invited paper)

Richard Gordon
Director, Computer Department for Health Sciences
Associate Professor, Pathology and Radiology Departments
University of Manitoba
Winnipeg, Canada

It is a great pleasure for me to address this colloquium of radio astronomers on reconstruction from projections, especially since this opportunity allows me to combine an early interest in astronomy with my current work.

Some of you have been applying methods of reconstruction since the mid-1950's. Radio astronomers formed a cohesive society then and the methods reached fruition and matured in a natural progression. Reconstruction from strip scans and then interferometer data has been so fundamental to the formation of images in radio astronomy that one may well say that the field could not have progressed without them.

The situation in medicine has been quite different. Shortly after the discovery of x-rays in 1895 they were used in medicine. By the 1920's "tomographic" methods were discovered for obtaining cross sections of the body by blurring out images of unwanted planes. The human body contains high contrast objects (bones) which appear in exquisite detail in either ordinary radiographs or the "newer" tomograms. High contrast is also obtained by injecting radio-opaque substances into the blood stream (angiography) and the intestinal tract. Thus inferences can also be made about soft tissues. Perhaps because of the success of tomography and the lesser mathematical training of radiologists, the reconstruction problem never got formulated within that field in a mathematical way. The math was, of course, second nature to astronomers.

Reconstruction from projections was primarily thrust upon medicine from without. The earliest published attempt was a complete one: a group of mathematicians, programmers and engineers (Kalos, et al., 1961), trying to find defects in atomic reactors, rediscovered the mathematics of Johannes Radon (1917), put together a prototype, and designed an instrument in 1961 which was essentially identical to the first EMI scanner marketed in 1972 (Hounsfield, 1973). Unfortunately, although this group realized the potential impact on medicine, they could not gain the cooperation of physicians. Similar

C. van Schooneveld (ed.), Image Formation from Coherence Functions in Astronomy, 317-325.

sad tales were to be repeated for a number of other early starters, including myself, during the 1960's. Many manufacturers were approached with the idea and uniformly turned it down, including major manufacturers of medical x-ray equipment. It took EMI, which is better known for its involvement in military electronics and phonograph records, to bring the idea to commercial fruition.

The x-ray scanners, now marketed by roughly 20 companies, have gone through five distinct "generations":

1) Two motion, single detector instrument using a pencil beam which is translated parallel to itself, and then rotated slightly (usually 1 degree) for the next scan.

2) Two motion, multiple detector instrument using a narrow fan beam. The rotation is made in larger angular increments, corresponding to the angle of the fan beam.

3) Single motion, multiple detector instrument using a wide fan beam which is simply rotated about the patient with the detectors.

4) Single motion, wide fan beam instrument, with a full circle of stationary detectors. Only the x-ray source rotates about the patient.

5) No motion instrument with a number of stationary fan beam x-ray sources and corresponding detectors.

The only fundamental distinction in this sequence of instruments is that each generation is faster in its data collection than the previous one. The last generation, which is just now being prototyped, uses a burst of x-rays to stop the motion of the beating heart (cf. Gordon, et al., 1975). The convolution algorithm (based on Radon's formula) is almost exclusively used in commercial scanners because it is computationally fast and easy to design into special purpose and parallel computers (cf. Tasto, 1977). This algorithm is not optimal in terms of giving the best image for the least dose (Gordon, 1976).

Computed tomography, as reconstruction from projections has come to be known in medicine, has had a large and immediate impact because it allows soft tissues to be seen with clarity. Classical tomography could do little more than image bones. Computed tomography may be used for nearly any medical problem whose diagnosis would be aided by peering inside the body.

Classical tomography has not been completely displaced, however, because computed tomography, as now practiced, cannot achieve the same spatial resolution without giving the patient dangerous amounts of x-radiation (Gordon, 1976, 1978). This anomaly deserves the attention of astonomers, for they too do not have as much radiation available as they would like. By careful design of the reconstruction algorithm, better images can be obtained from the available radiation.

In order to illustrate this point, I would like to give an example of optimal use of the radiation available in a problem in medical reconstruction, which may have application in x-ray astronomy. I will demonstrate an algorithm (called ARTIST) which attempts to make maximal use of gamma ray coincidence pairs in positron tomography (Gordon, 1975). In this diagnostic method a patient is administered a positron emitting radionuclide. Each positron emitted wanders a short distance within the patient (1 to 6 mm) before meeting an electron. The two annihilate with production of a pair of gamma rays which travel in opposite directions along a common straight line. If this line is in the plane of a ring of detectors, the two gamma rays will arrive nearly simultaneously ("coincidentally") at two of the detectors.

When a coincidence pair is detected, we know that the annihilation event occurred somewhere along a strip between two particular detectors. The reconstruction problem is to estimate the most likely point of origin of the photon pair. I will describe the ARTIST algorithm as it operates on each consecutively received coincidence pair, by assuming that it has already been operating for some time. Thus, imagine that we have an ordinary storage cathode ray tube, with a point displayed for each annihilation event whose location has already been estimated. We observe the screen through a slit, which is aligned with the coincidence line and opened slowly. The amount of light coming through the slit will increase in steps. We stop opening the slit either when we have reached a preset threshold count, or when we have reached some maximum slit width. The light coming through the slit is then scanned one dimensionally, resulting in a function which is regarded as a probability distribution. A random number is used to locate a single point along the coincidence line, chosen by this probability distribution. This point on the screen is then made to glow, and we are ready for the next coincidence event.

The ordinary approach to this reconstruction problem is to bin the coincidence counts in parallel strips between pairs of detectors. One may then use any ordinary reconstruction algorithm which handles data in parallel strips. However, fluctuations in the number of counts in each bin become noise in the data which is carried into the reconstructed image. In order to reduce this noise, the bin widths are typically set at 1 cm (Phelps, et al., 1977), which can be ten times the resolution limit set by the wandering distance of the positron! The algorithm I have presented, on the other hand, makes use of every single event and can do so without throwing away resolution during data collection. It thus suggests a different design of positron scanners, one with a large number of very small detectors.

The ARTIST algorithm may be a good replacement for the ART

algorithm (Gordon, et al., 1970) now in use in x-ray astronomy (Charles and Culhane, 1977). There are two points that its existence raises which may be germane to radio astronomers: 1) a reconstruction algorithm should be designed around the physics of the problem being investigated; 2) instrument design should come after such an algorithm is designed and tested. This apparently logical sequence of events is almost never followed in medicine.

I would like to point out some other areas in which astronomers are or could be using reconstruction from projections:

1) A number of gamma ray bursts of cosmic origin have been discovered recently. An attempt is being made to obtain their precise celestial positions and reconstruct images of these sources from data collected by a satellite camera with collimators consisting of multiple slits oriented in a number of directions (Gorenstein, Helmken and Gursky, 1975, 1976). Here we have another case in which reconstruction must be done from very few, individually recorded photons.

2) The solar corona is an x-ray and light emitting plasma whose structure may be reconstructed in three dimensions. The major difficulty is to obtain more than one view in a short enough time. This may best be done by having a number of satellites circling the sun simultaneously. Lacking such multiple views, solar astronomers make the assumption that the structure of the corona does not vary much from day to day as the sun rotates (Altschuler and Newkirk, 1969; Levinson, et al., 1975). Thus we may obtain a number of views of the corona from the earth.

3) When a star explodes in a supernova, it throws off a shell of gas and filaments which glow by emitted or reflected light. If the shell is rotating relative to our line of sight, it may be possible to reconstruct inhomogeneities in it from images recorded a few decades apart. This would be the first example of reconstructing the three dimensional structure of an object outside our solar system.

4) When a planet occults a star, we obtain an integral of the absorption of light at different wavelengths along the refracted path through the planet's atmosphere. Such data has been used to reconstruct atmospheres, assuming radial symmetry (cf. Colin, 1972 and Veverka et al., 1974). Ionospheres may also be reconstructed from occultations of radio sources (Fomalont, personal communication).

5) There is a new method for exploring inhomogeneities in the surface layers of the earth that could be used on the moon and the planets on which we can land a craft. First one drills two parallel holes into the ground. Then by placing transmitting and receiving antennas at various depths, the attenuation and phase of the signal may be measured for a number of paths between the two holes (Lager and Lytle, 1975). A reconstruction of the material between the two holes may then be calculated. A similar method has been used for reconstructing

an atmospheric profile using a pair of satellites (Liu, 1976).

6) Reconstruction from projections is just beginning to have an impact on nondestructive testing, and thus may find many applications in the testing of components of telescopes of all kinds.

7) The construction of large parabolic mirrors for visual astronomy is a major undertaking. In order to obtain larger apertures than currently available, a "strip telescope" has been proposed (Murata and Baba, 1975). The strip telescope would scan an object whose fine structure is desired, at many angles, and then reconstruct its image (Baba and Murata, 1975, 1977).

8) For the extraterrestrial and science fiction enthusiasts reconstruction from projections provides many exciting possibilities. For instance, a proposal has been made to image the interior of planets by observing the neutrinos passing through them on their way from the sun (Clement, 1970; cf. Saenz, et al., 1977). Certainly the rotation of the planet would provide multiple views for reconstruction. In reports of encounters of the third kind people sometimes tell of an "eye" which rotates about their body (J. Musgrave, personal communication). Perhaps it is a form of scanner. Science fiction cartoons have suggested that multiple "views" of a person be taken for transmitting a set of projections and materially reconstructing their bodies. In any case, since transmission of the projections of an image may be more efficient than transmission of a whole picture (Gordon and Herman, 1971; Wee and Hsieh, 1976), we might expect that we'll have to reconstruct extraterrestrial signals to make them intelligible!

I would like to end by discussing reconstruction from a mathematical and perhaps philosophical point of view. Reconstruction from projections is as fundamental to science as calculus. It is a battery of methods for unfolding a structure from observations. The data is an intricate convolution of our means of measurement with the object's structure. Indeed, this is generally true of all physical measurements. We are rarely interested in direct measurements of a system, but rather use them to infer an underlying structure.

A fundamental limitation has been discovered in reconstruction using parallel x-ray beams. This is the Indeterminacy Theorem (Smith, Solmon and Wagner, 1977) which states that "finitely many views give no information about an object in its interior". As I will show, this theorem has important implications for radio astronomers. Its proof is complex, requiring use of a fundamental theorem in differential equations which was only proved in 1956. I will give a heuristic "proof" which will make the result seem reasonable, even though it is counter-intuitive.

Let us suppose the object is contained within a connected two dimensional region A and consider any subset B of A. If we have N views or sets of parallel projections of the object in A then we may construct a 2N-sided polygon anywhere within A whose sides are parallel to the projections (Gordon and Herman, 1971). If we add some density at one vertex of the polygon, the same amount may be subtracted from an adjacent vertex, so that the ray sum along their shared edge is still the same as before. Alternate additions and subtractions around the whole polygon may be made, so that all N projections are left undisturbed. Now if one vertex of the polygon is in subset B and the rest are in its complement A-B, we may make an arbitrary change at a point in B and compensate for this change at points in A-B. By repeating the operation an uncountable number of times the reconstruction can be changed into any picture whatsoever in the region B (Gordon, 1979). (This is the mathematically weak step of this alternate "proof".) Since the size of B can be nearly the whole of A, we may say loosely that any picture is compatible with any finite set of projection data.

The Indeterminacy Theorem is directly applicable to radio astronomy because interference data is a subset of parallel projection data. This may be seen by taking the Fourier transform of a single visibility reading, which represents a component of the Fourier expansion for some projection.

The Indeterminacy Theorem is perhaps the most important result of pure mathematics in this area since Radon's original inversion formula. No matter how much data we collect, and no matter how precisely it is measured, we cannot narrow down what the reconstruction should look like. This is a sorry state of affairs, forcing us to re-examine the whole operation of reconstruction. The authors of the theorem draw one positive lesson, which perhaps indicates the way we must proceed: in light of the Indeterminacy Theorem it is not merely an improvement, but rather it is essential that we use a priori information to restrict the range of solutions to a given reconstruction problem. We must take a hard and long look at our sources of a priori information, ranging from the positivity constraint to the maximization of certain functionals, such as entropy, and to the use of "human judgement" (cf. Gordon, 1973). These are, of course, exactly the approaches being discussed in much of this conference. However, we have a limited understanding of the effects of a priori information. Thus we have not as yet mastered the fundamentals of reconstruction from projections.

REFERENCES

Altschuler, M. D. and Newkirk, G., Jr.: 1969, Magnetic fields and the structure of the solar corona I: methods of calculating coronal fields, "Solar Physics" 9, pp. 131-149.

Baba, N. and Murata, K.: 1975, Optical and computer simulations of a strip telescope, Digest of Papers, Optical Society of America Topical

Meeting on Image Processing for 2-D and 3-D Reconstruction from Projections: Theory and Practice in Medicine and the Physical Sciences.

Baba, N. and Murata, K.: 1977, Filtering for image reconstruction from projections, "J. Opt. Soc. Am." 67, 662-668.

Charles, P. A. and Culhane, J. L.: 1977, Extended soft x-ray emission from the Crab Nebula, "Astrophys. J." 211, pp. L23-L27.

Clement, H.: 1970, Star light, "Analog. Sci. Fict. Sci. Fact" 85, p. 55.

Colin, L., editor: 1972, Mathematics of profile inversion, NASA Tech. Memo X-62, 150.

Gordon, R., Bender, R. and Herman, G. T.: 1970, Algebraic reconstruction techniques (ART) for three-dimensional electron microscopy and x-ray photography, "J. Theor. Biol." 29, pp. 471-481.

Gordon, R. and Herman, G. T.: 1971, Reconstruction of pictures from their projections, "Comm. A. C. M." 14, pp. 759-768. (Figs. 6 and 7 are interchanged.)

Gordon, R.: 1973, Artifacts in reconstructions made from a few projections, Proceedings of the First International Joint Conference on Pattern Recognition, IEEE Computer Society, pp. 275-285.

Gordon, R.: 1975, Maximal use of single photons and particles in reconstruction from projections by algebraic reconstruction techniques intended for storage tubes (ARTIST), Digest of Papers, Optical Society of America Topical Meeting on Image Processing for 2-D and 3-D Reconstruction from Projections: Theory and Practice in Medicine and the Physical Sciences.

Gordon, R., Herman, G. T. and Johnson, S.A.: 1975, Image reconstruction from projections, "Scientific American" 233(4), pp. 56-68.

Gordon, R.: 1976, Dose reduction in computerized tomography, "Invest. Radiol." 11, pp. 508-517.

Gordon, R.: 1978, Higher-resolution tomography, "Physics Today" 31(7), p. 46.

Gordon, R.: 1979, A Treatise on Reconstruction from Projections and Computed Tomography, Plenum Press, New York, in preparation.

Gorenstein, P., Helmken, H. and Gursky, H.: 1975, Wide-field camera for the localization of gamma-ray bursts, Digest of Papers, Optical

Society of America Topical Meeting on Image Processing for 2-D and 3-D Reconstruction from Projections: Theory and Practice in Medicine and the Physical Sciences.

Gorenstein, P., Helmken, H. and Gursky, H.: 1976, Localization of gamma-ray bursts with wide field multiple pinhole camera system in near earth orbit, "Astrophys. Space Sci." 42, pp. 89-97.

Hounsfield, G. N.: 1973, Computerized transverse axial scanning (tomography): part I; description of system. "British J. Radiology" 46, pp. 1016-1022.

Kalos, M. H., Davis, S. A., Mittelman, P. S., Mastras, P. and Hutton, J. H.: 1961, Conceptual design of a vapor volume fraction instrument, Nuclear Development Corporation of America, White Plains, New York, NDA 2131-34, Contract AT(30-1)-2303(IX), United States Atomic Energy Commission, 31pp.

Lager, D. L. and Lytle, R. J.: 1975, Computer algorithms useful for determining a subsurface electrical profile via high frequency probing, Lawrence Livermore Laboratory, University of California, UCRL-51748.

Levinson, R. A., Nolte, J. T. and Sandler, S. S.: 1975, Application of algebraic reconstruction techniques to solar coronal x-ray emission features, Digest of Papers, Optical Society of America Topical Meeting on Image Processing for 2-D and 3-D Reconstruction from Projections: Theory and Practice in Medicine and the Physical Sciences.

Liu, A.: 1976, Vertical refractivity profiles of the terrestrial ionosphere and atmosphere from the GEOS-3/ATS6 satellite-to-satellite tracking data. "Eos" p. 235.

Murata, K. and Baba, N.: 1975, Measurement of the intensity distribution of an object by a strip telescope, "Japan J. Appl. Phys. Suppl." 14-1, pp. 1319-1338.

Phelps, M. E., Hoffman, E. J., Mullani, N. A., Higgins, C. S. and Ter-Pogossian, M. M.: 1977, Some performance and design characteristics of PETT III, In: Reconstruction Tomography in Diagnostic Radiology and Nuclear Medicine, Baltimore: University Park Press, pp. 371-392.

Radon, J.: 1917, Ueber die Bestimmung von Funktionen durch ihre integralwerte laengs gewisser Mannigfaltigkeiten, "Berichte Saechsische Akadamie der Wissenschaften (Leipzig) Mathematische-Physiche Klasse" 69, pp. 262-277. (English translation in Gordon, 1979.)

Saenz, A. W., Ueberall, H., Kelly, F. J., Padgett, D. W. and

Seeman, N.: 1977, Telecommunication with neutrino beams, "Science" 198, pp. 295-297.

Smith, K. T., Solmon, D. C. and Wagner, S. L.: 1977, Practical and mathematical aspects of reconstructing objects from radiographs, "Bull. Amer. Math. Soc." pp. 1227-1270.

Tasto, M.: 1977, Parallel array processors for digital image processing, "Optica Acta" 24, pp. 391-406.

Veverka, J., Wasserman, L. H., Elliot, J. and Sagan, C.: 1974, The occultation of Beta Scorpii by Jupiter, I; the structure of the Jovian upper atmosphere, "Astron. J." 79, pp. 73-84.

Wee, W. G. and Hsieh, T.-T.: 1976, An application of the projection transform technique in image transmission, "IEEE Trans. Systems Man Cybernetics" SMC-6, pp. 486-493.

SUBJECT INDEX

NAME INDEX

ASTROPHYSICS AND SPACE SCIENCE LIBRARY

Edited by

J. E. Blamont, R. L. F. Boyd, L. Goldberg, C. de Jager, Z. Kopal, G. H. Ludwig, R. Lüst, B. M. McCormac, H. E. Newell, L. I. Sedov, Z. Švestka, and W. de Graaff

1. C. de Jager (ed.), *The Solar Spectrum, Proceedings of the Symposium held at the University of Utrecht, 26–31 August, 1963.* 1965, XIV + 417 pp.
2. J. Ortner and H. Maseland (eds.), *Introduction to Solar Terrestrial Relations, Proceedings of the Summer School in Space Physics held in Alpbach, Austria, July 15–August 10, 1963 and Organized by the European Preparatory Commission for Space Research.* 1965, IX + 506 pp.
3. C. C. Chang and S. S. Huang (eds.), *Proceedings of the Plasma Space Science Symposium, held at the Catholic University of America, Washington, D.C., June 11–14, 1963.* 1965, IX + 377 pp.
4. Zdeněk Kopal, *An Introduction to the Study of the Moon.* 1966, XII + 464 pp.
5. B. M. McCormac (ed.), *Radiation Trapped in the Earth's Magnetic Field. Proceedings of the Advanced Study Institute, held at the Chr. Michelsen Institute, Bergen, Norway, August 16–September 3, 1965.* 1966, XII + 901 pp.
6. A. B. Underhill, *The Early Type Stars.* 1966, XII + 282 pp.
7. Jean Kovalevsky, *Introduction to Celestial Mechanics.* 1967, VIII + 427 pp.
8. Zdeněk Kopal and Constantine L. Goudas (eds.), *Measure of the Moon. Proceedings of the 2nd International Conference on Selenodesy and Lunar Topography, held in the University of Manchester, England, May 30–June 4, 1966.* 1967, XVIII + 479 pp.
9. J. G. Emming (ed.), *Electromagnetic Radiation in Space. Proceedings of the 3rd ESRO Summer School in Space Physics, held in Alpbach, Austria, from 19 July to 13 August, 1965.* 1968, VIII + 307 pp.
10. R. L. Carovillano, John F. McClay, and Henry R. Radoski (eds.), *Physics of the Magnetosphere, Based upon the Proceedings of the Conference held at Boston College, June 19–28, 1967.* 1968, X + 686 pp.
11. Syun-Ichi Akasofu, *Polar and Magnetospheric Substorms.* 1968, XVIII + 280 pp.
12. Peter M. Millman (ed.), *Meteorite Research. Proceedings of a Symposium on Meteorite Research, held in Vienna, Austria, 7–13 August, 1968.* 1969, XV + 941 pp.
13. Margherita Hack (ed.), *Mass Loss from Stars. Proceedings of the 2nd Trieste Colloquium on Astrophysics, 12–17 September, 1968.* 1969, XII + 345 pp.
14. N. D'Angelo (ed.), *Low-Frequency Waves and Irregularities in the Ionosphere. Proceedings of the 2nd ESRIN-ESLAB Symposium, held in Frascati, Italy, 23–27 September, 1968.* 1969, VII + 218 pp.
15. G. A. Partel (ed.), *Space Engineering. Proceedings of the 2nd International Conference on Space Engineering, held at the Fondazione Giorgio Cini, Isola di San Giorgio, Venice, Italy, May 7–10, 1969.* 1970, XI + 728 pp.
16. S. Fred Singer (ed.), *Manned Laboratories in Space. Second International Orbital Laboratory Symposium.* 1969, XIII + 133 pp.
17. B. M. McCormac (ed.), *Particles and Fields in the Magnetosphere. Symposium Organized by the Summer Advanced Study Institute, held at the University of California, Santa Barbara, Calif., August 4–15, 1969.* 1970, XI + 450 pp.
18. Jean-Claude Pecker, *Experimental Astronomy.* 1970, X + 105 pp.
19. V. Manno and D. E. Page (eds.), *Intercorrelated Satellite Observations related to Solar Events. Proceedings of the 3rd ESLAB/ESRIN Symposium held in Noordwijk, The Netherlands, September 16–19, 1969.* 1970, XVI + 627 pp.
20. L. Mansinha, D. E. Smylie, and A. E. Beck, *Earthquake Displacement Fields and the Rotation of the Earth, A NATO Advanced Study Institute Conference Organized by the Department of Geophysics, University of Western Ontario, London, Canada, June 22–28, 1969.* 1970, XI + 308 pp.
21. Jean-Claude Pecker, *Space Observatories.* 1970, XI + 120 pp.
22. L. N. Mavridis (ed.), *Structure and Evolution of the Galaxy. Proceedings of the NATO Advanced Study Institute, held in Athens, September 8–19, 1969.* 1971, VII + 312 pp.
23. A. Muller (ed.), *The Magellanic Clouds. A European Southern Observatory Presentation: Principal Prospects, Current Observational and Theoretical Approaches, and Prospects for Future Research, Based on the Symposium on the Magellanic Clouds, held in Santiago de Chile, March 1969, on the Occasion of the Dedication of the European Southern Observatory.* 1971, XII + 189 pp.

24. B. M. McCormac (ed.), *The Radiating Atmosphere. Proceedings of a Symposium Organized by the Summer Advanced Study Institute, held at Queen's University, Kingston, Ontario, August 3–14, 1970.* 1971, XI + 455 pp.
25. G. Fiocco (ed.), *Mesospheric Models and Related Experiments. Proceedings of the 4th ESRIN-ESLAB Symposium, held at Frascati, Italy, July 6–10, 1970.* 1971, VIII + 298 pp.
26. I. Atanasijević, *Selected Exercises in Galactic Astronomy.* 1971, XII + 144 pp.
27. C. J. Macris (ed.), *Physics of the Solar Corona. Proceedings of the NATO Advanced Study Institute on Physics of the Solar Corona, held at Cavouri-Vouliagmeni, Athens, Greece, 6–17 September 1970.* 1971, XII + 345 pp.
28. F. Delobeau, *The Environment of the Earth.* 1971, IX + 113 pp.
29. E. R. Dyer (general ed.), *Solar-Terrestrial Physics/1970. Proceedings of the International Symposium on Solar-Terrestrial Physics, held in Leningrad, U.S.S.R., 12–19 May 1970.* 1972, VIII + 938 pp.
30. V. Manno and J. Ring (eds.), *Infrared Detection Techniques for Space Research. Proceedings of the 5th ESLAB-ESRIN Symposium, held in Noordwijk, The Netherlands, June 8–11, 1971.* 1972, XII + 344 pp.
31. M. Lecar (ed.), *Gravitational N-Body Problem. Proceedings of IAU Colloquium No. 10, held in Cambridge, England, August 12–15, 1970.* 1972, XI + 441 pp.
32. B. M. McCormac (ed.), *Earth's Magnetospheric Processes. Proceedings of a Symposium Organized by the Summer Advanced Study Institute and Ninth ESRO Summer School, held in Cortina, Italy, August 30–September 10, 1971.* 1972, VIII + 417 pp.
33. Antonin Rükl, *Maps of Lunar Hemispheres.* 1972, V + 24 pp.
34. V. Kourganoff, *Introduction to the Physics of Stellar Interiors.* 1973, XI + 115 pp.
35. B. M. McCormac (ed.), *Physics and Chemistry of Upper Atmospheres. Proceedings of a Symposium Organized by the Summer Advanced Study Institute, held at the University of Orléans, France, July 31–August 11, 1972.* 1973, VIII + 389 pp.
36. J. D. Fernie (ed.), *Variable Stars in Globular Clusters and in Related Systems. Proceedings of the IAU Colloquium No. 21, held at the University of Toronto, Toronto, Canada, August 29–31, 1972.* 1973, IX + 234 pp.
37. R. J. L. Grard (ed.), *Photon and Particle Interaction with Surfaces in Space. Proceedings of the 6th ESLAB Symposium, held at Noordwijk, The Netherlands, 26–29 September, 1972.* 1973, XV + 577 pp.
38. Werner Israel (ed.), *Relativity, Astrophysics and Cosmology. Proceedings of the Summer School, held 14–26 August, 1972, at the BANFF Centre, BANFF, Alberta, Canada.* 1973, IX + 323 pp.
39. B. D. Tapley and V. Szebehely (eds.), *Recent Advances in Dynamical Astronomy. Proceedings of the NATO Advanced Study Institute in Dynamical Astronomy, held in Cortina d'Ampezzo, Italy, August 9–12, 1972.* 1973, XIII + 468 pp.
40. A. G. W. Cameron (ed.), *Cosmochemistry. Proceedings of the Symposium on Cosmochemistry, held at the Smithsonian Astrophysical Observatory, Cambridge, Mass., August 14–16, 1972.* 1973, X + 173 pp.
41. M. Golay, *Introduction to Astronomical Photometry.* 1974, IX + 364 pp.
42. D. E. Page (ed.), *Correlated Interplanetary and Magnetospheric Observations. Proceedings of the 7th ESLAB Symposium, held at Saulgau, W. Germany, 22–25 May, 1973.* 1974, XIV + 662 pp.
43. Riccardo Giacconi and Herbert Gursky (eds.), *X-Ray Astronomy.* 1974, X + 450 pp.
44. B. M. McCormac (ed.), *Magnetospheric Physics. Proceedings of the Advanced Summer Institute, held in Sheffield, U.K., August 1973.* 1974, VII + 399 pp.
45. C. B. Cosmovici (ed.), *Supernovae and Supernova Remnants. Proceedings of the International Conference on Supernovae, held in Lecce, Italy, May 7–11, 1973.* 1974, XVII + 387 pp.
46. A. P. Mitra, *Ionospheric Effects of Solar Flares.* 1974, XI + 294 pp.
47. S.-I. Akasofu, *Physics of Magnetospheric Substorms.* 1977, XVIII + 599 pp.
48. H. Gursky and R. Ruffini (eds.), *Neutron Stars, Black Holes and Binary X-Ray Sources.* 1975, XII + 441 pp.
49. Z. Švestka and P. Simon (eds.), *Catalog of Solar Particle Events 1955–1969. Prepared under the Auspices of Working Group 2 of the Inter-Union Commission on Solar-Terrestrial Physics.* 1975, IX + 428 pp.
50. Zdeněk Kopal and Robert W. Carder, *Mapping of the Moon.* 1974, VIII + 237 pp.
51. B. M. McCormac (ed.), *Atmospheres of Earth and the Planets. Proceedings of the Summer Advanced Study Institute, held at the University of Liège, Belgium, July 29–August 8, 1974.* 1975, VII + 454 pp.
52. V. Formisano (ed.), *The Magnetospheres of the Earth and Jupiter. Proceedings of the Neil Brice Memorial Symposium, held in Frascati, May 28–June 1, 1974.* 1975, XI + 485 pp.

53. R. Grant Athay, *The Solar Chromosphere and Corona: Quiet Sun.* 1976, XI + 504 pp.
54. C. de Jager and H. Nieuwenhuijzen (eds.), *Image Processing Techniques in Astronomy. Proceedings of a Conference, held in Utrecht on March 25–27, 1975,* XI + 418 pp.
55. N. C. Wickramasinghe and D. J. Morgan (eds.), *Solid State Astrophysics. Proceedings of a Symposium, held at the University College, Cardiff, Wales, 9–12 July 1974.* 1976, XII + 314 pp.
56. John Meaburn, *Detection and Spectrometry of Faint Light.* 1976, IX + 270 pp.
57. K. Knott and B. Battrick (eds.), *The Scientific Satellite Programme during the International Magnetospheric Study. Proceedings of the 10th ESLAB Symposium, held at Vienna, Austria, 10–13 June 1975.* 1976, XV + 464 pp.
58. B. M. McCormac (ed.), *Magnetospheric Particles and Fields. Proceedings of the Summer Advanced Study School, held in Graz, Austria, August 4–15, 1975.* 1976, VII + 331 pp.
59. B. S. P. Shen and M. Merker (eds.), *Spallation Nuclear Reactions and Their Applications.* 1976, VIII + 235 pp.
60. Walter S. Fitch (ed.), *Multiple Periodic Variable Stars. Proceedings of the International Astronomical Union Colloquium No. 29, Held at Budapest, Hungary, 1–5 September 1975.* 1976, XIV + 348 pp.
61. J. J. Burger, A. Pedersen, and B. Battrick (eds.), *Atmospheric Physics from Spacelab. Proceedings of the 11th ESLAB Symposium, Organized by the Space Science Department of the European Space Agency, held at Frascati, Italy, 11–14 May 1976.* 1976, XX + 409 pp.
62. J. Derral Mulholland (ed.), *Scientific Applications of Lunar Laser Ranging. Proceedings of a Symposium held in Austin, Tex., U.S.A., 8–10 June, 1976.* 1977, XVII + 302 pp.
63. Giovanni G. Fazio (ed.), *Infrared and Submillimeter Astronomy. Proceedings of a Symposium held in Philadelphia, Penn., U.S.A., 8-10 June, 1976.* 1977, X+226 pp.
64. C. Jaschek and G. A. Wilkins (eds.), *Compilation, Critical Evaluation and Distribution of Stellar Data. Proceedings of the International Astronomical Union Colloquium No. 35, held at Strasbourg, France, 19-21 August, 1976.* 1977, XIV+316 pp.
65. M. Friedjung (ed.), *Novae and Related Stars. Proceedings of an International Conference held by the Institut d'Astrophysique, Paris, France, 7-9 September, 1976.* 1977, XIV+228 pp.
66. David N. Schramm (ed.), *Supernovae. Proceedings of a Special IAU Session on Supernovae held in Grenoble, France, 1 September, 1976.* 1977, X+192 pp.
67. Jean Audouze (ed.), *CNO Isotopes in Astrophysics. Proceedings of a Special IAU Session held in Grenoble, France, 30 August, 1976.* 1977, XIII+195 pp.
68. Z. Kopal, *Dynamics of Close Binary Systems,* xiii + 510 pp.
69. A. Bruzek and C. J. Durrant (eds.), *Illustrated Glossary for Solar and Solar-Terrestrial Physics.* 1977, xviii + 204 pp.
70. H. van Woerden (ed.), *Topics in Interstellar Matter.* 1977, VIII + 295 pp.
71. M. A. Shea, D. F. Smart, and T. S. Wu (eds.), *Study of Travelling Interplanetary Phenomena.* 1977, XII+439 pp.
72. V. Szebehely (ed.), *Dynamics of Planets and Satellites and Theories of Their Motion. Proceedings of IAU Colloquium No. 41, held in Cambridge, England, 17-19 August 1976.* 1978, xii + 375 pp.
73. James R. Wertz (ed.), *Spacecraft Attitude Determination and Control.* 1978, xvi + 858 pp.

GPSR Compliance

The European Union's (EU) General Product Safety Regulation (GPSR) is a set of rules that requires consumer products to be safe and our obligations to ensure this.

If you have any concerns about our products, you can contact us on ProductSafety@springernature.com

In case Publisher is established outside the EU, the EU authorized representative is:

Springer Nature Customer Service Center GmbH
Europaplatz 3
69115 Heidelberg, Germany

Batch number: 10809439

Printed by Printforce, the Netherlands